Thor Heyerdahl's *Tigris* adventure began in the Garden of Eden on the bank of a river that flows from Ararat, where Noah's legendary ark once came to rest. It took him and his companions from nine nations in search of sea routes which he was sure must have been used by the ancient Sumerians 5,000 years ago on vessels like his own. His voyage down the Tigris, through the Gulf and eventually the Indian Ocean led to many discoveries and included many hazards. Modern shipping, bandits, reefs, and politics dogged *The Tigris Expedition*. Finally with permission to land only in the tiny republic of Djibouti and unable to continue the voyage for political reasons, *Tigris* was ceremonially burnt in protest against the intervention of major powers in African disputes which resulted in death and misery for millions of local people.

Thor Heyerdahl was born in Norway in 1914. After studying at Oslo University he married and went with his wife to the Marquesas Islands in the Pacific on field research. For over a year they lived on remote Fatu-Hiva Island, where rock carvings and other remains suggested to him that the first settlers might have come from America. *Fatu-Hiva* is a reminiscence of this adventure. Specialising in the Polynesian race back in Norway and in the United States, he began to formulate his theory of Pacific crossings on balsa-wood rafts or great double-canoes over a thousand years ago. His work was interrupted by two wars which he served in the free Norwegian Air Force and as an officer in a special parachute unit operating in Arctic Norway.

He went back to research after the war, and his Kon-Tiki expedition was undertaken to prove the possibility of the migration theory. The *Kon-Tiki Expedition* is an account of this voyage. He later published *American Indians in the Pacific* in which he marshalled the full evidence.

Subsequently, in 1953 and 1955, Thor Heyerdahl visited the Galapagos Islands and Easter Island. His original research into the origins and methods of erection of the great stone statues of Easter Island is described by him in *Aku-Aku*. He has also written *The Archaeology of Easter Island* and *The Art of Easter Island*.

In 1969 and 1970 Thor Heyerdahl embarked from Morocco on another ambitious voyage, this time in a boat made of papyrus, to test the theory that the earliest cultured people of the Mediterranean world could have crossed the Atlantic and set up communities from Mexico to Peru. *The Ra Expeditions* is the account of those voyages. His most recent voyage, *The Tigris Expedition*, took him into the ~~Indian~~ strate that the Sumerians used

By the same author

AKU–AKU
AMERICAN INDIANS IN THE PACIFIC
ARCHAEOLOGY OF EASTER ISLAND
ART OF EASTER ISLAND
EARLY MAN AND THE OCEAN
FATU–HIVA
KON–TIKI EXPEDITION
RA EXPEDITIONS

THE TIGRIS EXPEDITION

In Search of Our Beginnings

THOR HEYERDAHL

London
UNWIN PAPERBACKS
Boston   Sydney

First published in Great Britain by George Allen & Unwin 1980
First published in Unwin Paperbacks 1982

Unwin® Paperbacks
40 Museum Street, London WC1A 1LU, UK

Unwin Paperbacks
Park Lane, Hemel Hempstead, Herts HP2 4TE, UK

George Allen & Unwin Australia Pty Ltd,
8 Napier Street, North Sydney, NSW 2060, Australia

© Thor Heyerdahl, 1980, 1982

British Library Cataloguing in Publication Data

Heyerdahl, Thor
　　The Tigris expedition.
1. Tigris (*Ship*)　2. Indian ocean
I. Title
910'.09165　　　　G530.T/

ISBN 0-04-572024-X

Set in 11 on 12 point Bembo
and printed in Great Britain
by Hazell Watson & Viney Ltd, Aylesbury, Bucks

CONTENTS

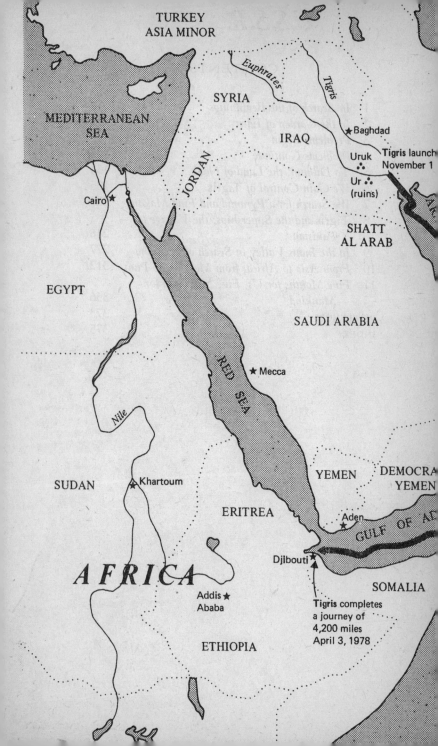

U.S.S.R.

AFGHANISTAN

Indus

IRAN

PAKISTAN

STRAIT OF
HORMUZ

Mohenjo Daro
(ruin)

GULF OF
OMAN

Karachi

P

Ancient
copper
mines

Muscat

OMAN

INDIA

★Bombay

ARABIAN SEA

SOCOTRA

0 Nautical Miles 300

SIMPLIFIED AND ADAPTED FROM A MAP
DESIGNED BY DOROTHY MICHELE NOVICK
DRAWN BY SNEJINKA STEFANOFF
COMPILED BY GUNARS J. RUTINS
© 1978 NATIONAL GEOGRAPHICAL SOCIETY

ILLUSTRATIONS

Plates 1 and 7 by David Graham. All other photographs by Carlo Mauri and the expedition crew.

CHAPTER 1

In Search of the Beginnings

THE BEGINNING. The real beginning.

This was the place.

This was where written history began. This was where mythology began. This was the source of three of the mightiest religions in human history. Two billion Christians, Jews and Moslems all over the world are taught by their sacred books that this was the spot chosen by God to give life to mankind.

Here two large rivers, the Euphrates and Tigris, drift slowly together, and their meeting-place is shown on every world map. Yet it is not a spectacular scene. Silent as the rivers when they meet are the narrow rows of date palms lining the banks, while sun and moon, passing over the barren desert, are reflected day and night on the calm waters. A rare canoe glides by, with men casting nets.

This, most of mankind believes, was the cradle of *homo sapiens*, paradise lost.

A narrow point of green land is drawn out between the two rivers as they meet and greet each other with slow whirls, forming a single river, the Shatt-al-Arab, which quickly hides from view behind a palm-lined bend. Between the rivers, at the very point of the land, a little resthouse was once built and subsequently half abandoned. With its three guest rooms, large hall and still bigger terrace, facing the sunrise over the river Tigris, the modest building bears an impressive name in big letters above the door: THE GARDEN OF EDEN RESTHOUSE.

This boastful name is justified if we are to take a nearby noticeboard literally. Hardly a stone's throw from the resthouse, and separated from it by a greensward large enough for building a boat, are a couple of inconspicuous green trees leaning over the Tigris. Between them lies a thick, short stump. This was part of a fallen tree of undefined species, but venerated by modest candles and solemnly fenced in as a very simple sanctuary. Old men from the nearby town of Qurna sometimes come here to sit and meditate. A placard with text in Arabic and English tells the rare passer-by that this was the abode of Adam and Eve. Abraham, it says, had come here to pray. Indeed, according to the scriptures, Abraham was born at Ur, a few hours away.

The long-lost branches of this aged tree certainly had never carried apples. And Abraham had probably never venerated that pleasant riverside spot, since the ground level has risen an estimated six metres in the last few millennia and must have altered the original water-course. Nevertheless, the meeting place of the two rivers, the whole locality, merits the humble meditation of the passer-by, for something began in these surroundings. Something of importance to you and me and most of mankind.

As I moved with my luggage into the resthouse and leaned over the terrace fence to watch the silent rings made by fish as they broke the surface, the sun slipped away behind me and drew a red curtain over the sky, causing the black silhouettes of date palms on the other side to be reflected for a while in a river that seemed as if turned to blood.

There was adventure in the air. How could it be otherwise? Here was the homeland of the *Thousand and One Nights*, of Aladdin's Lamp, the Flying Carpet, and Sindbad the Sailor. Ali Baba and the Forty Thieves belonged to these riverbanks. Downstream, the waters drifted past Sindbad Island, named after the great yarn-spinning sailor of Arab folk-tales. Upstream it had its source near the foot of the soaring cone of Mount Ararat, where Hebrew records have it that Noah grounded his ark. Near the banks, modest road signs still point to time-honoured ghost cities like Babylon and Nineveh, whose Biblical brick walls still seem to shake off rubble and dust in their attempt to reach the sky. Jetliners roar into timeless Baghdad, where modern cranes and concrete buildings crowd between golden domes and minarets.

Located, as it were, to the east of the West and to the west of the East, people of all types and buildings of all epochs blend naturally in this Arab republic known today as Iraq. Mesopotamia, the 'Land between the rivers', was the descriptive name the conquering Greeks once gave to this same territory which men of antiquity had regarded with awe, wonder and admiration and knew under many different names. Best known and first among them was Sumer; subsequent were Babylonia and Assyria. Since the generation of

Mohammed the Prophet this has been the Arab's eastern frontier.

Today oil flows through pipelines across the sterile deserts which were once productive pastures and irrigated fields. In the withered landscape temple-pyramids, minarets and oil drills stand side by side as symbols of the changing civilisations. Near the former banks of the Euphrates, south of Baghdad and half way to the distant Garden of Eden, the road passes a huge, shapeless pile of crumbled bricks. A signpost in the rubble brings to mind the classic tale of man's first attempt to build higher than what he himself could control. The sign reads: THE TOWER OF BABEL.

The road continues in the direction of the midday sun. It runs through endless desert plains, with Arab towns so colourful and picturesque that Sindbad and Ali Baba seem to sit on the doorsteps or move in the throngs in the market places; and one passes from the former domain of Babylon into the coastal area where everything began – Southern Iraq, formerly known as Sumer.

The naked deserts, flat as a farmer's field, continue to the open gulf. Only in the flooded area where the twin rivers slowly converge are vast stretches of green marshes full of birds and fishes. Here a truly unique culture has survived since Biblical times, hidden in a world of canes and reeds that grow tall and dense as a jungle. More Sumerian blood probably runs in the veins of these marsh dwellers than in any other Arab tribe. The rivers from Noah's distant mountain seem to ignore the scorched landscape of former Mesopotamia throughout its length to overflow in joy at encountering the timeless Marsh Arabs. They alone of Noah's descendants seem to have been blessed with eternal life, while all the great city-states and kingdoms around them have followed one another in collapse.

The desert, encroaching upon the spring-green marshes from all sides, has swallowed up the former Sumerian homeland and all that it contained. Bright dunes of arid sand have rolled over colossal temple-pyramids erected to forgotten gods, and covered abandoned cities ruled by kings themselves reduced to dust. The landscape which once throbbed with life is today as silent and lifeless as the North Pole. Like the snow-filled crevasses of the Polar ice, endless irrigation systems and former shipping canals run from horizon to horizon, though not a drop of water enters them and not a green leaf grows.

This is the cemetery of an entire civilisation, the oldest ancestor to our own. No one who wants to know our own beginnings can ignore what archaeological detectives have extracted from the local sand. For in this dead and buried world human words live on like ghosts, ready to speak again whenever recovered from the grip of the soil. Among the excavated palaces and dwellings scientists have uncovered real 'libraries', with tens of thousands of baked clay tablets incised with the earliest script known. The original Sumerian script was composed of hieroglyphic symbols, but these were soon replaced by cuneiform characters easier to cut into clay.

In tombs and temples the archaeologists have also come across incredible art treasures of gold and silver, witnessing to a standard of taste and a level of civilisation so outstanding that simpler minds of our own day have been led to suspect that these long vanished people must have come from outer space. The learned publications of the archaeologists rarely reach the common reader and the market has been free for stories about extra-terrestrial visitors landing to build pyramids and bring civilisation to the barbarians on our planet. Such entertaining books and films have spread like wildfire all over the world in recent decades while men from Earth have set foot on the moon. Intellectuals smile and shrug their shoulders, but millions believe in them. They satisfy modern man's growing desire for an immediate answer to a question science has only slowly and meticulously begun to disentangle: How did it all begin?

What could the first Sumerians have told us if they had returned to life? They were supposedly there to receive the civilised space-men, if they did not directly descend from them.

The Sumerians do not have to come back to witness to their origins. Their words are still with us. They left their written testimony. Their tablets record how they came and from where. It was not by spacecraft. They came by ship. They came sailing in through the gulf, and in their earliest works of art they illustrated the kind of watercraft that brought them. They came as mariners to the coast of the twin river valley where they founded the civilisation which during the ensuing millennia was to affect in one way or another every corner of our world.

Their written testimony seems at first to open the door to another mystery. Where is the eastern land of 'Dilmun' from which the seafaring Sumerians said they came?

It was to get acquainted with these original testimonies and to obtain practical lessons from people still living in the marshes that I returned to Iraq time and again in my search for a tiny piece missing from a great puzzle. The real puzzle was that human history has no known beginning. As it stands it begins with civilised mariners coming in by sea. This is no real beginning. This is the continuation of something lost somewhere in the mist. Is it still hidden under desert sand, as was Sumerian civilisation itself, remaining unknown to science until discovered and excavated in southern Iraq in the last century? Was it buried by volcanic eruption, as was the great Mediterranean civilisation on the island of Santorini, unknown until discovered in our own time under fifty feet of ashes? Or could it possibly be submerged in the ocean that covers two-thirds of our restless planet, as suggested by the hard-dying legend of Atlantis?

If we are to believe the Sumerians, who ought to know, their merchant mariners returned to Dilmun many times. In their own days, at least, their ancestral land was neither sunk in the sea nor buried by volcanic ash. It was within reach of Sumerian ships from Sumerian ports. One little piece missing from the big puzzle is that nobody knows the range of a Sumerian ship. Their seagoing qualities were forgotten with the men who built them and sailed them, their range lost with their wakes.

Practical research into ancient types of watercraft leads one upon many untrodden trails. It led me to remote islands in Polynesia, lakes in the Andes and central Africa, and rivers and coasts of all the continents. Lastly it brought me to what was formerly Sumer, today the home of the Marsh Arabs. There began my quest for human history beyond the zero hour. There began also a voyage that brought me and my companions into adventures far from those of the astronauts, back to the remote days and nights when our planet was still big. So big it was in those times that unknown and unforeseen worlds, alien to the voyager, beckoned beyond every horizon. Worlds with plants and animals never imagined. Peoples, buildings, and living manners so distinct from those at home, as if pertaining to another planet under a different sun. Such worlds once existed side by side, separated by barriers of wilderness and united by the open sea.

The sea roads between them were in use before the Sumerians came to settle in Sumer. Their tablets speak of navigating kings and merchant mariners coming from or going to lands overseas, and they give long lists of cargo imported from or exported to foreign ports. A few even speak of shipwrecks and maritime disasters. Such records reflect the hazards always involved in a marine enterprise even when the vessel is built with the experience of a whole nation and manned by a crew at home with the craft. In reading the tablets such dramas come to life. You can almost hear the cry: 'All hands on deck!'

How often such shouts of warning and despair must have been drowned by the thunder of surf against reefs or by a roaring ocean which in fury tried to devour a tiny vessel fighting to resist an unexpected gale.

'All hands on deck!'

This time the warning was for me. For me and my sleeping companions inside the tiny bamboo cabin. It was Norman's voice. A roaring noise filled the darkness. This was reality. In my sleep my body had bounced about so violently that I had dreamed I was riding in a car with one wheel off the road. Instead I found myself clinging to a bamboo post to stay put in a strange bedstead where water trickled down my face.

We were in trouble. Out.

'Out!' I shouted, and kicked the sleeping-bag away.

Others were already crawling over my legs, flashlights in their hands, heading through the tiny door opening.

No time to dress. Just to tie the safety ropes around our waists. We were all needed on deck. This was the real thing, no bad dream. A gale had suddenly overtaken us during the night. Hard to stand upright for the wind and rolling. Spray and rain whipped the skin. Seeing nothing, we fumbled from stay to stay or clung to the bamboo wall trying to locate the threatened sail and rigging with our flashlights.

'Tie yourselves on!' I shouted. A wild sea sent our ship bouncing like an antelope. Sea and air were in uproar, the noise of waves and screaming woodwork was terrifying. The storm howled and whistled in ropes and bamboo. The kerosene lamps had all blown out except for one swinging like a maddened firefly high in the mast top, shedding no light on deck.

'Get Norman to reef the sail!' It was Yuri at the rudder oar, yelling to Carlo on the cabin roof. Suggestions, orders, questions, violent exclamations in many languages were swallowed by the din before they reached the ears they were intended for, though the voice of Norman, our sailing master, cut through from somewhere with overtones of despair. We all knew that our rigging was in danger.

The mast, rather than the sail, was our Achilles' heel. The sail might split, but it could be repaired. The feet of the straddle-mast were set in wooden shoes lashed to the reed-bundle ship with rope. We feared that the reeds or the ropes might rip away and all our rigging with sail and masts would disappear in the wind. Whilst we all hung upon stays and halyard to press the mast legs down and to lower the maddened mainsail, we heard a sharp report overhead, followed by a terrible crash. The vessel leaned over; everything to cling to leaned over, as I looked up and tried with my feeble flashlight to discern what had happened up there.

In fact, the solid topmast had broken into splinters. The main piece dangled upside down in front of the mainsail with the sharp clawlike end of the fracture threatening to rip the canvas which fluttered and battered like a huge kite above our heads, pulling the vessel ever further over on to its starboard side. A chasing wave caught at the edge of the sail and refused to let go. Everybody with a free hand gathered to try to drag the drowning canvas back to the ship. We feared the worst. Any normal vessel would have been in the utmost peril of capsizing or springing a leak.

But this was not a normal vessel in the eyes of a modern sailor. This was a kind of ship once used by the earliest Sumerians, a kind of ship regarded by modern scholars as a mere river boat. As wave upon wave thundered aboard the uncontrolled vessel while the ocean held the corner of our mainsail in its grip, it seemed for a while that anything could happen. Whatever did happen would give the answer to one of the questions that had led to the experiment: how seaworthy was such a pre-European vessel? It was for a lesson like this that we had come on board. To know the Sumerian hardships and pleasures, and to feel on our own skins their practical problems at sea.

I thought of the warm, dry ground between the firm date palms in the Garden of Eden. Would I have started this if I had known that a moment like this was awaiting us? I believe so. I believe we all realised that one cannot go to sea in an unfamiliar kind of ship without at some point running into hardship and trouble.

This was my third ocean voyage on board a reed-ship, but it was the first in which we did not simply drift with the elements. This time there was no marine conveyor belt to carry us to our destination. Yet our previous experience of the Egyptian-type papyrus ships gave us great advantages, for in ship design the earliest Sumerians and the earliest Egyptians have a common heritage. Scholars had even pointed out that the oldest hieroglyphic sign for 'ship' in Sumer was the same as the one for 'marine' in the earliest hieroglyphic script of Egypt.[1] It depicted a sickle-shaped reed-boat with crosswise lashings around the vessel and reeds spreading at bow and stern. Could this mean that the earliest scribes in these two countries had inherited the idea of a script from a now lost common source? Or does it mean only that both people had inherited the same kind of reed-ship?

The reed-ships of ancient Egypt had been illustrated by the artists of the Pharaohs in such detail that I had been able to copy the vessel, including its steering gear and rigging, when I built the reed-ships *Ra I* and *Ra II* for testing in the open ocean. Such minute details are not shown in Mesopotamian art, but the reliefs on a slab wall brought to the British Museum from a royal palace in Nineveh depict a realistic battle between reed vessels of Assyria and Babylonia. The ships are big enough to reveal double rows of Assyrian soldiers on deck as they embark to massacre fleeing men and women, with the victims thrown overboard to the fish and crabs. These large reliefs show in great clarity that the reed-ships of the twin-river country were built like those on the Nile.

No motif is more common in the miniature art of the Mesopotamian cylinder seals than the reed-boats used by the legendary heroes in the period of settlement. In all principles these boats are the same as those shown in greater detail in Nineveh and Egypt. Yet there was one fundamental difference. The Egyptian reed-boat builders had access to papyrus. In former times papyrus grew in abundance all along the banks of the Nile from its sources to its delta. The Sumerians had no papyrus; instead, the marshes of Mesopotamia offered another tall fresh-water reed, locally known as *berdi*.

After our two experiments with an Egyptian style vessel in 1969 and 1970 I knew that a correctly built papyrus ship could cross a world ocean. *Ra I*, built by Central African Budumas, had almost crossed the Atlantic when the lashings broke. In the following year, with *Ra II* built by South American Aymara Indians, we sailed all the way from Africa to America. But berdi differs markedly from papyrus both in form and in substance. And what is worse, science had decided that berdi is very water absorbent. There was only one authority on ancient Mesopotamian watercraft in our time, the Finnish scholar Armas Salonen. In his learned and thorough study of all types of vessels used in the twin river country in former times he has nothing to say about the *elep urbati*, the reed vessels, except a reference to the general belief that they quickly absorbed water 'and unquestionably had to be brought ashore to dry out after use'.[2]

There was agreement on this point in the very sparse literature touching on this subject. The Sumerian reed-ships, accordingly, could have served only as river boats.

How could this modern verdict be reconciled with the ancient texts and illustrations? This was the question that had led me to move into the Garden of Eden Resthouse. I came back to Sumerian territory to try to disentangle the theoretical controversy by a practical test. I wanted to see how long a berdi ship would float, and to attempt to retrace some of the obscure itineraries recorded in the old tablets with their references to Dilmun, Makan, Meluhha and other disguised and long forgotten lands.

I had ended up beside Adam's tree by mere accident. I had searched for a safe and suitable place to build a reed-ship in the marshlands and along the river banks. Down by the gulf, modern cities and expanding industry had occupied all convenient terrain, and upstream quagmire, mud flats, date plantations or steep escarpments made boat building and launching difficult. Then the representatives of Iraq's Ministry of Information showed me the empty garden plot by the side of Adam's tree, and generously offered me the Garden of Eden Resthouse as assembly site for the expedition. The endless reed marshes began a few minutes from the door, and from the terrace we could sail down the river straight to the open sea. Even with the help of Aladdin's Lamp I could not have been offered a better solution.

I had been amply warned that Iraq was not the easiest country in which to mount an expedition at that moment. The republic was in the melting-pot after the overthrow of the monarchy and the casting-off of British influence. The leaders of the pan-Arab Bath party had won the most recent revolution and closed the borders to tourists, though they welcomed constructive projects in science and industry. The only two available hotels in Baghdad were crowded with businessmen and engineers from all the industrialised nations of East and West. All seemed eager to bring the ripe fruits of modern civilisation back to the scorched soil where the first seeds had once germinated. The oil pipelines were the veins of modern Iraq, just as the irrigation trenches had been in earlier days.

The new republic was in a truly explosive development period where the future was far more important than the past. But a solid core of scientists at the National Museum in Baghdad were well aware of the fact that the more we know of the past the better we can plan the future. Man cannot know where he is going unless he can see his tracks and know the direction from which he has come. No situation is entirely new under the sun, and unless we seek to learn by our own errors we must learn by the mistakes of others.

Dr Fuad Safar called together his colleagues and collaborators at the National Museum, with its well-stocked library, and together we discussed the project I had in mind. Did the Sumerians cover their reed-ships with bitumen to make them waterproof and buoyant? They certainly had access to natural asphalt that came to the surface in open springs near Ur and in several other localities higher up the river, and they used it to waterproof receptacles and roofs. On the museum shelves there were reed-boat models five thousand years old, thickly covered with asphalt.

Or, since asphalt was heavy, did they coat the bigger ships with shark oil, as was still the custom among many of the fishermen in the gulf, who used it on the planks of their wooden vessels? There was even a very early tablet that spoke of a famous hero who mixed six measures of pitch with three of asphalt and three of oil when building a huge reed-ship. Was this mixture intended for impregnating the water-absorbent berdi reeds?

Where was the ancestral Dilmun so often visited by Sumerian merchant mariners? Most scholars now believe it was the island of Bahrain, where recent archaeological work has uncovered extensive towns, tombs and temples which in part even antedate Sumerian time. But Bahrain was far out in the gulf, so perhaps Dilmun was the smaller island of Failaka, which lay just off the Sumerian coast?

I felt I had derived great benefit from the Museum meeting, but I was also very uncertain and bewildered. What should I use to impregnate my reed-ship? Anything at all? After weeks among the Baghdad Museum artifacts and translations of the tablet texts I had notebooks full of facts and theories. But one thing was clear: the early Sumerians were shipbuilders and mariners. Their civilisation was based on the import of copper, timber and other raw materials from foreign lands, and their growth into a dominant power at the mouth of the rivers was due to cities like Ur and Uruk being major ports and centres of very extensive trade. They had no access to copper and little chance of profitable trade in the immediate vicinity, so it seemed obvious that their sailors must have gone very far.

Only a practical life-like test could give the answer. I decided to carry out my project, and the Museum staff convinced their Ministry that to reconstruct a prehistoric vessel such as I had planned was a sensible thing to do. I was then granted permission to harvest reeds in the marshes, to import equipment free of customs, to assemble the expedition's crew in Iraq irrespective of nationality, and we would all be the guests of the country until we sailed away.

I lost no time in returning to the marshes. An interpreter from the museum was sent with me. From our rooms at the Garden of Eden

Resthouse we could walk through the town of Qurna, where the main asphalt road from Baghdad passed full of thundering traffic. No camels on this road. Not even bicycles. Huge transporters, articulated tankers and army trucks rolled by to the gulf ports. Immediately on the other side of the highway the marshes began, and for mile after mile they led ever deeper into a world of their own, unlike anything I had imagined. These marshes are about six thousand square miles in extent.

As we reached the first water channel, two tall marshmen in flowing Arab gowns were waiting for us, each with a long punt-pole of cane. One held back a long black canoe with his big bare foot as they welcomed us and signed to us to step on board. This was their usual *mashhuf*, the slender, flat-bottomed long-boat built to standard lines by all Marsh Arabs today. While formerly built of their own reeds, they are now pegged together from imported wood and covered, like their reed prototypes, with a smooth coating of black asphalt. Prow and stern soar in a high curve like the Viking ships, following the five-thousand-year lines of their Sumerian forerunners.

I stepped on board the unsteady vessel and sat down on a pillow at the bottom, the two slim marshmen standing erect at either end, punting with expert strokes, long and slow, against the shallow bottom. The water was crystal clear; plants grew on the bottom; I saw fishes and there were long garlands of water-crowfoot floating on the surface. We slid silently away from the green turf and slipped in between two high walls of canes and bulrushes. As these tall water plants closed in about us and shut the green door behind us we left the bustling, rumbling modern world and felt as if transported with the speed of spacecraft into the past. With each calm punt-stroke by the two silent marshmen I sensed that I was travelling back through time, not into savagery and insecurity, but into a culture as remote from barbarism as ours and yet incredibly simple and uncomplicated. My interpreter from the Ministry of Information had not been with me on my earlier visit, and as we reached the first floating villages he was as fascinated as I was.

On that first visit five years earlier I had the feeling that the authorities in Baghdad had been a bit reluctant to let me into the marshes. Not that the people living there were dangerous. They were just not yet in step with modern Iraq. The surrounding desert Arabs in a slightly humiliating way referred to them as Madans, the keepers of buffaloes instead of camels. Or, rather, instead of cars, for even camels are now out of fashion.

The water in the marshes is too deep for wheeled traffic and too shallow for normal boats. With its boggy bottom it has always kept horsemen and camel riders away. A pedestrian would be totally lost if wading among the tall reeds. Only the Madans knew the hidden labyrinths of narrow and shallow canoe passages through their bulrush jungle, and for this reason have been left to lead their own lives. But the enthusiasm of the British explorers Wilfrid Thesiger, Gavin Young, and the few others including myself who had been inside the marshes and emerged full of admiration, had begun to affect the attitudes of Baghdad. This time I was even encouraged to use a film camera and to bring as many Madans as I needed out of the marshes to my ship-building site near the resthouse.

Only a few columns of smoke far apart revealed to us that people were living in the marshes. We did not see a single trace of human waste. Not a roof disclosed the whereabouts of the villages until we came within a spear's throw of a building. No elevation, not a stone to step on is found to permit a view above the canes and bulrushes that stand compact and much taller than a man's eyes on boggy ground that yields to the foot like a mattress. Geese and ducks and other waterfowl of all colours and sizes abound, as if guns had not been invented. An occasional eagle sails in from the surrounding shores, and kingfishers, and an endless variety of little birds, some of brilliant colour, sit and sway everywhere in the reeds, especially in the migratory season. Tall white herons and red-beaked storks stand like sentinels between the stalks, and stout pelicans scoop up fish with their big bucket-like beaks.

With luck one may catch a glimpse of a shaggy black boar as it ploughs its way in heavy bounces between swinging reeds. Only when approaching the hidden habitations do we see huge water buffalo wading lazily in our way or climbing up into the reeds, their broad black bodies shining like wet sealskin in the sun. They stop to watch us with their friendly bovine eyes as we pass, flapping their broad ears and flicking their slim tails patiently to shake off flies as they imperturbably continue to chew the last of the green sedge that hangs down from their jaws.

Suddenly the village is before us. What a revelation! What perfect harmony with nature! The vaulted reed houses are as much at one with the environment as are the birds' nests that hang among the canes. Some are small and scarcely more than shelters to creep under, but most are big and roomy. They are hidden simply because we ourselves travel behind high and unbroken screens of greenery. The tallest houses are big enough almost to resemble hangars, with walls and roof arching in perfect symmetry from side to side, usually with one end open. Some have both ends open, like a railway tunnel. No wood, no metal goes into these big structures. A skeleton of thick, arching bundles of cane is covered with reed mats lashed on with bulrush fibres.

In its elegant perfection, the architecture is as impressive as the result is astonishingly beautiful, each dwelling recalling a little temple with its golden-grey vault outlined majestically against a perpetually cloudless sky extending from the surrounding desert. Some are mirrored in the water together with the blue sky.

This was pure Sumerian architecture. The industrious people who first handed the art of writing down to our ancestors had lived in such houses. In their regular cities they had built walls from enduring bricks, but in the marshes they had constructed their houses entirely from reeds. They are realistically illustrated in Sumerian art five thousand years old, carved in stone and incised on seals, just as their present boats are identical in line to the small models in silver or asphalt-covered reeds, found as Sumerian temple offerings. Both have proved themselves perfectly adapted to the environment and to local needs.

As we jumped 'ashore' the ground under our feet swayed like a hammock and my friend, unprepared for this, tottered and grabbed for an arm. We were walking on a floating mattress of reeds. The few steps from the water's edge to the tunnel-like entrance to the big house brought us on to a thicker, more solid foundation. A middle-aged man welcomed me with both hands and then touched his chest close to his heart: '*Salam alaikum*, peace be with you!'

I was back among friends. This was where I had been five years earlier when I had met an old man I could never forget. This was his son.

'Friend, how are you? How is your family?'

'Praise be to God. And you? Your children? And your old father?'

'He is alive, praise be to God. But he is in hospital in Basra. He is more than a hundred years old now.'

I was sorry, for in a way it was this old man who had brought me back to the marshes. Another old man with a long white beard appeared in the cool shade of the reed tunnel and for a while we all continued in Arab fashion, asking each other how we felt and how each member of the family felt, praising Allah for his generosity to our households. We slipped off our shoes and sat down in polite, traditional silence on the clean oriental mats on the floor, our hosts with their legs folded under the long gowns once they had stuffed piles of colourful cushions behind their two guests, who were expected to lean comfortably against the soft reed wall.

I looked around me and recognised with pleasure this big airy building, the guest house of the old man now in hospital. I could not have reached the ceiling even with a fishing rod. Seven stout cane bundles thicker than a human body arched like parallel ribs holding up the tight skin of braided reed mats. It gave me the feeling of sharing the biblical adventure of Jonah in the stomach of the whale. But this whale had its mouth wide open at both ends, leaving a double view of a perfectly blue sky, blue water, green reeds, and a couple of fringe-leafed date palms.

Only a few of the Madan villages can muster date palms. Most of them are built on entirely artificial islands formed by untold generations of rotting reeds and buffalo dung. Quite often these islands are actually afloat and rest on the bottom only in the dry season. New top layers of reeds have to be added annually as the bottom layers disintegrate. To prevent the edges from being washed away by the slowly moving water, they are fenced in with tight pallisades of canes stuck into the bog-bottom below. While the islands with the reed-houses rise and sink within their pallisades according to season, the canals between them permit the passage of the slender canoes and make up a village complex in the pattern of Venice.

A Marsh Arab can rarely walk more than a couple of steps before he has to enter his canoe. Some of the floating islands are so small that with the traditionally big house or buffalo stable on top they look like house-boats or some sort of Noah's Ark with barely enough foothold to walk around the walls. In the lake areas deep inside the marshes, the floating Madan families bob up and down on swaying reed carpets with their ducks, hens, water buffaloes and canoes, and the big buffaloes have to dive in with the ducks and swim for the reed fields every morning when their owners unfasten the mat barriers of their vaulted reed stable.

Our caftan-covered host stirred up the embers of a small fire on a mud patch in the middle of the floor. Then from an elegant teapot small silver-framed glasses were filled and we were offered drink, the perfumed steam strong in our nostrils. More marshmen, covered but for their ruddy faces in caftans and flowing gowns, came silently in, saluted in the name of Allah and sat down in the shade by the reed columns.

'The berdi has to be cut in August.'

The old man broke the silence by repeating the sentence I had now heard everywhere in the marshes.

'Why?' I asked, repeating a question I had asked a hundred times. By now I knew the answer all too well: if cut in any other month the reeds would absorb water and lose buoyancy. Only if cut in August would berdi float for a long time. Some said for a year. Some said two, three, or even four years. Some said they did not know why they cut in August; it was the custom.

'In August there is something inside the stalk that keeps water away,' said the old man. 'We have to harvest our reeds then and let them dry for two or three weeks before we use them.'

The first time I had heard this statement was on the banks of Shatt-al-Arab at the village of Gurmat Ali, between the marshes and the gulf. It was my first visit to Iraq after my two experiments with Egyptian-type papyrus ships, and I had been quite astonished at seeing half a dozen huge reed rafts moored like floating wharfs along the river bank. I had measured one that was 112 feet long, 16.5 feet wide and about 10 feet deep. One third of the depth was below the river surface. I jumped on board this raft and had tea with a marshman who lived on it in a tiny makeshift shelter, also made of reeds. A patch of mud on the reed floor prevented the huge raft from catching fire, as all he burned to heat his pot were short bits of the same dry reeds. I asked how long Matug had lived on his big reed-raft. Matug had lived on his *gáré* for only two months so far. He had spent one of them floating it down from Sueb in the marshes. And how far had it sunk into the water in that time? Nothing, he said. Matug had cut his reeds in August. He had come here to sell the reeds of his gáré to a small factory that made them into cardboard for modern building insulation.

Two months! After one month on our papyrus ships our reeds were already completely waterlogged and for the rest of the voyage we had floated with our deck at surface level and the waves breaking over all cargo not kept high above the reed bundles. Matug even told me that in the previous year he had to sit for nine months on his gáré before he could sell it, as the factory was full.

The seemingly unsinkable gáré gave me no rest. We had harvested the reeds for both *Ra I* and *Ra II* in December to have the ships ready built for departure in the spring. Why had my experienced Buduma reed-boat builders from Lake Chad not told me that we had cut the reeds in the wrong season? And why had the Ethiopian monks at Lake Tana, who harvested the reeds for me, said in reply to my express inquiry that any month was equally good for cutting reeds for boat building? The answer was simple. The African reed-boats on Lake Chad and Lake Tana were only used for a day or two at a time, after which they were either beached or carried ashore, and had no chance to become waterlogged. The Madans of Iraq, however, spent their lives on top of their reeds.

At once it became clear to me that the Marsh Arabs could still teach me lessons not taught in any faculty, nor found in any scholarly books. A police car gave me a lift to Qurna, where the rivers met, and from there the Sheriff drove me on a narrow dirt road to the ferry-point across from Madina, a major town on solid sand built up by the Euphrates some fifteen miles inside the marshes. I was housed by the Sheikh and served a breakfast I shall never forget: coffee, tea, fresh milk, yogurt, eggs, lamb, chicken, fish, figs, dates, Arab bread, white bread, pastries, compotes. I could hardly stagger down to the banks and press myself into the straight mashhuf the Sheikh had waiting to take me to Om-el-Shuekh, further inside the marshes.

My proud hosts in Madina had shown me their willingness to share everything they had, but they could not give me what I had come for: information about reed-boats. They had all seen gáré, such as I had myself seen down the river, but these were just loose cross-piles of berdi temporarily stacked on top of each other to facilitate their transport to the paper mill. The only real boats they knew inside the marshes were the various types of wooden mashhuf: *tarada*, *mataur* and *zaima*, all coated with asphalt, and the broader *balam* used for the transport of mats and canes. Reed-boats belonged to the past. I had to talk to some really old men to learn about them.

Thus the Sheikh sent me to Om-el-Shuekh, where lived the oldest man they knew. He was said to be a hundred years old.

I did not expect much memory from a man of that age, and was no more optimistic when the reed screens opened and we silently slid forward to his bank. Both ashore and reflected in the water I saw old chief Hagi Suelem in his white gown and with a long white beard sitting in the opening of his own reed hangar and looking like an image of Methuselah. But when old Hagi rose to meet me and wish me peace I looked into a pair of friendly and alert eyes that made the whole man become big and powerful. He was obviously looked upon with great respect by the men who gradually assembled around us and sat down with us in two rows, facing each other across the hall. Like myself, they all listened to old Hagi's wisdom and humour with interest and approbation. The tea tray was soon there beside the flickering fire, and some big, broad fish, split open like giant butterflies, were balanced on edge close to the flames without a pan. Crisply toasted, but white and juicy inside, rolled up in oven-warm Arab bread baked broad and thin like pancakes, the fish was so delicious that I ate as if the Sheikh of Madina had forgotten to give me breakfast. The old man watched me attentively and saw to it that the man at my side dug out the best pieces of the fish with expert fingers and fed me like a royal baby. True to custom, before and after the meal, soap and towel were passed around by a man who went from person to person pouring hot water as we rubbed our hands in the jet over his swill-pail.

Hagi apologised for the simple meal, as if he had not seen my appetite, and assured me a real treat if I promised to come back. This I promised. I had to come back. I had lived with so-called primitive peoples in Polynesia, America and Africa, but these marshmen were not primitive in any sense of the word. They were civilised, but differently from us. They lacked the push-button services and took the direct shortcut to food and joy provided at the source. Their culture had been proved viable and sound by persisting while the Assyrian, Persian, Greek and Roman civilisations progressed, culminated and collapsed. In this stability through untold ages is reflected something the rest of us lack: respect for their progenitors and confidence in the future.

'We are not poor,' said old Hagi, as if he had read my mind. 'Our pride is our wealth and no marshman is hungry.' He had once been to Baghdad, but could hardly wait to get back to the peace and security of Allah's marshes. The city in his opinion breeds greed, competition, jealousy and theft. Here in the marshes nobody stole. They all had what they needed and nobody had anything to lose, praise be to God. There was plenty of fodder for the buffaloes, there was plenty of fish to spear, there were fowls, and there were boat-loads of watermelon and braided reed-mats to trade for flour and tea in Madina. Moreover, and here the old sage raised his hand, there were beautiful women. He himself had four wives. All along the walls there was approving laughter at his virility.

The marsh women are indeed beautiful. That is probably why they were never permitted to eat with us or even to serve the tea. They were wrapped in black from head to bare feet, and as black shadows they glided by between the reed screens, feeding their chicken or baking flat bread clapped vertically on to the inner walls of cylindrical clay ovens open at the top. Their profiles were sharp and fine like their men's. Their sparkling eyes and white teeth shone like stars if one got a glimpse of them before they shyly turned their heads or pulled the black cloth over their noses. Like the men, they were fabulous paddlers and punters. I saw women alone, punting huge balams loaded with mats while herding a flock of swimming buffaloes. But only when they were tiny little girls or old crones could they stand with boys and grown men and laugh and wave at us as we slid by their abodes in our canoe.

A surprising number of the people were red-haired. Especially among the bareheaded little girls red hair seemed almost as common as black. Among the Madans in the ancient boat-building village of Huwair I had seen more red-haired people than in any town in Europe, so many that it could not be due to foreign intermixture, especially since the death penalty for unchaste behaviour or adultery is still the unwritten law among the marshmen. Hagi could confirm that during the British administration few foreign soldiers had ventured into the marshes and none to visit Arab women.

Hagi was well aware of the fact that we in the cities would not survive without a culture that was dependent upon automobiles and electricity. But he was not at all sure that his people would be made happier by projects to bring electricity cables into the marshes and bricks for building houses. When people are happy they smile, he said. Nobody had smiled at him in the streets of Baghdad.

'There are too many people in a city,' I explained. 'One cannot smile at everybody there.'

But Hagi had also walked in streets where there were few people and had seen no difference there. I could not protest. It could not be mere coincidence that the marsh people came out of their houses waving to us with broad smiles as we paddled by, their children racing to the water's edge with happy shouts and laughter. A faint smile from us and the whole assembly laughed happily back. Not so, I had to admit, when we walked in the poorer sections of a city. Old Hagi was right: the marshmen were not poor people.

The dignity of Hagi was not that of a poor man. With his manners and appearance he could have been a powerful oil sheikh, a former statesman, a retired scholar. But in his attire he looked more like a wise prophet or patriarch out of the Bible, timeless as the Sumerian reed-house above us.

Hagi did remember reed-boats. There were three types when he was young. Two were hollow like canoes or receptacles, asphalt-coated inside and out. These were the beautiful *jillabie* and *guffa* I had personally seen still in use higher up the Euphrates, above Babylon. The jillabie was like a slender canoe or mashhuf; the guffa was perfectly circular and looked precisely like a giant rubber tyre, but with a bottom, so steady that it did not tilt when I sat on the edge. I had never seen plant stalks worked with greater perfection in symmetry and detail, except in the reed-boats on Lake Titicaca, and now in the huge arches of cane that held the lofty ceiling above us. They too represented perfection. All equally spaced and identical to a fraction of an inch, beginning as columns thicker than a man at base on either side and gradually narrowing towards the apex of the roof, all of one piece from base to base. Hagi pointed to these stout and strong arches. The third kind of reed-vessel he had seen was built like these, except that the bundles got narrower towards either end instead of thicker. In that way many bundles lashed together would create a compact watercraft raised and pointed at either end. Berdi had to be used for the bundles, *kassab* only for the shelter on top. Berdi was a reed with a spongy pulp inside, but kassab was a hollow cane that would crack and fill with water. He had a boy fetch a stalk of each kind and showed me how the tender lower part of the berdi could be eaten, just as I had seen with the same part of the young papyrus. It was crisp and tasty.

To make the berdi-boat float for a long time, Hagi added, each bundle should be pressed as tight as two men could pull the rope ring around it; it should be as hard as a log. I felt once more the arches of his house. I could not press a finger in; it felt like touching a tree. In reply to my direct question, Hagi answered that he had never known of asphalt or other impregnation used on this kind of bundle-boat.

I had seen a single photograph of the kind of reed-boat Hagi now described to me. It had been published in the *Daily Sketch* of 3 March 1916, during the First World War, and the faded caption read: 'This is the kind of boat our men in Mesopotamia are constantly seeing.' It was strikingly similar to the reed-boats I had seen on Easter Island and on Lake Titicaca in South America, except for the Arab on board.

But for the information provided by Hagi's memory I had come to the former Sumerian territory half a century too late; Hagi was there as a bridge to the past. Looking at him I caught myself thinking of Abraham. In fact, he could very well be a direct descendant. All Arabs, like all Jews, begin their pedigree with Abraham, and after all, Hagi lived close beside Ur, where Abraham was born. In these Biblical surroundings even Abraham could not be overlooked by one who wanted to trace the beginnings, for he not only began both Moslem and Hebrew history but through him we have one of the earliest recorded descriptions of how the Mesopotamians of antiquity built their boats.

Abraham is recognised today as an historical personage who lived in Mesopotamia about 1800 BC. According to the Old Testament he was born in Ur where he left his kinsmen and followed his father's tribe and their livestock on their migration from the fringe of the marshes northwards to Harran in Assyrian territory, then across to Mediterranean lands. Although born in Ur, he went even as far as Egypt before he turned and decided to settle for good in his chosen land, leaving us an example of recorded overland contact between Mesopotamia and the Nile Valley in early antiquity. Although today we think that to man of antiquity Mesopotamia and Egypt must have been two worlds apart, they were not so remote from each other but that Abraham might claim that his descendants had been promised all the land 'from the River of Egypt to the Great River, the river Euphrates . . .'³

Today the river Euphrates and the green marshes have withdrawn half a dozen miles from the buried ruins of Ur, but the gigantic Sumerian temple-pyramid still rises out of the dust against the blue desert sky as a breath-taking monument to human enterprise and impermanence. This lofty stepped pyramid has been rebuilt time and again by successive cultures, but was already age-old when Abraham played around its base and bathed in the nearby river that for centuries had made Ur a port of paramount importance. In Ur's bustling harbour Abraham had come face to face with merchant mariners from foreign lands, and in the shade of the pyramid temple scribes and elders had shared with succeeding generations their knowledge of the past and their recipe for a happy after-life. From them he must have received the long history of his ancestors, which in turn he passed on to his own descendants until it was recorded in the Old Testament. He probably saw the boat models, some of silver and some of asphalt-covered reeds, which the priests buried as temple offerings from prudent sailors, and he must have been familiar with the kind of ships that docked along the local wharfs and river banks.

As Hagi sat there on the floor of his reed-house and described the building principle of the jillabie, its ribs clad with reeds waterproofed with bitumen, it sounded like a miniature of a famous vessel described in the Old Testament. Mention Noah's Ark and people will smile with happy childhood memories of the naïve story of a bulky house-boat and a gangway packed with pairs of elephants, camels, giraffes, monkeys, lions, tigers and other beasts and birds of all kinds, herded by a friendly old man with a long beard. As a boy, when I played with wooden animals parading into a wooden Ark, I never dreamt there was anything to learn from the old tale, still less that I should come to the homeland of the legend or study learned volumes which attempted to trace its origin.

The Hebrew version of Noah and the Flood, as known to us from the Old Testament, might have survived as oral tradition until recorded in Hebrew characters centuries before the time of Christ. It seems equally possible that a patriarch like Abraham was not illiterate, since he came from a city where script in varying forms had been in use for more than a millennium. However this may be, the Noah myth dates back to a remote period in human history, antedating the spread of civilisation from the Middle East into Europe. The ark described was not a European ship but a Mesopotamian watercraft. The story is one of joint Judaic and Moslem belief and allegedly came with Abraham, who grew up in Ur. The migrating tribe of Abraham could not have avoided passing through the Assyrian kingdom in northern Mesopotamia, and there is even reason to believe that they spent some time there.

By then the Assyrians, too, were well familiar with the story of the flood that had destroyed the majority of mankind. The vast library of the Assyrian King Assurbanipal, which consisted of tens of thousands of inscribed tablets, was found in 1872 to include a detailed version of the Universal Deluge. It is so similar to the younger Hebrew version that both must clearly have had a common origin. Since the Assyrians acquired their writing system as well as their mythology from the Sumerians, and since the Hebrews claim to have come from the former Sumerian capital, Ur, it would be reasonable to suspect that Sumer would be the common source of the Assyrian and Hebrew Deluge stories.

In the Assyrian text the old man who built the ship to save mankind is referred to as King Utu-nipishtim rather than Noah, but both names are probably allegorical. The ocean-god Enki also takes the place of the monotheistic Jahve of the Hebrews. The Assyrians pretend that the story of the Universal Deluge was told to one of their ancestors by the boat-building King Utu-nipishtim himself while he was still alive in Dilmun. He claimed that the ocean-god took a liking to him and revealed the secret conspiracy of the other gods to drown all mankind. It was the ocean-god who had told him to build a large ship and to take on board his family, his attendants, and his livestock. King Utu-nipishtim followed this advice and built the ship:

'On the seventh day the ship was ready. As the launching was heavy, rollers were used . . . With all my property I loaded the ship, with all my silver I loaded it, with all my gold I loaded it, with all my living seed I loaded it. All my family and servants I brought on board, the livestock, the beasts of the field, all the craftsmen I brought on board . . . To the master of the ship, the captain Puzur-Amurri, I entrusted the large structure and its cargo.'[4]

For six days and seven nights the flood raged over the land, and on the seventh day the big ship grounded on a mountain top in upper Kurdistan, the region where the Hebrews had Noah landing on Mount Ararat. The Assyrian text says that a dove and a swallow were in turn sent out, but returned, and only when a raven was let loose and never came back did the King realise that the waters had abated and he was on safe ground with his followers and live-stock. They all disembarked and offered sacrifices to the gods, who promised never again to punish all mankind for the sins of some.

It is interesting that in this Assyrian text the survivors from the big ship were told to go and 'dwell in the distance, at the mouth of the rivers'. To the Assyrians this meant the mouth of the Euphrates and Tigris, the former Sumerian territory. In other words, the Assyrians recorded that the gulf coast and marsh area of southern Iraq was the part of their world first resettled by the new genera-tions of mankind.

It would therefore be particularly interesting to know what the older Sumerians themselves had to say in this connection. Their version of the same event was discovered subsequently by a team of archaeologists from the University of Pennsylvania. In the general area where the rivers meet they were excavating the enormous Sumerian ziggurat of Nippur, one of the sun-oriented and stepped pyramids with temple at the top which was the main feature of all early Mesopotamian cities, when they hit upon another well-stocked library. At the foot of the pyramid they found a collection of 35,000 inscribed tablets, and one of them contained the original Sumerian version of the Universal Deluge.

This Sumerian record, unlike the younger Assyrian, and the still younger Hebrew writings, does not say that the survivors of the flood landed on any inland mountain top, but that after the flood mankind first settled in Dilmun, somewhere across the sea towards the sunrise. Later the gods led them to their present abode at the mouth of the rivers.

A clear indication that the Assyrian text was only borrowed from this older Sumerian original is seen in the fact that both refer to the Sumerian ocean-god Enki and give him the credit for having saved mankind. Also in the Sumerian original Enki's divine choice fell on a pious, god-fearing and humble king of an unidentified kingdom, but in the Sumerian language he was referred to as Ziusudra. Here, too, the god 'advised him to save himself by building a very large boat'. The part of the tablet describing how Ziusudra built this large boat is unfortunately destroyed, but at least it was big enough to carry his livestock in addition to his family. Once they were all aboard, the deluge raged over the surface of the earth for seven days and seven nights. 'And the huge boat had been tossed about on the great waters', when finally Utu, the Sumerian sun-god, came forth and shed light on heaven and earth. Then Ziusudra 'opened a window of the huge boat' and prostrated himself to the sun-god. He then sacrificed an ox and a sheep, which indicates that he must have had more than one pair of each kind on board, as distinct from Noah. But then again he carried no wild beasts. In short, during the millennia, the original version of survivors, who only carried their domesticated animals on board, had been slightly embellished until Noah also saved the beasts of the wilderness.

The essence of all three versions is their reference to big ships. They all speak of domesticated animals and even refer to the existence of cities and kingdoms before the flood, but none of them suggests that a tall pyramid or tower saved mankind and his herds from the inundation. Five thousand years ago scribes put on record what today is the oldest known attempt at written history. It begins with families and livestock, after some catastrophic event, landing with a big ship at a place called Dilmun and from there reaching Ur in Mesopotamia, also by sea.

It is regrettable that the tablet is broken where the building of the big vessel is described, but since Ziusudra and Utu-nipishtim are clearly two names for the same royal ship-builder, we may draw some inference from the Assyrian version. In the earliest epic ever rediscovered, the Assyrian poet sent his hero, King Gilgamesh, by boat to the ancestral land of Dilmun, where the long-living King Utu-nipishtim tells his own story of the flood. He first introduces himself as the son of King Ubara-Tutu, who ruled in Shuruppak before the universal disaster, then he alludes in poetic terms to the words of the ocean-god who told him how to build the ship:

'Reed-house, reed-house. Wall, wall. Reed-house, listen! Wall listen! Man from Shuruppak, son of Ubara-Tutu. Tear down your home and build a ship!'

Obviously, by tearing down a reed-house one could only build a reed-ship. This is also in conformity with the Hebrew version. The instruction to Noah was: 'Make yourself an ark with ribs of cypress; cover it with reeds, and cover it inside and out with pitch.'[5]

Even the Assyrian epic hints at their boat-building king using some covering on his reed-ship. It was none less than Utu-nipishtim who said, after tearing down his royal reed-house: 'six *sar* of pitch I poured into the melting-pot, three *sar* of asphalt I added. Three *sar* of oil were brought by the mixing crew, apart from one *sar* kept in the hold and two *sar* hidden by the captain.'

The boat-building principles accredited to Noah were indeed those used in miniature in the building of the jillabie described by Hagi and seen by me with six men on board near Babylon. Naturally, the dimensions of the royal vessels of antiquity would have been in proportion to the sizes of the other structures built in the days of the totalitarian kingdoms. The Assyrian epic gives the dimensions of King Utu-nipishtim's ship as one '*iku*', a field measure which equals the ground plan given for the Tower of Babel.[6] Although this measure is hardly to be taken literally, it would nevertheless have been easier to build a structure that big out of long reed-bundles freely harvested in the vast marshes than out of small bricks each moulded and baked in brick ovens.

The Hebrews were rather more modest in their measurements; they recorded that the vessel was 450 feet long, 75 feet wide, and 45 feet deep, which was only four times longer than Matug's makeshift reed-gáré which I had measured myself.

The Assyrians probably added something for the benefit of readers when they wrote that Utu-nipishtim's ship was built with nine inner compartments and 'six superimposed decks'. Again the Hebrews were more modest: the Ark was supposed to have 'three decks, upper, middle and lower'.

Clearly the Assyrians, and also the Hebrews before they left Mesopotamia, had seen big ships. Otherwise they would not even have been familiar with the concept of vessels with more than one deck. Nor should we underestimate the ability of the Sumerians to build giant structures of reeds, when we know that they built real mountains of sun-baked and oven-baked bricks, so huge in fact that we would most definitely have thought it impossible in those days but that the structures are still there to stupefy us, like the pyramids in Egypt. That civilised societies in the Middle East were familiar with extremely advanced ship-building five thousand years ago should no longer surprise us after the discovery of Pharaoh Cheops' truly astonishing vessel, one that was much larger than any Viking ship and had been built a thousand years before Abraham came to Egypt. In fact, if Abraham and Sarah had seen it in its hidden crypt at the foot of the Great Pyramid, it would have been as old to them as the Viking ships are to tourists in Norway today.

The extensive Danish excavations on the gulf island of Bahrain have a direct bearing on the original flood myths. The cities uncovered were interpreted as the first concrete confirmation of Bahrain being the Dilmun of the Sumerian merchant records and the alleged land where the Sumerian ancestors settled after the flood. The prominent Danish archaeologist P. V. Glob,[7] in summarising the results of the first fifteen years during which he led the Bahrain excavations, supports a widely held view as to the origin of the Sumerian flood legend. Beneath Ur, the royal city of the Sumerians, Leonard Woolley found in 1929 a layer of homogeneous mud, ten to thirteen feet thick, of a type deposited by water. Under this again were discovered the ruins of the first city which was there when some gigantic flood wave had buried all lower Mesopotamia under twenty-five feet of water until the flood subsided. To the few survivors this would have seemed to be the destruction of the world, and the memory of it would have continued until recorded on Sumerian tablets. Glob assumed that the few survivors might have saved their lives by climbing the highest walls of the inner city. But why, I thought, why, when the city was a port and probably full of reed-ships?

I looked around me. Reed-house, reed-house. Walls, walls. My thoughts had wandered. Certainly Hagi could have made a nice ship by tearing down the big reed-house we sat in. Actually, if turned upside down, this hall could become the hull of a roomy ship ready made with solid reed ribs; Hagi would only have to cover it with pitch or asphalt, inside and out, the moment both ends were closed.

Nobody planned to tear down Hagi's reed-house. I was delighted to find it there, as good as ever, when I came back after an interval of five years, although I missed the old reincarnation of Abraham. But his sons were there and gave me a royal reception. The Biblical setting was still there, the descendants of Terah, Abraham's father, kept up most of the old traditions.

Sha-lan, Hagi's oldest son, and all the men in his house, got excited when I told them I had come back to find people in the marshes who could help me harvest berdi and build a bundle-boat like the ones old Hagi had described to us five years earlier. I needed twenty men. Sha-lan immediately assured me that he would choose them himself. No problem. After a short discussion among those present, Gatae was thought to be the best man to lead the work. Gatae was a master reed-house builder and therefore would know how to make perfectly tapering bundles.

Gatae was fetched in a mashhuf and proved to be a fine elderly marshman with a humorous twinkle in his eyes. Tall and slim, he stood in the canoe as straight as a mast, his checkered caftan fluttering. With his dignity and his trim white beard, he reminded me of my late English publisher, Sir Stanley Unwin. We met as if we had been friends for a lifetime. Gatae was not in the least surprised that I wanted to build a reed-ship and sail away into the sea. He went straight to the point: How much reed did we need?

I paced the floor of the hall. *Ra I* had been 50 feet long, *Ra II* only 39. This time I wanted a larger crew and thus a ship 60 feet long, just about the length of the one-room reed-house we were in. But I had learnt from Hagi that the spongy reeds had to be compressed, so I needed more reeds than the final volume of the ship.

'We had better cut twice as much berdi as what would be needed to fill this house from floor to ceiling,' I estimated, and we all looked to the vault high above our heads. Gatae was not impressed. He would make the bundles any size I wanted.

We agreed that twenty men under Gatae's leadership should come to Adam's tree and begin the building in September, but I should first return in August and see that the reeds were cut and properly dried by other Madans in the village of Al Gassar, closer to the building site. There the government had built an elevated dirt road right to the edge of the marshes so that canes could be delivered directly from the mashhufs to the trucks that brought them to a new paper-mill on the river Tigris. From that point I could later bring the sun-dried reeds to the building site on the river Tigris, next to the Garden of Eden Resthouse.

CHAPTER 2

In the Garden of Eden

AUGUST CAME, and I was back in the marshes again. August is the hottest month in southern Iraq. The thermometer wavered between 40° and 50°C (105°–120°F) in the shade, but there was no shade anywhere in the open swamp-land where we cut the reeds. The Marsh Arabs advanced with curved machetes into the reed thickets with the speed and energy of a band of warriors and the long green stalks fell like slaughtered troops. The heat was so great that I soon became exhausted merely watching the battle from the canoe, and as my marshman interpreter, with a vocabulary of a few dozen English words, assured me that there were no longer any bilharzia in the water, I jumped into the canal and joined the Madans, who were waist deep and fully dressed. From Lake Chad and the Nile I had learnt to dread the little bilharzia worm that lives on snails in the reeds and drills its way through the human skin in a few seconds to multiply inside the body. I enjoyed the slowly running water until a beautiful snail shell came floating by. I picked it up and hesitantly showed it to my reed-cutting informant.

'That?' he said. 'That is only the *house* of the bilharzia.'

I was back in the waterproofed mashhuf in one leap. Better to sweat than wade in a stream with bilharzia worms.

I lost all count of the number of mashhufs towering with green berdi which the men and women of Al Gassar punted through the channels to leave on the banks of the marshes to dry. It looked as if I was planning to build a ship every bit as big as Noah's.

In the meantime I had to return to Europe for a few weeks to organise the expedition; at the Garden of Eden Resthouse I could organise nothing. In Baghdad I had managed to grab an ivory-coloured telephone in my hotel room and speak to Oslo, Tokyo and Sydney in a few minutes. But the telephone in my Eden was a leftover from an early administration, a relic to be cranked like an old Ford. If I finally got through and thought I had London, the whole resthouse joined me to listen and shout into the mouthpiece until we learnt that the faint squeak we heard was from the poor operator in Qurna across the street, who desperately tried to tell us that the line to Basra was broken, so nothing doing today. An English engineer came all the way from Basra to comfort me with the news that he was entrusted by Baghdad to put up a modern line. Ready next year. Good news! – for those who were to come long after we had gone.

The oil boom which had initiated a building explosion and the massive importation of all kinds of goods from the outside world, also enabled all imported products to be bought up as soon as they arrived, and the vast fleet of tankers and cargo ships that entered the river mouth were far too many for the port facilities. Ships of all nations were anchored in the open bay for two, three, or even four months, waiting for a turn to come up the river and deliver their cargo. Everything from Indian timber and bamboo to Danish butter or American frozen chicken would be swept away from the lumber yards or grocers' shelves before I could lay a hand on it. If I ordered anything to be sent to Iraq by ship it might get stuck at anchor in the gulf. I saw only one solution. Everything I needed for the expedition had to be assembled in one place and from there sent overland to Iraq by chartered road transport.

So I chose Hamburg. In three days German friends helped complete the purchase of everything required, apart from the berdi reeds. A ropery set aside its orders for nylon cables and twisted many miles of assorted hemp-rope for our bundles and rigging. Two thirty-foot ash legs for our straddle-mast were hewn to shape by a genuine boat-builder of the old school, who also hand-carved two twenty-five foot rudder-oars and a dozen rowing-oars with extra long shafts that could be sawn progressively shorter as the tall reed-ship settled deeper in the water. An equally genuine and conscientious sail-maker hand-sewed two square sails from Egyptian cotton canvas; they tapered from top to bottom as in pre-European times. One was bigger and thinner than the other, intended for good weather only. We needed bamboo for the super-structure. A rain collector was also required, flags and signal lights, kerosene lamps for illumination, primus-stoves, and pots and pans for the kitchen. Also fishing gear. And a tiny inflatable rubber dingy with a 6-hp outboard motor for the cameraman to film us at sea.

The shopping tour ended with food and all daily requirements, including jerrycans with drinking water for eleven men for three months, trusting that additional supplies could be obtained during the journey. We had tried, on an earlier occasion, to eat only what people of antiquity could have stored on board their vessels. On the *Ra* voyage we bought only such food as could be stored in ceramic vessels and baskets, and all our water was kept in jars and goatskin bags. In this way we had survived the voyage without any dietary problem, so it was unnecessary to repeat the experiment. Even so, on a reed-ship without electricity for a fridge there was a very definite limit to what we could store on deck. Fresh meat, fruits and vegetables would not keep, and even most canned foods would go bad in temperatures such as we could expect before the start and after. Nevertheless, there were tons of provisions and equipment, sufficient to fill a forty-foot transport trailer, sealed after approbation by the Iraqi Embassy in Bonn and prepared for the two weeks' drive from the freeport of Hamburg to the very doors of the Garden of Eden Resthouse in southern Iraq.

From Hamburg I flew to London to meet the representatives of an international consortium of TV companies improvised for the occasion by the BBC. After much brainwork, typing and retyping, a thirty-one page contract was signed, obliging six television organisations in Great Britain, France, Germany, Japan, Sweden and the USA to finance a reed-ship expedition by buying four one-hour television programmes not yet filmed of an expedition not yet undertaken. The contract was made more difficult by the fact that I could only say where the voyage would start but not where it would go nor how long it would last, as these were the questions we sought to answer ourselves. This obstacle was overcome by the wording that we should sail as far as the vessel could be navigated or kept afloat above water. The American member of the Consortium, the National Geographical Society and their television producers, WQED, further insisted on sending with us their own cameraman with a special camera, who should be free to record everything and anything done and said on board. He would have no duties but to film, even if the vessel sank. I agreed. We all signed.

With reluctance I also went with the BBC representative to the University of Southampton, where a six-foot plastic model of the reed-ship had been built according to my own drawings. It was to be tested and filmed in a wind-tunnel as well as in the sea. It was beautiful to see the big yellow model bobbing in the waves while the nautical experts from the university pressed buttons for long distance control that made the rudder-oars twist and the little vessel turn and roll sideways to the waves. It would even cut into the waves when the total area of the oar-blades was increased, and achieve a real tack.

The lesson of the experiment was that the bigger the sail and the more or the bigger the oar-blades put into the water, the better the twin-bundled raft-ship tacked into the wind. The final result from the wind-tunnel studies would be mailed to me later. There were only a few slight hitches: the model was made of plastic and not reeds; no one knew how deep the reed-ship would sit in the water, nor how fast the changing buoyancy would decrease her original freeboard. Besides, the delivery of the model had been delayed and by the time we got the answers from the wind tunnel it would be too late to change the measurements of the sails and oars already cut to size to meet the transport deadline from Hamburg.

Just outside Southampton was the beautiful Broadlands estate, the home of Admiral of the Fleet, the Earl Mountbatten of Burma, where I was expected for lunch. Common interests in the sea and other means of bridge-building between nations had made us friends in recent years; his enthusiasm had led the old Sea Lord to make me honorary Vice-President of the United World Colleges, of which he was a very active President. As a member of the Royal Family and former Viceroy of India he had friends and contacts all over the world and had succeeded in getting students even from China enrolled at Atlantic College in Wales, one of the units I had recently visited. Apart from bringing together bright boys and girls from all nations, the United World Colleges put a good deal of stress on marine life-saving and on boating as a sport.

I had promised Lord Mountbatten to bear in mind some of the graduate students if I ever again had plans of maritime experiment. I had now just fulfilled my promise. I wanted a truly multinational

1. In a Marsh Arab reed-house. *(opposite)*
2. Floating reed islands and river banks of southern Iraq. The boat is still to the Marsh Arabs what the camel once was to their neighbours. *(p. 42)*
3. Women bring berdi reeds for our reed-ship. *(p. 42)*
4. A wooden jig was built to give correct shape to the ship and to hold the reed-mat made by South American Indians to envelop the forty-four bundles prepared by the Marsh Arabs. *(p. 43)*
5–7. Ship building on the banks of the river Tigris as in the days of the pyramid builders. *(pp. 43–5)*

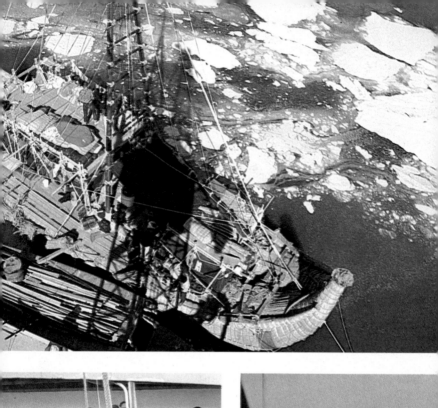

crew on my planned voyage from Iraq. The late Secretary-General of the United Nations, U-Thant, had granted me the right to sail *Ra I* and *Ra II* under the UN flag, and his successor, Kurt Waldheim, had kindly repeated the permission for my forthcoming expedition. I had informed the United World Colleges that I wanted to give priority to any of the crew of my previous reed-boat experiments who cared to come along again, but I was prepared to draw reserves from the United World Colleges. The headmasters of the various colleges had therefore sent out a circular to former students, enclosing a copy of the press release concerning my plans that had just been circulated by the BBC Consortium. The teachers would send me a list of recommendations selected from those graduate students who applied.

It was great to be back at Broadlands and enjoy a meal in the bright dining-room facing the park and its old giant trees. A tiny tree, barely visible above the grass, I had planted myself on my last visit, in accordance with the Admiral's own tradition.

'Thor, would you believe that you have got me into international trouble?' Lord Mountbatten looked at me sternly as we took our seats at table in the company of his adjutant and were served melon by a butler in a naval uniform.

'Of course I believe that,' I replied, laughing to show that of course I did not believe.

'You have brought the anger of the Shah upon me,' my host continued calmly.

I laughed again and enjoyed my cool melon. 'Of course, of course!'

Lord Mountbatten stopped eating and looked at me: 'What can I do to make you realise that I am not joking?' He sent his adjutant to fetch a letter from the Imperial Court in Teheran. It was a long and sharp protest against the wording of the circular from the United

8. The reed-ship *Tigris* sails down the Shatt-al-Arab on its way to the sea. *(p. 46)*
9. The legendary Garden of Eden seen from the building site where the rivers Euphrates and Tigris meet. *(p. 47)*
10–12. Chemical pollution from a paper mill threatens to dissolve the reeds of *Tigris* before reaching the open sea. *(p. 47 & opposite)*
13–14. Entering the gulf at the river's mouth as the wind died down, we were left adrift among ships of all nations, until a strong onshore wind blew up and forced us to sail towards Kuwait. *(opposite)*

World Colleges concerning my planned reed-ship expedition. The Iranian Ambassador to Great Britain had telephoned Lord Mountbatten from London before the letter arrived, immediately the press release from the BBC had been broadcast on radio and television. I was quoted as planning to sail down the river Shatt-al-Arab into the 'Arabian Gulf'. How could an international college institution headed by the Admiral of the Fleet use such a fictitious geographical term? Was this the result of the growing tendency to flatter the Arabs? The true and only name for this body of water was the 'Persian Gulf', and it was indeed the British Admiralty that had originally given the gulf this proper name.

I was sorry. This was an unforeseen problem for a reed-boat voyager who had to test an ancient vessel in a modern world. I explained to Lord Mountbatten that I had indeed originally used the name the 'Persian Gulf', which I had learnt in school. But officials in Baghdad had corrected me and my message and made it abundantly clear that if I wanted to sail anywhere from Iraq it had to be into the 'Arabian Gulf'.

Lord Mountbatten saw my problem but objected even to the term 'the Gulf', used diplomatically by many shipping people to distinguish this Old World gulf from the Gulf of Mexico. His ultimate solution was that the place should not be referred to at all. However, though I did not want to insult anybody, I could not sail down the river into nowhere, so I contacted the Norwegian Foreign Office. Their general practice, they told me, was to speak of the 'Persian Gulf' when referring to a port on the Iranian side and to the 'Arabian Gulf' when the port was on the coast of some Arab nation. However, since our experiment was to sail in open water I could not refer to ports, so I contacted a public relations officer of the United Nations.

'It is a considerable problem,' he admitted. 'All the nations around the gulf you are sailing into are to have a meeting against pollution of their common waters, but we cannot agree on a word to explain where it is!' This was a forewarning that should have told me that the twentieth century, with all its radar and lighthouses, is not the easiest one for avoiding hidden reefs.

I had hoped for some truly exotic candidate from the United World Colleges, and was a bit disillusioned to find two young Scandinavian candidates heading the list: a Norwegian medical student and a Danish student of mathematics. As a former sergeant in the Engineers, the Norwegian had the rare advantage of being specialised in rope-work and bridge-building, which was just what

I needed, for the ship would be all ropes and reeds. Even the masts and the cabins would be lashed on without a nail. Moreover, a bridge-like structure of sticks and poles was immediately required as a cradle to facilitate the assembly of the reeds during the building, and permit the final launching into the river. I telephoned the young applicant in Norway, and Hans Peter Böhn, who insisted on being called HP, joined me in Rome with a rucksack and camera, and together we flew back to Baghdad.

After seven hours by car we reached the Garden of Eden Rest-house a day later than anticipated. I was to have a rendezvous with the two truck-drivers from Hamburg any day now and wanted to be on the spot when all the expedition's food arrived, so as to get it quickly into the shade of the resthouse before it was spoiled by the baking sun.

At the steps of the resthouse we met to my surprise a young European with his face red from blisters, carrying an armful of huge German smoked sausages, just like those I had bought in Hamburg, which he dumped into the river Tigris. His eyes blinded by perspiration, he hardly noticed us as I introduced myself and asked what he was doing.

'Disaster!' he said. 'All the food is spoiled. The truck has returned to Hamburg without me and I am alone to dump it all into the river!'

In one jump HP was over the terrace rail and came out of the water with a giant sausage in his arms like a baby. The last one. All the others were gone with the current and there was none left in the storage room, where the young perspiring German was already back to dig out a smoked ham and a carton of cans of Norwegian mackerel to throw in after the German salami.

I realised we had come at the eleventh hour. Like a robot, the stranger went on with his dumping until we made him sit down with us on some of his unladen rope coils to explain the great idea of bringing food from Hamburg to throw into the river Tigris.

'The customs,' he said. 'The customs.'

Ali, the friendly boy of the resthouse, had to run for a cool can of beer before we could get any sense out of his wild story. He was a special envoy of the excellent Montan Transport in Hamburg, sent along with the two truck drivers to see that they had no customs problems and to ensure a speedy unloading directly into our shaded store room in the resthouse. It was hot even in Europe, and the truck had no cooler, so he made the drivers race to cut down on the two weeks estimated driving time. In this he had had more success than expected. In southern Turkey, Kurds had ambushed them and started to shoot. They had accelerated. There were bullet-holes in the trailer when they caught up with a whole convoy of transport trucks driving desperately to reach Iraq. In the confusion they had not got clearance from the Iraqi border customs, who were awaiting them with special orders from Baghdad to break the seals. Instead, they had rushed on southwards into ever hotter conditions down the whole river country, past Baghdad, past the Garden of Eden, ending up beside the ships in the over-crowded port of Basra, where nobody had orders to help them. A policeman showed them the road until they suddenly found themselves locked up in the big customs yard of the harbour. There was no place to park in the shade. It was 45°C (113°F) outside the trailer and 70°C inside, according to the shattered envoy. The three of them were almost dead by the third day in the customs yard; even the big bamboo they brought for me had started to burst from the heat and it was like sitting on a cargo of firecrackers. Then a friendly soul who spoke English had helped them telephone to the Ministry in Baghdad, and this had ended the confusion and cleared them and their cargo from their prison.

As he spoke we were startled several times by violent explosions around us: the thick bamboo was still drying up and cracking with the sound of gun-shots.

No wonder the two truck-drivers had hurried home and left the special envoy to clear the rest of the mess, which he did by dumping all our comestibles into the great river. The cheese had gone first: 40kg of select Norwegian Cheddar, 24kg Edam, and a variety of other types estimated to resist modest heat. Leaking cans of liquid soap and melted butter followed, and assorted smoked meats. He had barely had time to liquidate 20kg of specially prepared polony salami when we arrived.

HP sat with a knife shaving the fresh beard of mould off the polony he had saved, and savoured a piece with delight. In one leap he was back into the river. Nothing more to be found. We were never to taste another one like it. But with the one he saved, throughout the building period, we and all our guests at the resthouse had the best sandwiches of the sort any of us had ever tasted.

The cans, however, scared even us. They were no longer cylindrical, but round like cannon-balls, normally a sure sign of poisonous contents. I filled a sack with one can of each kind and sent them with the envoy by air from Baghdad to Hamburg for a laboratory test before we would consent to discard all this costly expedition food. A few days later our entire collection of cannon-balls had resumed the shape of normal cans; they had merely swollen in the immense heat. Those poor chaps from Hamburg had been sitting on top of them while they expanded!

We had to replace all the lost supplies, and shortage of time now forced me to order shipment by air. The shippers promised this time to send it to the airport in Baghdad, and inform the Ministry of the waybill number and hour of arrival, for the Ministry had offered to forward it direct to the Garden of Eden by road. The day of arrival came, the heat was still with us, and the shipment disappeared. Hamburg confirmed that the food had been sent, Baghdad airport reported that it had never arrived. After a week of worries I got the crazy idea of sending a messenger to the same magic customs people in the harbour city of Basra. Three days later the indefatigable emissary came back in triumph; he had located the lost replacements.

HP and I occupied two of the three nice big bedrooms in the resthouse, and the many friendly servants were all at our disposal. Ali and Mohammed were excellent, and even knew some English. Good Arabian meals were served us alone in an empty restaurant hall, so large that we planned to build the reed-bundles indoors if the heat continued into the next month. We were up the moment the red desert sun arose across the Tigris, and the morning air and first shower in the bathroom were refreshingly cool. On the first morning I lay down in my tub and let it overflow with the coldest water I could get; until two long brown antennae began to waver out of the overflow, and two expressionless eyes peeped at me. A huge cockroach dived out of the hole into my bathwater, followed by another, and another, and another: they were twenty-two in all. With this swimming armada I was out of the water in a second and tried to flush them all down the drain, but they were too big and I had to scoop them with a glass into the toilet. I woke HP and suggested he should take a cool bath. I waited a while and then heard a yell through the wall. 'You know what happened to me?' he asked as we sat down to a grand egg and cheese breakfast on the lawn. 'Yes,' I replied. 'How many?'

There was plenty of work ahead to prepare the building site for the arrival of all the groups of helpers: Marsh Arabs, South American Indians, *dhow* sailors from Bombay, and expedition members from all continents. The three bedrooms would not suffice, but two rooms could be made serviceable upstairs under the roof, and mattresses and camp beds could be put up everywhere.

With HP I began to prepare the ground for the combined jig and building scaffold. I had the idea of digging two deep and broad trenches side by side in the garden, so that the boat-builders could walk underneath the double-body of the vessel when the thick spiral rope was to be wound around the two final bundles. Half a dozen Arab workmen from Qurna came with picks and shovels and they soon began uncovering big, square, yellow Sumerian bricks which they unperturbedly carried away in baskets on their heads to dump. My archaeological conscience made me stop the work in the Garden of Eden immediately, until HP stooped down and picked up a tiny living tortoise among the débris. Then I too noted something: between the truly ancient bricks lay the neck sherd of a beer bottle, showing that others had ploughed up the whole area before and filled in irregularities to level the ground for the building of the resthouse.

The digging was resumed, but on the third day, just as we began to see a satisfactory result, a committee of solemn gentlemen in European dress came and to my surprise began to measure our trench. Soon afterwards one of the workmen spoke in Arabic to Kais, my young interpreter, who sat with me in the pleasant shade of a date palm, wondering what was going on.

'They say we should dig two metres closer to the road,' said Kais, translating the message.

'No,' I protested. 'We should keep as close as we can to the river.'

The workman left and the digging was resumed, but soon he was with us again. 'They say our marks are not in line with the resthouse!'

'Nonsense,' I explained. 'They don't have to be. You are digging just right.'

The workman left again, but as I saw gesticulations around the trench I went over to calm everybody down and clear up the obvious misunderstanding. A little friendly man with a prominent nose, who knew English well, introduced himself and his companions and with both hands he showed me in a very friendly way where he wanted the trench to go.

'It makes little difference,' I admitted, 'but the closer to the banks the easier will be the launching.'

'Launching?' he said, and he looked at me as if I had escaped from an asylum.

'Of course,' I laughed. 'You don't expect me to leave the ship ashore?'

'Ship?' Now he really showed big eyes and an open mouth. 'This isn't going to be a ship!'

It was my turn to suspect that the little man in front of me had crawled through the fence of some institution. 'Call it a haystack if you wish, but to me it will be a ship,' I said.

The little man stepped back and looked at me with profound suspicion: 'You make fun of me. Sorry, sir, but I have my orders from the Ministry of Information!'

'That is the Ministry that granted me the right to build here,' I answered, and began to suspect some real confusion. The little man now looked really unhappy: 'Sir, please, I have the masons and the carpenters all ready to start tomorrow, we shall add twenty-five bedrooms to the resthouse just here.'

Obviously we had dealt with two different offices in the same Ministry. We had to disentangle this problem.

'If you build your house now,' I replied, 'I cannot build my ship here afterwards. But if I build my ship now we will sail away and you can begin building your rooms two months from now.'

For a moment the little engineer looked at me in despair, then he pointed to the top of the palm tree above our heads: 'Do you see that date palm? You will find me hanging with a rope around my neck just up there if I go back before I have done my job!'

We all began to laugh as he capitulated, and as Mr Ramsey had come with all his luggage I agreed that he could stay in the resthouse where the easy-going manager gave him a spare bed in his room. Thus, in a sense, an engineer I had not asked for had half-joined the expedition as a local consultant whom we should discover we really needed.

About this time another extremely friendly and polite middle-aged man turned up, also with his suitcase. Mr Shaker al Turkey, appointed by the Museum authorities to be my local guide and liaison officer. This was just when the second shipment of food had gone astray, so, happy suddenly to have two interpreters, I sent Shaker to Basra, his own home town, to search for the lost consignment. No sooner had he left before Baghdad managed to get through on the phone: the Ministry had learnt that I now had two interpreters, so Kais was immediately ordered back to the capital.

Kais hated what he had termed the wilderness and left that same night. Thus I was left with no interpreter at all when an army of Arabic-speaking truck drivers and marshmen knocked at my door next morning. Not even Ali or Mohammed was around. The Arabs showed me that the road outside was lined with lorries laden sky high with dry golden berdi. Here they all were, all waiting for me to explain what to do. I ran and opened the iron gate to the Garden of Eden where the trench was dug. The lorries could barely pass between the brick pillars flanking the entrance. I have never seen so many trucks, trailers and bulldozers as in Iraq; the government imported them by thousands at a time, and when the Mayor of Qurna got orders from Baghdad to help me fetch the berdi from the banks of the marshes near Al Gassar, he sent a battalion of trucks so as to have the job done in one day. As they all began their shuttle operation, I had up to nine lorries at a time around me inside the narrow fence, trying to get rid of their loads, all blocking the passage for each other, and in a struggle to get in and out driving over the brittle stalks unloaded by others. I ran between the drivers, who shouted angrily to each other in Arabic and smiled happily at me, ignoring with good conscience all the orders I gave in various European languages that were all Greek to them. When the evening came I was exhausted from waving and pointing and stumbling among the reeds. At sunset I finally found myself alone with the silent river, beside Adam's tree, both gate-posts having been broken by the trucks and the whole garden a dense chaos of reed piles, leaving no place to build a ship or even place a foot.

No sooner was all the sun-dried berdi there before Gatae and his chosen men showed up next morning as if by magic. Gatae was a bright personality, and language problems were no obstacle. With Mohammed as a sort of interpreter, I had the marshmen make me the first tightly-bound reed bundle sixty feet long, which they did with astonishing dexterity and speed. But to my surprise the result, very much thicker than a man could embrace, was so heavy that Gatae estimated that possibly eighty men would be needed to lift it and carry it to the intended scaffold. We clearly needed many more but much thinner bundles to make up the final ship.

When Shaker came back from his successful mission we set all men to work for two days assembling the berdi within convenient reach of the building site, piling it into parallel stacks as high as a man could reach and with ample space to walk between. All the broken stalks were thrown on the banks and those that were not carried away by old women for kitchen fires were in no time turned into a flotilla of reed rafts filling the river with jubilant boys and girls.

Material to build a wooden jig, a temporary cradle for the ship, was not easily located in Iraq. Through the earliest works of art and inscribed tablets we know that Sumerian territory was originally covered with forest, but that these were gradually destroyed by man in antiquity so that timber became a major import, judging by the cuneiform records listing ships' cargoes. The price of imported timber is today so high that we were delighted to find a modest lumber yard near the Basra docks where long natural poles and rods from forests in the northern mountain regions were available. We needed them in hundreds and selected those with least bends and twists. HP, as expert army bridge builder, succeeded in raising the barkless poles and sticks into a sturdy crisscross framework, measured and designed to give size and shape to the reed-ship when the bundles were assembled inside it. A serious problem was the lack of gang-planks for the high scaffold. The main body of the final ship should consist of two compact reed cylinders each ten feet in diameter amidship, getting narrower as they curved up in bow and stern to the height of about twenty feet. We had to reach this height by means of the combined jig and scaffold.

Aladdin's Lamp must still have been working for us, for two truck-loads of used scaffold planks and crate boards were dumped outside the fence for our use just when they were needed most. Indeed, strange rumours had spread up and down the river to other Europeans temporarily at work in the country. The gossip was that we were erecting building scaffolds in the resthouse garden and had already bought tons of reeds, so we were certainly about to set up another paper mill! Far up the river was a German paper mill actually under construction, and the Germans who came down to look at us learnt that we were about to build a reed-ship to sail away with one of their own countrymen on board. Far down the river, below Basra, was a barely finished Danish cement factory, and the Danes who came up learnt that one of their compatriots was also to come with us. The result of the German and Danish visits were two stacks of boards and planks, which are worth their weight in gold in Iraq today, even as in Sumerian times.

HP's wooden jig was a master construction that merited permanence for its architectonic perfection and elegant lines. We were just about to turn our attention to the actual reed work with the berdi beautifully stacked within easy reach, when another committee of European-looking Arabs marched in with tape-measures and began pacing about between our stacks.

I smelt new problems and approached the party politely. Indeed, they were about to build a fountain, and the berdi had to be moved, since we had placed them just where the fountain was to stand.

'A fountain?' I said, 'but can't the fountain wait until we have finished the ship?'

No, the fountain was needed now.

'But there is a big fountain out of service between the trees just across the road,' I said. 'Can't you use that?' I pointed to a large wreck of a structure, dry as a bone, with rusty tubes and spouts fifty yards from our fence.

No, the fountain had to be right here, and they generously gave us a week to clear the centre of the garden. Our stout friend, the Mayor of Qurna, explained to us in a friendly way that the fountain was something important.

We spent another day carrying the brittle reeds again in all directions, away from the building platform and up nearer to Adam's tree, so as to leave a clearing in the middle. Two old Qurna workmen then came with pick and shovel and began to dig a big circular hole in the ground, knee deep and the size of a little swimming pool. Nothing further happened to it as long as we were in the Garden of Eden, but our own poor workmen who had to bring loads of berdi on their heads from the stacks near Adam's tree kept stumbling into the new hole and had to climb up again to get across to the building site with their burdens.

The calendar now showed that the day was very near when the Aymara Indians from South America would come and turn the bundles made by the Arabs into a boat. Without them we would never be able to obtain a sickle-shaped ship that would neither capsize nor lose its shape in the ocean waves. The marshmen were still masters in reed-work of all kinds, but ship-building from reed bundles was to them a lost art, just as it was forgotten in modern Egypt and the many other areas of early civilisation where it formerly existed. There was one marked exception: the intricate construction system still survived in perfection in the region around the ruins of South America's most spectacular prehistoric civilisation, Tiahuanaco. There, on Lake Titicaca, high in the Andes, the Aymara, Quechua and Uru Indians still build watercraft identical with those of ancient Egypt and Mesopotamia. When the Spaniards reached this area after the discovery of America, the Indians around the lake told them that their ancestors had first seen such boats built by strangers who had come to their land and erected the colossal stepped temple-pyramid and the giant stone statues at Tiahuanaco, before they were driven away by hostile tribes and moved on to the coast to sail away into the Pacific.

The Indians told the Spaniards that the leader of these foreign visitors was a divine priest-king known to them as Kon-Tiki, with the Quechua suffix Viracocha, which means sea-foam. He claimed descent from the sun and had his divine ancestors depicted in pottery and stone reliefs as human beings with birds' heads and wings. Kon-Tiki and his men, however, were human beings; they were tall, white and bearded like the Spaniards, but differently dressed, for they wore long loose robes to the ankles, with a belt and sandals.

When the Spaniards found the old stone statues and golden figurines depicting Kon-Tiki-Viracocha just as the Indians described him, they suspected that some apostle from the Holy Land had crossed the Atlantic before them, as they had heard precisely the same story as soon as they landed in Mexico. Surrounded by colossal stepped temple-pyramids and giant stone sculptures, the Aztecs had welcomed Cortez in Mexico just as the Incas had welcomed Pizarro in Peru. They told the Europeans that white and bearded men in long gowns had come across the Atlantic long before the Spaniards and instructed the ancestors of the Aztecs and the Mayas in pyramid-building for sun-worship, in hieroglyphic writing, and in all the other aspects of civilisation unknown to the primitive masses of American Indians north of Mexico and south of the restricted Andean area of South America. Everywhere within this continuous belt ruins of some lost civilisation were found among different tribes of Indians who invariably ascribed them to white and bearded foreigners whose leader descended from the sun. In each country he had a different name. The Aztecs called him Quetzalcoatl, the Mayas Kukulkan, and the Incas Viracocha. The Spaniards were at first so confused that in Mexico the monks mistook Quetzalcoatl for St Thomas, and in Peru they formed a St Bartholomew order to venerate a large bearded statue of Kon-Tiki-Viracocha at Kana, north of Lake Titicaca.

On an island in the lake itself, later called the Island of the Sun, the sun-king and his escorts of white and bearded men were said to have intermarried with local women. Then they had set forth in a flotilla of reed-boats to civilise the local Uru Indians and build the impressive cult and culture centre of Tiahuanaco.

Since that time the art of reed-boat building had survived among all the Indians around the lake where reeds, stone and mud are the only building materials. The Uru Indians, who formerly dominated the area all the way down to the Pacific coast, not only build reed-boats; they also live in reed houses on floating reed islands, just like the marshmen of southern Iraq.

I had taken four Aymara Indians with their interpreter from Lake Titicaca to Africa to build the papyrus-ship *Ra II*, and with the experience of their stormy mountain lake they had built a reed-ship that stood up to the test and crossed the Atlantic Ocean without a reed lost. But when *Ra II* was brought afterwards to the Kon-Tiki Museum in Oslo and the swollen, waterlogged reed-bundles gradually dried out, the proud vessel sank like a coat without a hanger, and neither boatmen nor scientists were able to make it stand and give it back its sickle shape. So ingenious was the ancient building technique that, even though we had seen it done, the Museum Board had to invite the same four Indians and their interpreter to Oslo to restore the vessel. Since nobody else could do the job like them, I would try to get them across the Atlantic for a third time, this time to Iraq.

It would be preferable to launch and even be out of the entire gulf area before the winter rains began, so the Titicaca Indians were needed as soon as possible after the reeds cut in August had dried. But September was still burning hot in the lowland marshes of southern Iraq, while the Aymara Indians lived in the cold, thin air of their stormy lake 12,000 feet above sea level, with the snow remaining on the highest neighbouring peaks all the year round. For them such a sudden change of climate could be disastrous, so I had to delay their arrival until the September heat was over, though it would still be necessary at both ends to modify the climatic shock. From their barren island in the mountain lake the Indians were escorted by their interpreter on a journey down into the Bolivian jungle at the sources of the Amazon. There they all ate and slept and got used to the heat for two weeks before they flew from La Paz to Baghdad with my Mexican friend Gherman Carrasco, who was later to join the expedition.

On the Iraqi side the Ministry of Information had generously offered to install air conditioning in one of the two rooms we had prepared for the Indians in the upper part of the resthouse. The cage-like apparatus was torn loose from a window downstairs and soon became to me a nightmare. For days it just hung there as dead and useless as if the canary had escaped; then it was suddenly filled with wild and snarling tigers, while sometimes it began to shudder like a roaring helicopter failing to take off, until the miracle happened and the ugly monster began to spew an Arctic breeze into the empty room just as the Indians arrived.

But by this time the peaceful river-house had within a few days become like an overcrowded seaside resort, if not a madhouse. Loaded front and back with cameras and tripods, a five-man Arab television team from Baghdad had tumbled through the door and begun at once to shoot at us from every corner. They needed Shaker's room and he squeezed in with the manager and the engineer. They were hardly unpacked before a five-man British television team, sent by the BBC, conquered the house with 103 cases, boxes, crates and bags that filled all the stairs and corridors until their Arab colleagues kindly left them Shaker's room and moved into the attic beside the 'helicopter room', which I was keeping for the Aymaras. The next to arrive was a gentle, soft-spoken young Arab who spoke flawless English as he introduced himself: Rashad Nazir Salim, art student from Baghdad, recommended by the Norwegian consulate as expedition member to represent Iraq. At first impression he looked too sophisticated for a rough sea adventure, but we put him in with HP. Then tired, perspiring persons unknown to me dropped in one by one, seeking beer, bath and bed after a seven-hour drive in a Baghdad taxi through a sun-scorched landscape: a German reporter, a press photographer from the USA, a reporter from Baghdad, a Swedish journalist, two Norwegian journalists, and another German. Then I lost track of the newcomers. The peace-loving manager welcomed them all and some were squeezed into a windowless laundry-closet on mattresses between towering piles of linen of all sorts that menaced to collapse and bury them if they snored.

The Garden of Eden Resthouse was virtually bursting from the inside before the key people had arrived. Apart from the South American Indians, we expected three Asiatic dhow-sailors from Bombay, and all the expedition members due from Asia, America and Europe. In panic I got through to the Ministry in Baghdad and learnt that the country was still closed to tourists and journalists, but according to their promise they let anybody in who said he was coming to me. They promised to look at my own list from now on, but added that other reporters were already on the way.

The mountain Indians were supposed to be driven through the hot area by night. The sun was burning from the zenith, however, when a station-wagon rolled up in front of the overcrowded building and five short and broad men in heavy ponchos of llama wool and woven caps with ear-flaps tumbled out and embraced me in silence, giving me the greeting reserved for chiefs, just as they had done to the King of Norway. They then shook the others' hands and followed me upstairs, each with a little bag in which I knew they brought the round water-worn stone and wooden hook, all they needed for working the ropes and reeds. They all posed in a row in front of the confounded air-conditioner that now sounded like the cage of a snarling polar bear, and as they cooled off I recognised five great friends who began to smile from ear to ear: the three brothers Juan, José and Demetrio Limachi and Paulino Esteban, all Aymaras from Suriki Island in Lake Titicaca, and their Bolivian interpreter, Luis Zeballos Miranda from the Tiahuanaco Museum in La Paz.

Our stoic Aymara boat-builders showed no sign of surprise when they entered the Garden of Eden and saw the endless stacks of reeds. But their eyes grew bigger for a moment, and they tore off their woollen caps to expose their raven-black hair, when Adam's tree was pointed out. Then they turned again to the berdi, which they tore to bits with their hands and calmly condemned as no good for boat-building. Gatae and our marshmen stood curiously in a circle around us and looked at the short Indians as if they were creatures from the moon. They were amazed that this tribe of South American reed-boat builders called themselves Aymara, the more so when I could add that their neighbours on the lake were called Uru and lived on floating reed islands like the Marsh Arabs. Our Arab builders were surprised, for the nearby town of Amara and the ruins of Ur were for them two of the most important names around the marshes.

Gatae asked me to explain to the sceptical Indians that the reeds were brittle as paper now, but once they were wet they would be tough and as flexible as rope. This the Aymara Indians knew well, for their own totora reeds at Lake Titicaca had that same property. But still berdi was no good for them. These were not simple stalks like totora or the papyrus we had provided for building *Ra II* in Africa. Papyrus was even bigger and better than totora. But this plant fanned into thin branches like grass with no real stalk, and they did not know how to handle it.

The Aymaras hurried back to their own cool room and I was afraid they wanted to go home.

I had to agree that there was a big difference. For us this would be a completely new experiment. Only the sweet smell and the fluffy inside pulp seemed the same. In the totora and the papyrus this airy pulp was completely surrounded by a thin, watertight skin, and the straight stalk was like a rod with a rounded triangular cross section all the way from the root to the bushy flower on top. The berdi, however, had many separate layers of skin and pulp rolled up inside each other like an onion, and the oval stem at the base gradually opened up into long, sharp, separate leaves. The skin was waxy and surely as watertight as the skin of papyrus, and since water penetrated only at the cut ends berdi had at least one advantage: there was nothing to cut off at the top, so that water could only enter from the root section, whereas the truncated flower stalks of the other reeds drew water from either end.

As it became cooler towards evening, the four Aymaras came down to take a second look. With Señor Zeballos as experienced mediator, Gatae and I succeeded in convincing them that if they would just show us how to combine the bundles into a boat, the Marsh Arabs would make the bundles themselves to any measure the Aymaras ordered.

Nothing on the whole expedition was more pleasing to observe than the spontaneous friendship and mutual respect between the Indians and the Marsh Arabs as they sat down together and began handling the reeds. The eyes of Señor Zeballos and the four Aymaras reflected astonishment and approbation the moment they saw the marshmen select the best reeds and throw them together with loops around until they became bundles as compact and smooth as if made of the best totora. The Aymaras concluded that the Arabs of Iraq were superior to the Arabs of Morocco, who could not work reeds like this. Obviously the Arabs here descended from Adam.

The conversation began with the Indians speaking to Señor Zeballos in the Aymara language, which he translated to me in Spanish, and I to Mr Shaker in English, who then told the marshmen in Arabic what the Aymaras wanted. The system was cumbersome, but it did not last long. When I emerged next morning, I found the Aymaras in their caps and ponchos and the marshmen in their caftans and long gowns squatting around a long mat they had already produced together. They were talking to each other, nodding and smiling, asking for strings and reeds and handing each other what was wanted as if they all had a fluent knowledge of Esperanto. At first I stood behind a palm to make sure I was not mistaken, then I ventured closer to hear what language they had in common. I found that I did not understand a single word. Zeballos and Shaker came and could testify that the Aymaras spoke Aymara and the Arabs spoke Arabic and the two languages were as different as English and Chinese. But these people had the reeds in common and were equally earthbound and alert. With such fine people as those of the Iraqi marshes, the Titicaca group declared themselves willing to make a boat of any size. And Gatae, beaming with satisfaction, said that his men had never worked with more pleasant and able persons than these South American gentlemen. One more day, and I found the tall, dignified Gatae in a short brown poncho with llama cap, and Zeballos and his square-built Indians all hidden like five white ghosts in long Arab gowns and headgear. The change of attire was just too comical; the Indians kept stumbling about in the too long sarks, and Gatae suffered from heat and itching. In Basra we tried to find some straw hats for the Indians; they were worn for one day only and then the woollen caps with ear flaps were on their heads again.

What the Aymaras had taught the marshmen on the first day and without interpreter was how to tie together the very special mats that would be folded around each half of the twin-bodied ship like a tight sausage skin. These smooth mats were as long as the entire ship and hand-woven in such a way that all reeds pointed one way and not a single reed end would jut outwards. This was important for two reasons: to increase the speed of the ship through the water and to decrease absorption. When several mats had been made in strips about 60 feet long by 3 feet wide, they were carefully carried one by one into the cradle-shaped jig by thirty men. The giant sausage skins were now ready to be filled with the compact bundles. I had worked out that thirty-eight bundles two feet in diameter would be needed to give the ship the desired proportions. Open spaces between them would be filled with thin bunches of reeds. The marshmen and Aymaras worked so fast that they averaged three of these large bundles each day instead of one as estimated. The bundles had to be much longer than the total length of the ship, since they must curve upwards, sickle-shape, at each end.

A difficult decision was whether or not to use asphalt. Scientists argued that the Sumerian boat models from Ur were thickly covered with asphalt. Clergymen reminded me that Moses started his life in an asphalt-covered reed basket on the Nile, and that Noah had saved the lives of his companions by coating his reed-ship in the same way. But Gatae agreed with what old Hagi had said: the bundles would float well enough just as they were. We had two identical twelve-foot test bundles prepared, and one of them was brought to the boat-building village of Huwair, where a red-haired, blue-eyed marshman spent his life coating mashhufs in the same manner and with the same kind of wooden spatula and rolling-pin as was used in this trade five thousand years ago. His speed and precision were admirable. The asphalt came from a natural well up river. He coated our roll in the same manner, and we noted that an estimated burden of sixty kilos of the black bitumen was added to the little bundle to cover the reeds well. The two identical test rolls, one asphalted and one not, were now launched side by side in the river Tigris and anchored to the bottom with heavy burdens of bricks and scrap iron. They were to remain submerged for six weeks or so until our vessel was fully built.

Parallel with this experiment HP began his own. He filled my room with truncated and transparent plastic bottles holding bits of berdi set on end, some in fresh water, some in salt water. Some had their cut base end tightly bound with string. They all floated so well that even a complete ten-foot reed set vertically in a bottle of water remained floating upright without touching the bottom. But the results were confusing: after a few days some of the reeds to our surprise started to rise higher in the water, probably because their bases had swollen. HP became optimistic and suggested that we might end flying across the ocean like a Zeppelin. As the weeks passed all the reeds in fresh water sank quite insignificantly; those in salt water not at all.

But as the Aymaras started the ship-building we ran into a no less dramatic water problem on dry ground.

'*Maku mai!* There is no water!'

This was the first Arabic I learnt. We heard it every day. Then: '*Aku mai!* There is water!' This was the standard phrase of our little Arab engineer, Mr Ramsey, who happily shouted back to us from the resthouse roof.

Without water the berdi was as brittle as a match and broke if we bent it. Since green berdi was worse still, the reeds had to be sun-dried first, but then drenched on their outside to become pliant before they could be tied into mats and bundles. With Baghdad and other major cities upstream, the river Tigris was probably so polluted that it might affect the dried reeds if we daily poured bucket-loads of river water over them. This we reluctantly did at the beginning. But Mr Ramsey solved the pollution problem: he had two big tanks brought from Basra and, after endless problems, installed on the roof. They were pumped full of filtered drinking water from Qurna, and Sr Zeballos could spray the reeds and bundles all day long. But the pipelines of the resthouse passed through the same tanks, and the busy kitchen department and crowded guest-rooms competed with Zeballos and his thin rubber hose.

'*Maku mai!*' Zeballos and all the rest of us learnt to yell in despair as the bone-dry reeds cracked under our feet. '*Aku mai!*' we heard a moment later from the little man wielding the big pliers on the roof. His happy message was not infrequently followed by an angry roar from some soap-covered television man or journalist under a shower that had run dry. I was so afraid of losing Mr Ramsey that on one occasion I dragged him out of his car as he tried to escape for a day's visit to Basra.

Before we moved into the Garden of Eden Resthouse, the Ministry had generously offered to close it to all but men of the expedition group. This I refused to accept as I knew that the big restaurant hall and the adjacent riverside terrace were the favourite meeting places for local people. The Mayor and other officials of Qurna, as well as the incredibly large number of school teachers from the marsh area, used to come here in the evenings and on Moslem holidays to enjoy a cup of tea or a cool Iraqi beer, and I knew the country was closed to tourists anyhow. But I had not counted on the tens of thousands of foreigners who were already inside Iraq on government contracts to develop the industries of the Baghdad area and down near Basra and the oil fields. When local television had shown that South American Indians were working in the Garden of Eden, they all flocked to our fence, and while Russians, Japanese, Germans and Poles were invading the resthouse for cool beer, Ali and Mohammed turned the water to the toilets and Ramsey turned it back to our hose in an endless internal battle.

'*Maku mai!*' '*Aku mai!*' '*Aku maku!*'

During this chaos the ship took shape under a burning sun. Bundle after bundle was carried on the shoulders of thirty men, and, winding like sixty-foot Chinese dragons between the date palms and stacks of reed, they were carried up on to the feeble scaffolding and down into the huge sausage skin.

The air cooled slightly. Three weeks had passed since the Aymaras arrived at the beginning of October, and the big body of the vessel was now ready in two parallel halves. Each half was separated from the other by a wide passage where the backbone of the peculiar vessel was to go: the invisible third bundle that was the professional secret of the ingenious building system.

At this stage I began to feel in desperate need of the three Indian dhow sailors who should have arrived from Bombay long before. Without them I could not start preparing the special dhow-type sail for which I had brought extra lengths of spare canvas. The dhows, a type of sailing vessel peculiar to the local waters since time immemorial, had a sail that looked like an ancient Egyptian sail set at a slant. Undoubtedly it was the surviving transition form between the earliest prehistoric type and the modern lateen sail. Such a sail would be of great value for our experiment, but the sailmaker in Hamburg did not know how to make one.

I had been down the banks of the river all the way to the gulf, and had even visited Kuwait in search of dhow-sailors who could help us first to rig the vessel and then to guide us through the chaos of reefs and tankers in the gulf until we reached the open ocean. But there was not a single Arab left in the area who had not sawn down the mast of his dhow and installed a motor, for fuel was now as cheap as the wind in the gulf countries. Everywhere I got the same reply: today only dhows from Pakistan and India still use sails.

Twenty years ago the Shatt-al-Arab and the Tigris as well were full of white sails hoisted on open dhows bringing dates from the plantations to Basra. Today these proud sailing-dhows can only be seen as trade-marks decorating the box-lids of stoneless Iraqi dates.

When I first came to Iraq, Indian sailing dhows used to ply regularly between Bombay and Basra, but due to smuggling this traffic had come to a temporary standstill. The Indian consul in Basra had therefore promised to get me three professional dhow-sailors through the Seamen's Union in Bombay. However, before he could complete the transactions he was transferred from Iraq and I was planning to go to Bombay myself and hand-pick the men, but too many other tasks held me down to the boat-building site. The BBC then promised on my behalf to locate dhow-sailors through a seamen's agency in Bombay: the requirements were three men thoroughly familiar with sailing in the gulf and at least one with some knowledge of English.

Cables from London confirmed that the three men had been hired and were on their way by plane from Bombay. Two weeks later the men were still missing and a new cable confirmed that they were temporarily lost in New Delhi, where they had gone to get their promised Iraqi visas.

While the BBC and the Bombay agency tried to relocate the lost dhow-sailors, the expedition members started to arrive to assist with the launching and the final superstructure. HP flew home to Norway to cool off and rest before the voyage started, and a new acquaintance, Detlef Soitzek, a young captain in the German merchant navy, came to take HP's place as my right-hand man in the shipyard. Then came three of the experienced reed-boat sailors who had been with me on both *Ra I* and *Ra II*: the expedition navigator and second in command, Norman Baker from the USA; the expedition doctor, Yuri Senkevitch from the Soviet Union; and the

Italian mountain climber Carlo Mauri. The new men followed a few days later: Toru Suzuki, underwater cameraman from Japan; my Mexican globe-trotting friend Gherman Carrasco, who had been back home since escorting the Aymaras from Bolivia to Iraq; Asbjörn Damhus, the young Dane from the United World Colleges; and the mystery man Norris Brock, the American film photographer sent to us by the National Geographical Society as a sort of independent outsider.

We were seated at a greatly extended table in the big hall eating an excellent supper on 2 November when Ali came in with the happy news that he had found the lost Indian dhow-sailors: they were standing in reception with their duffle-bags. We all left our plates in sheer excitement to welcome our lost companions, who were to join us for at least the first leg of the expedition. There they were: three real, darkly tanned Indian dhow-sailors. I felt I had known them all my life and introduced them to my friends: Saleman Taiyab Changda, Ibrahim Harun Sodha and Abdul Alim Vasta. We hardly gave them time to wash their hands before we dragged them into the restaurant hall, extended our table and seated them beside us to enjoy all the delicacies Ali and Mohammed could pile on to their plates. At first embarrassed and then delighted at our obviously unexpected comradeship, they grabbed half a fried chicken each and poured down one beer after the other. Saleman spoke English. He translated to his friends.

With Indian added to Aymara, Arabic, Japanese, Russian, and all the West European languages, someone suggested that we should forget the ship and rebuild the nearby Tower of Babel instead. Once it was restored we could place an Esperanto school on top. The multinational spirit was high, and Mohammed kept on carrying loads of full beer cans in from the kitchen and empty ones out on to the terrace. Curious, I followed him, and saw that he dumped all the empties into the river.

'Mohammed,' I said, 'you people dump all your rubbish into the Tigris. Where do you think all these beer cans will go?'

Mohammed's face lit up in a happy smile: 'To America?'

The three newcomers were tired from the journey and it was already dark, so they had to wait till next morning to see the vessel. All our rooms and even the lobby and its adjacent soft-drinks bar were full of beds and mattresses, so the three had to squeeze in on camp-beds in our storage room between popping bamboo and coils of rope.

Next morning at sunrise I woke them up. In pyjamas they

followed me up the stairs to the flat roof, where we had a magnificent view of the river and the garden. Our crescent-shaped vessel looked marvellous in the half-light; it was as if a golden new moon had landed on the banks of the river Tigris and was ready to take off again. In recent days the Aymaras had organised the most difficult job; they had managed to pull the two big halves together to form one complete ship. Each half still retained its circular cross-section with the many big core-bundles now all nicely wrapped inside the woven reed skin. The open space which until recently separated the two half-vessels and made them appear like two parallel canoes, had now been filled in with a slim bundle serving as a sort of common backbone. This backbone-bundle was tied to each of the huge side-bundles by a half-inch rope hundreds of yards long, wound in continuous spiral from bow to stern. First the backbone was bound in with one side of the ship, then with the other, by separate ropes which ran in complementary spirals. When these two spiral ropes were pulled tight by all our men, aided by blocks and pulleys, each of the two half-bodies moved slowly in towards the central backbone, until this was literally squeezed into the two main sections and became completely invisible between them. The result was a sort of a compact catamaran with no gap between the twin hulls.

From the roof I let the three sleepy Indian dhow-sailors look down upon the beautiful vessel from which only cabins and rigging were missing. For a long while they seemed to admire the ship in silence. Then Saleman said slowly:

'And where is the engine?'

'Engine,' I said. 'There will be no engine!'

'But how will it move?' Saleman was curious.

'By sail, of course. Aren't you dhow-sailors?' I asked and looked at the three chocolate-coloured men who gazed at the reed-ship.

'We are dhow-sailors. But our dhows go with engines,' said Saleman calmly.

None of them knew how to sail, or how to make a dhow-sail! I was horrified. But at least they could serve us as pilots through the gulf. How many times had they made the voyage between Bombay and Basra?

Basra? Saleman looked at Ibrahim and Ibrahim looked at Abdul. None of them had been to Basra. None of them had seen the gulf.

I gave up. We had to send them back to Bombay. They were visibly relieved at this decision and showed clear signs of horror when they took their last look at the reed-ship. But before I sent them on their long and costly journey home I had the brilliant idea that we might still be able to use Saleman because he spoke English. We had seen several big Indian dhows at anchor in the river at Basra, but all attempts to get any sense out of the idle crews failed because none of them spoke any European language and not even Arabic. If some of them could sail, I thought, Saleman could be interpreter.

That evening the expedition members were going to Basra by minibus to be guests of honour at a dinner offered by the President of Basra University. But before the party, Norman, as our navigator and sailing master, was willing to accompany the three Indian dhow-sailors to Basra harbour and use them to get some information out of their countrymen at anchor in the river. Yuri had just borrowed a beautiful new Russian car from the Soviet Consul-General, and our Museum interpreter Shaker offered to drive.

Norman and his four companions never showed up at the University in time for dinner, and in the end we had to begin. We must have enjoyed a dozen or so savoury Arab dishes, every one fabulously good, when a man all wrapped up mummy-fashion in white was shown in by a nervous waiter and stood immobile like a ghost inside the door. A red nose and a red ear were all we could see. But the red was from scratches! It was Norman! All in bandages, waiting to be introduced.

'What has happened?' I exclaimed, horrified.

'The car rolled around three times off the road,' Norman mumbled.

'And where are the dhow-sailors?'

'In hospital for a check-up.'

'And Shaker?'

'At the police station. He drove.'

'And the new car of the Soviet Consul-General?'

'Wheels up, a total wreck.'

Norman had only surface wounds and there was room enough between the bandages on his face to prove that his famous appetite remained. But when the three unfortunate dhow-sailors came limping in, one with an injured leg, one with an injured head, and one with an injured back, they had no appetite at all and begged me to send them back to India by the first plane. This I did, as fast as I could get Baghdad on the telephone. They flew from Iraq, with course for India, but were lost on the way and have never shown up at the agency in Bombay.

We were in a real fix. In a few days we had to get the ship into the water; the winter was approaching and there had already been a few drops of rain, a month earlier than normal. Norman was so impressed by the model testing report from Southampton University that he insisted on a bigger sail and bigger blades on the rudder oars than what had been made for me in Hamburg. We had spare canvas for a dhow-sail, but no one to make it for us, so he suggested we should forget the dhow-sail and use the spare canvas to make our own sail much larger instead. 'An easy job,' said Norman, and fixed his eyes on our best sail and a pair of scissors.

We had only one thick solid sail of about seventy square yards, with rope reinforcements sewn in along the edges, together with 'cringles', lateral eyes for guide ropes to be used when tacking. Our second sail was merely a larger substitute to gain more speed without fear of capsizing in the event of feeble, following wind. It was also of Egyptian cotton canvas, but much thinner, and measured some ninety-five square yards.

We succeeded in locating a couple of old sailmakers among the boat-builders in Huwair, and Norman split our best sail along the central seams to add the extra canvas in the middle and thus leave the fine edges with their reinforcements and cringles intact. Once the good sail was spread in separate bits all over the dining-room floor, the sailmakers went to work for half a day and then declared themselves unable to add the new pieces to the old because they no longer had the proper tools. Too bad, I said, then you have to put the sail back as it was.

But in spite of all their attempts, the Arab sailmakers, Norman, and HP, who returned from Norway with sailmakers' needles and gloves, were unable to put the severed sail together again in its original shape. The rain had started, there was no time to send the pieces back to Hamburg, and we had to start our voyage with nothing but a thin down-wind sail. Fortunately for us, in the winter season there would be a north wind, blowing from Iraq down the gulf. We would therefore have a steady following wind all the way to the island of Bahrain, and everybody assured us that there we could find any number of sailmakers to fix our sail and even dhow-sailors to take the place of the three departed Indians.

Norman had more luck when he wanted to enlarge the blades of the rudder oars: we got unexpected help from a truly professional carpenter. But this time we worked like the inmates of a zoo. Every day hundreds of workers from the industrial areas came in bus-loads to gaze through the iron fence at the peculiar vessel on the river bank; Russians, East and West Germans, Poles, Japanese,

Americans and Scandinavians mostly, apart from local Arabs. A never-failing spectator was Josef Czillich, foreman carpenter of Polensky & Zöllner, the West German contractors building a huge mill upstream near Amara, intended for making paper from marsh canes. Before anybody knew it, we found Czillich busy working inside with us every Friday. With a master's hand he helped Norman to peg and glue lateral boards to the blades of the rudder oars until they were so huge that thicker shafts were needed. By fitting extra wood in a superb fashion even the shafts became colossal, though oval in cross-section instead of round. The masters behind these spectacular wooden monsters vigorously defended their highly unorthodox products: that the shafts were oval did not matter so long as they were to rotate in an open fork and not in a circular hole. In any case, there was no way to slim down the shafts unless we also reduced the dimensions of the blades, and this, as Norman pointed out, would nullify the value of the Southampton model test. And Norman was right: why the test if we ignored the results.

For the time being we had much more pressing problems to worry about. Norman, my right-hand man in the rigging, took to bed with a temperature of over 40°C (104°F), and so did our second navigator Detlef, and HP, and all the other men one by one, except Carlo and myself, and Yuri who suffered from pains but ran from bed to bed with his special medical kit designed for Russian astronauts. Even the camera teams tumbled into bed, and for a while the outlook was very serious. A cholera epidemic had struck the surrounding Arab world and was reportedly raging in villages down river, but no cases were reported from the Qurna region.

While nearly all the men were in bed the rain came. The real rain. It poured down as described in the story of Noah. The Marsh Arabs kept to their reed houses and waited for their islands to lift and float. The sandy clay in the Garden of Eden turned to sticky mud. With a couple of men I tried to cover our vessel with reeds to shed the worst of the water, for we could see that the bundles were gathering weight. I was terrified when I saw some of the vertical poles of our perfect jig begin to lean outwards. Tons of water must have been added to the weight of the vessel and the jig was ready to collapse. We hurried to hammer slanting props into the ground to support the endangered structure, and Mohammed ran to Qurna and came back with green plastic sheeting to cover the entire ship. Ugly though this material was, it looked now like a Noah's Ark with its gabled roof streaming with rain, while pools of water grew in size and number on the ground. It was exciting to see if the woodwork of the launching-jig would resist the pressure. It did. And as the clouds drifted away and the sun shone as before on the Garden of Eden, the mud dried up, the ugly plastic was removed, and Noah's Ark lay as dry as if it had landed on Mount Ararat.

It was time to decide whether to cover our golden ship with black asphalt. We pulled the two test bundles up from the bottom of the river. To our horror they both lay much deeper in the water and were obviously waterlogged. We dragged them ashore and with a saw cut them off at mid-length. They were soaked right through, both of them. The asphalt had cracked and done nothing but add extra weight to the bundle.

I looked at Gatae and his marshmen, who stood speechless around the waterlogged pieces. They were as bewildered as I was disappointed. Gatae had no explanation except that we had forced the bundles down to the bottom instead of letting them float on the surface. But, as I explained, our ship too would be submerged, partly by tons of cargo and partly by towering waves. Gatae had no comment to make and went stooping away with his silent men. I remained alone with the dissected test bundles. The pieces of berdi in the glasses of drinking water in my room did not behave like this, especially the samples standing in salt water. Could it be that the polluted water of the river Tigris had caused the reeds to rot? That, I recalled, was precisely what happened when the Papyrus Institute in Cairo had left their test reeds to putrefy in the stagnant water of a bathtub.

We got rid of the waterlogged bundles that left a gloomy impression on all of us, and had to concentrate on getting our ship launched before the next rain, which would be more than a passing shower. After I had first chosen the building site the water in the river had gone far down. The Tigris runs like a drainage ditch throughout the length of Iraq, receiving the downpour only in Turkey, and it therefore shrinks through the summer and autumn when there is no more snow to melt in the mountains around Ararat. But this was a minor problem. We could always tilt the reed-ship down to the water. The real obstacle was a high and solid concrete wall which the river authorities had built from the rest-house terrace to well beyond Adam's tree in the short time I was away in Europe to organise the expedition. The mayor told me it was no problem, the wall could be broken through wherever we needed a gap for our launching.

The estimated weight of our tight-packed bundle-ship was about thirty-three tons, and the system I had devised for the launching had to be extended to pass the new wall and reach the river. The vessel was built inside a wooden jig that would remain ashore when the reed-ship itself was launched; but it rested upon an iron sledge that would pull the vessel into the river and then sink free when the vessel began to float. The runners of the sledge consisted of steel beams with I-shaped cross sections set on edge into rails of the same kind of beams but laid on their sides to provide channels. They had to be extended another two ships' lengths to reach the water's edge.

First the wall had to go, and this required permission from the river authorities. I could visualise papers circulating up and down the river until the rainy season started, so we took a short-cut by directly approaching the two bulky building contractors who had put up the wall in such a hurry, offering them a reasonable payment to knock it down and build it up again.

They accepted and guaranteed with a double handshake to have the wall down by noon the day before the launching. They even gladly agreed to half pay if the wall was not down on time. They came as promised, and asked to borrow our picks, but an hour later I saw their broad backs as they calmly walked away. Shaker came running and said they refused to do the job because the wall was too hard. I hurried to catch up with them and told them that since they had made the wall they should be able to knock it down. But they explained that they could build, but they lacked the tools to destroy. Goodbye.

I tried the Arab wall with a pickaxe and it only threw up sparks. I blessed the Danish cement factory downstream that had made it possible for these men to mix a concrete harder than mountain rock. Norman jumped out of bed, his eyes still red with fever. He was a Commander in the US Navy Reserve all right, but in his daily life he was a New York building contractor. With Shaker he drove to the village smith and had iron wedges and a huge sledge-hammer made, and in no time fever-bitten men tumbled out of the resthouse and took turns with the heavy sledgehammer; even the British and Arab cameramen joined in Operation Jericho. And the wall crumbled to provide a gap as wide as we needed.

The road was open into the river. The river ran into the long gulf that opened into the Indian Ocean, an ocean I had never been at grips with. A gateway to unknown adventures stood agape in the Garden of Eden, and Noah's Ark lay ready to float as new rain clouds gathered on the horizon.

CHAPTER 3

Problems Begin

ZERO HOUR. All flags up.

'Are we ready?'

'Ready!'

'OK! Let go!'

We had invited no one, but the Garden of Eden was packed with spectators who seemed as curious as we were to see the reed-ship enter the river. The atmosphere was that of a big theatre with the rumbling of countless voices that came to a sudden silence as we set to work with two small hand-jerks we had borrowed to pull the heavy vessel into the river Tigris.

The silence was broken by the roar of thousands of jubilant voices as the lofty reed colossus began to move and then to slide along the metal rails. Slower than a turtle, she moved in jerks towards the gap in the broken wall, with the flowing river below.

There would not have been room for an apple to fall to the ground in Adam's peaceful garden, the way the spectators now pressed forward for a better look, and the Iraqi police who had come along to protect the high officials from Baghdad were in difficulties to save the dignitaries, expedition members and work-men from being pushed beneath the advancing Sumerian curiosity. Even the river was full of Arabs and foreigners in mashhufs, balams, police boats and motor launches.

It was a great relief to see the monster moving out of our own home-made wooden jig and on to the improvised steel beams which an engineer from the upstream paper-mill had kindly welded together as rails to the water's edge. The corn-coloured vessel still carried blood marks on the bow that rose proudly like a swan's neck, but covered by red human handprints after the recent naming ceremony.

There had been some discussion about this ceremony. The Marsh Arabs had come to the launching site with six beautiful sheep. They wanted them sacrificed and I was supposed to hand-print their blood on the bow of the new vessel. This I refused to do. *Kon-Tiki* had been baptised in South American coconut milk, *Ra I* and *II* in milk from Berber goats. I could not even propose the use of buffalo milk which would seem an insult to the Marsh Arabs. They insisted on the local custom of animal sacrifice as adhered to since the days of Abraham. They always hand-printed blood on any new building, whether house or boat. The marshmen blankly refused to let the ship enter the river unless the proper sacrifices had been performed, and Gatae insisted they would do it themselves if I would not. They were so dead set on this rite that even our educated Baghdad friend Rashad refused to come along on the voyage unless the customary ceremony was performed.

We found a compromise. The marshmen should be allowed to carry out their own rite beforehand, but the naming ceremony should be the way I wanted it.

November 11 was the day that had been set for the launching. In the late morning I found Gatae by the ship in his spotless white kaftan with a blood-stained right palm. He was still stamping red hand-marks on the golden bow from the last sheep sacrificed, while his men sat happily squatting on the ground devouring roast lamb with no effort to conceal that this was to them the most important part of the ceremony.

By midday the dignitaries had arrived from Baghdad, bringing a white silk ribbon and scissors to be used by the Director-General of the Ministry in a proper bridge-opening ceremony in front of the bow. Then it was our turn; Gatae's beautiful little grand-daughter was to name the vessel. Gatae stepped up from the river's edge leading this tiny black-haired lady in colourful costume by the hand. Little Sekneh struggled to carry a traditional Marsh Arab bottle-gourd dripping full of river water that was to give the ship its name. With sparkling eyes she splashed it successfully on the reed bow, forgot all her lessons, and only those who stood close could hear her mumble 'Dídglé', the local name for Tigris. Her grandfather never let go her hand as he took over and declared with loud voice in Arab:

'This ship is to enter the water with the permission of God and the blessing of the Prophet, and will be called *Dídglé Tigris*.'

No sooner had these words been proclaimed than a low rumble of thunder was heard in the south. All heads were turned as if in surprise or awe; it had sounded as if approbation had come from the mightiest of all Mesopotamian gods, the sun-god who struck thunder and lightning with his hand-mace. Black clouds were approaching from the horizon. Even those of us who were not superstitious almost felt a shiver down the spine at this timely comment from the weather-god.

A screw on the borrowed 'come-along' system broke and had to be replaced by a piece of nail, so it took an hour to pull the newly-named *Tigris* through the broken wall to the critical hump where the slope began to run steeply down to the river. We had tried to ease this rather sharp incline by filling in more soil, but the current immediately carried the lower part away. As the *Tigris* reached the hump we all held our breath. Would the sixty-foot reed bundles and the thin steel sledge they rested on stand the strain when half the ship balanced unsupported on the hump?

To our relief the giant tilted over like one massive block and began to slide downhill by its own weight. The bow hit the water at the foot of the steep slope with a splash. Thousands of jubilant voices rose to a crescendo of triumph as the broad bow was lifted up by the water and began to float in the river, high as a rubber duck, while most of the ship was still up on dry land.

What uplift, what buoyancy! I ran close behind to ensure that all ropes were held tight to prevent the unrigged body from disappearing down the river. Then the stern suddenly stopped just at the moment when the bow rose on the water. I heard a terrible crash of breaking timber and saw the steel beams twist like spaghetti under the broad body of the ship. The jubilant sound of applause sank to a deep murmur of lament, mingled with screams of despair as the broad vessel slowly settled on the solid ground like a rebellious hippopotamus refusing to enter water.

There it sat. Firmly aground. Bow launched and afloat in the river and stern solidly planted up among the crowds in the Garden of Eden.

An army of volunteers ran to try to help push the vessel down into the river. In vain. Only a few could get their shoulders to the high stern that curved steeply into the air, and those who tried to push along the sides sank into loose earth and river mud.

We began to dig beneath the vessel to inspect the damage and try to let the water in under the stern. Our idea was close to that of the wise prophet and the mountain: if our ship did not want to come to the river, then the river had to come to our ship.

The dignitaries saluted politely and left with the police. One by one the crowd also melted away as evening approached and drizzle began to fall. Worst of all, the Aymara Indians had left for South America the previous day, because their excursion-rate return tickets to Bolivia would otherwise expire. Without them it would be a major problem to repair the ship if ropes and bundles had been damaged. But until we had access to the broad twin bottom we could have no idea of the extent of the damage caused by broken ends of timber and steel. As night came on an incredibly chill wind began to sweep the landscape. We had to run into the resthouse for more clothing. The black night sky was split by spectacular streaks of lightning.

We took turns at using the few shovels available, and the British cameramen stretched cables to the resthouse roof terrace to beam light from their lamps in our direction. David, a young Jewish assistant in the group, confided to me his last night's dream: he had seen a whole herd of sheep come on board and devour our reed-ship in revenge for their sacrificed relatives. Gatae and the Marsh Arabs also had their comment: six sheep was not enough for such a big vessel, we should have sacrificed a bull.

These pessimistic observations were hardly uttered before the rumble and strong headlights of a huge Russian truck made us drop the shovels. The unexpected visitor bumped in through the gate over broken timber and twisted rails and took up a position as if to push us all into the river. Two husky Russian drivers jumped out and, with Yuri as interpreter, we explained that though we needed a push, our ship was as brittle as crispbread until in contact with the water. They then helped us rig up thick reed fenders extended in front of their tall engine housing on wooden beams taken from the abandoned scaffold, thus saving the heavy truck from being launched with the ship. Suddenly in the dark the informal and most unconventional launching began with a steady thrust from the truck. The vessel moved and sank slowly into a porridge of thick mud where we had dug. As mud and broken timber floated away with the current we held the ship on tight ropes and made her fast to a floating reed-bundle mole we had prepared beside the resthouse terrace. In the light from the truck's headlamps we had seen the twisted sledge and rails all following the vessel into the river. The Russians saluted and left. We had to wait for daylight to judge the damage. *Tigris* did not seem to float with a perfectly even waterline all around. It seemed to tilt slightly towards the port side bow.

As daylight broke we were all on the spot. Our new ship looked magnificent. Bigger and stronger than any of the two *Ra*s, she rode very high on the river, and above the turbid water there was not a scratch. But another fear was confirmed: in spite of the precision of our craftsmen and the perfect symmetry of all the curvatures, the ship seemed to lie slightly deeper in the bow than in the stern, although not more than we could probably adjust when we stored the cargo. Someone suggested that this tilt was due to quick water absorption, for the bow had been launched hours before the stern! This pessimistic suggestion was rejected as an ominous joke. It could be, however, that part of the reed bundles under the bow had been ripped off by the broken sledge.

Detlef and Gherman swam under *Tigris* with goggles and seemed gone an eternity before they came up with their reports. Visibility was nil in the muddy water, but they could feel that steel and timber of the launching frame was still stuck to the bundles beneath the bow. No wonder there was a tilt. Fumbling with their hands, they could feel that a thick reed bundle tied to the bow as a temporary fender had been tied to the sledge as well. With a knife they cut this bumper free, and as the jumble of steel and wood sank to the bottom the bow of the vessel tipped into balance. Feeling their way all along the ship's bottom they were both left with the impression that every loop of the spiral lashings was still intact. The ship was undamaged. Jointly the thousands of berdi reeds remained afloat and had won the battle with modern steel that lay twisted on the bottom of the river.

Our two dissected test bundles had shown us that modern river pollution was harder on reeds than twisted metal, and we were in a hurry to get into clean salt water. But it took two weeks to rig and load the vessel. On the lawn in front of the resthouse our marshmen had prefabricated two huts of green *kassab* cane from the marshes. The glossy canes were braided in a decorative local pattern and tied to a framework of bamboo. The huts were tall enough for us to sit but not to stand upright under the vaulted cane ceiling. The largest measured about four yards by three and had barely space for eight men to sleep outstretched on the floor. The smaller was only half that size and intended for three men and camera equipment.

Both cabins were carried aboard, ready made, and tied to widely spaced planks which in turn were lashed across the main bundles. The main cabin was set aft with a window-like door-opening on either long wall facing the sea. The little cabin was set forward and crosswise to the first, with a single similar door opening, one metre square, facing the central deck and the main cabin behind. A roof platform, fenced by low bamboo uprights, added an upper terrace to either cabin. With some imagination, in a world of fish and waves, the reed-deck space between these two golden-green jungle dwellings appeared like a tiny village square. The huts were only intended as sleeping quarters and retreats in bad weather. Our daily living quarters was the open space between them.

This was also the place where we hoisted the colourful sail on its bipod mast. As on all reed sailing vessels of the Old and the New World, the yardarm holding the sail had to be hoisted on a double mast with its straddling legs resting one on each of the twin bundles. There was no hold for a mast along the centre line where the thick bundles barely met, so the thirty-three-foot ash masts were set into large wooden 'shoes' lashed on top of the bundles. Masts and shoes were held together by wooden 'knees' made from branches naturally curved at right angles. Such small but important details were copied from Egyptian frescoes and tomb models. Following these prototypes the straddle-mast was drilled where its legs met at the top to pass the halyard hoisting the yardarm. Cross-bars held the two legs together and formed a convenient ladder to the top.

If the reeds kept afloat our life for months ahead would revolve around this ladder between the two cabins. In the space between the ladder and the forward cabin we sewed nicely polished scaffold planks together with rope to form a long table with two benches, set crosswise to the vessel. Behind the ladder the roof and side walls of the main cabin were extended three feet forward to form an open alcove to serve as galley. This small shelter could hold four primus stoves and all our pots and pans.

On the last day tons of food and water were carried aboard and stored under table and benches, along the cabin walls and down in the deep angular trench that ran from bow to stern between the two big bundles. Clothing and personal property were stored together with film and vulnerable equipment in asphalt-coated boxes set together to form a sort of raised floor in the main cabin. This floor was our common bed.

Tigris was ship-shape and ready to sail the moment the two huge rudder oars were lashed astern, one on either side. A wooden steering platform or 'bridge', three feet wide and three feet above deck, followed the rear wall of the main cabin from side to side and permitted the two helmsmen to see over the roof. The sail, however, would inevitably limit their vision. The oar shafts slanted to the rear and were rotated on their axes by crosswise tillers near the upper end to make the blade turn like a rudder. The shafts were held by tight loops both at deck level and up on the railing of the bridge.

Tigris was ready to sail.

'Let go the moorings! Hoist the sail!'

I was filled with relief and pleasure as I shouted the orders and waved to the incredible new crowd of spectators that had once more gathered in the Garden of Eden. They must have come by intuition. After our experience of the launching we had told nobody of our intended departure, saying only that we would sail the moment we were ready.

'Hoist the sail!' I shouted again, desperately. Seconds counted. The moorings were gone and we were already moving as victim of the current, but the sail that should have given us steering control was still down. The only response to my cry were hundreds of Arabs ashore shouting unintelligible words and pointing into the air. I looked up, and there in the mast top, just where the yardarm was to go, hung our dear Mexican friend Gherman, almost upside down, with his inseparable movie camera perpetuating the great moment. Round and jovial but incredibly agile he almost fell from the mast as I sent up to him an unintelligible roar which in decent language would read: 'Please come down very, very fast, dear Gherman, and let us have room to hoist our sail, otherwise you may be shaken down on the bank the moment we make one more crash landing.'

In a second Gherman and the sail exchanged places. The current already had us in its grip, but the wind filled the sail and we took control of the situation. Norman was in charge of the sail, Detlef was hidden somewhere in front of the forward cabin awaiting orders to raise or lower the *guara*, our wooden centreboard, and Carlo and I were on the bridge with one rudder oar each. It was grand to see the banks of the Garden of Eden and the resthouse move away. It had been a great place, but it was high time to get into the ocean. To Adam's tree, farewell! Farewell Ali, Mohammed, Gatae, Kais, Shaker, Ramsey, and all our other Iraqi friends. You are disappearing now as a forest of waving arms, but we will often recall you in the days to come, among the waves.

We were gaining speed. We were heading for the other bank of the Tigris. Carlo and I turned the clumsy oars over and I yelled to Detlef to lower the forward guara. The vessel obeyed beautifully and we passed the green point of land where the twin rivers meet and become the Shatt-al-Arab. Behind us lay the Garden of Eden

Resthouse with its cluster of palms and trees; to its right was the Tigris and to the left the Euphrates opened into view with the Basra road bridge that had prevented us from building the reed ship among our friends in the marshes, since our mast would not have passed beneath it.

'Hurrah, we're sailing!' It was Norman's voice, full of joy and vitality for the first time since his violent fever. Nearly all the men had barely recovered, but every one was dead set on getting away and all eyes were bright with energy and excitement. The sail really invited rejoicing as it drew its breath and pulled ahead of the wind. For the first time we saw unfolded the tanned Egyptian canvas on which our Iraqi art student Rashad had painted a huge reddish sun rising behind a terracotta-coloured Mesopotamian pyramid.

The date-palms flanking the river must have signalled our arrival, for people lined the banks shouting and waving, and as we moved on more and more spectators emerged from huts and villages of reed and mud-brick. Everywhere men and boys began to run with us along the river banks as far as they could keep up with our speed. But we sailed faster than they could move on the uneven terrain, and passed with great speed through a changing landscape of barren wasteland and palm plantations. Everywhere, however, people seemed informed and some even shouted our names. There were high spirits on board. We would pass the down-river cities and industrial areas and reach the gulf in a few days. A motor-driven balam with ten Arabs reputed to know the river came along to pilot us through the hectic traffic and other modern obstacles in the lower part of the river. They had just disappeared around a very sharp right-hand bend when I noticed that our ship was out of control.

Carlo was a mountain climber and like me had never sailed on a river. Not familiar with local banks and shallows and with pilots who ran around as they pleased, we chose to keep an equal distance from both banks. But at this bend the river seemed to end straight ahead of us in green grassy banks that attracted the current like a magnet. No matter how much we turned the rudder oars and adjusted sail and guara, we were pulled sideways fast and forcibly towards the turf. We turned with the river and followed the bend, but ever closer to the banks, where another gathering of men and boys awaited us and ran enthusiastically along in our company until they started to scream and yell as they saw the reed-ship skidding too close to the shore.

The current ran at its fastest at this outer edge of the curve and we swept along so close to the shore that the broad blade of port side rudder oar began to dig up mud. All hands not fighting the land with our long punting poles joined Yuri and Carlo in trying to pull up the colossal port side rudder oar before it broke under almost forty tons of pressure from ship and cargo. But the oar was too heavy to lift and jammed in the double rope loops. At any moment we could expect a deafening crash from the shaft, which was as thick as a telephone pole.

We were now rushing along so close to land that the Arabs running with us ashore tried to push us away with bare hands while the oar blade began cutting up solid dirt along the edge of the turf. But we sailed faster than anyone could push while running, and neither they nor we on board with punting poles could do much to prevent what looked like disaster. For hundreds of yards we followed the curve of the river like a fast and highly effective plough, turning up the fat earth that would have been the envy of any farmer. At every second we expected the oar to break, but Norman and his master carpenter had done an amazing job. The oar held; instead the whole steering bridge to which it was fastened began to yield. With a horrible creaking and squeaking from rope and wood it began to lose shape. Carlo and I were ready to jump the moment the rope lashings burst and the bridge, perhaps the whole stern, was torn apart. Our pilots in the motor balam were on their way back, and some jumped ashore to help their running compatriots to push. But before they had gained a good grip, Norman got a new angle on the twisted sail and we rushed away from the port-side river bank like a bird taking off from a freshly-ridged potato field.

There was barely time to draw a deep breath of relief on the wobbly steering platform before we looked to the other bank and saw a solid forest of grey palm trunks coming rushing towards us. We had the green fringes of the leaves almost above our heads when quick manoeuvres with sail and oars helped us shoot back towards the naked banks we had just been ploughing. The motor balam now followed us from side to side like a drunken companion, trying to get in between us and the banks to serve as a fender. But it was always on the wrong side. Suddenly it turned around and disappeared upstream. It was gone for two hours before we learned on its return that they had been away looking for four men they had left ashore and forgotten up river at the place where they had been running along the banks to push.

In the meantime we had become masters of the situation. We began to know our new vessel. We were as alert to the invisible drag towards the outer bends as to the threatening shallows built up along the inner. Soon the Shatt-al-Arab began to float straight and even as an autostrada. Few houses. No traffic. A man with a ragged piece of canvas on a makeshift reed-raft sailed downstream at half our speed. Our huge and all too heavy rudder oars now hammered back and forth like colossal sledge hammers, and with each bang the bridge shook and squeaked and cracked; we clung to the cabin roof, which still seemed the firmer part of the structure.

The motor balam guided us past large herds of black water-buffaloes in the shallows beside Beit Wafi, a large and decorative reed-house village in marked contrast to the no less picturesque Arab adobe houses of home-made bricks or sun-baked mud. Brick kilns, as in the days of Abraham, were still functioning all down this section of the river and offered a spectacular sight. They rose like pyramids over the plain, and when operating could be mistaken for active volcanic cones, sending out vast columns of smoke from burning reeds and canes.

The influence of the high tide in the gulf could be noted more than a hundred miles up the river. As the rising sea blocked the outlet of the Shatt-al-Arab, the river water dammed up and even began to flow in the reverse direction. In the late afternoon the surface around us became as motionless as a lake, and before it began to flow the other way we asked our pilots to show us a safe anchorage for the night. They recommended the west bank near Shafi village, where we furled sail and threw out our two small anchors on their ropes. Our experienced river companions knew that anchors would come loose when the river began running the other way, and they thrust punt poles into the muddy bottom to keep us in one spot like the fenced-in floating islands in the marshes.

It was a great evening. Our first on board. The sun set red behind the smoking brick kilns and we felt as if we were in ancient Sumer. In fact we were. A cold wind began as the sun went down and suddenly we could feel the current beginning to drag up river. The boys lit the kerosene lamps and looked into the water as if they expected Mohammed's beer cans to come back from America. While in action up river, we had eaten only biscuit, but now Carlo got a primus going, and we all gathered on the benches along the deck table with our individual mealtime bags, from which we fished out bowls and forks ready for the steaming spaghetti. Exquisite. *Buonissimo. Wunderbar. Koroshii. Deilig.* Carlo received his well-deserved praise in many languages. The men were hungry and tired. But for the first time for many days we had a chance to relax.

The last days before sailing had been worse than a madhouse. Neither garden fence nor guards had been able to keep the curious crowds away. We had tried to rope off at least the small area around our floating reed mole so that we might have space for our carpentry work and free access to and from the reed-ship. But to no avail. We had to elbow our way through and with scant success, for people pulled us back with broad smiles and brutal determination. Pencils and bits of paper waved everywhere. Nobody realised that if they did not let us finish the ship our signatures would be those of a doomed expedition. The Arabs were scarcely interested in our scribblings at first. Most of them could read only Arab script anyhow. But when they saw how Russians and Japanese fought for autographs, they all wanted them before we disappeared. One of the Aymara Indians was literally dragged away from work to build a reed-boat model for a German journalist. Three Russian carpenters who volunteered to help us lash on the steering bridge had to sign as well. No matter what we signed. The smartest would request a dozen signatures for friends and family. We were not left room to move arms or feet until we had done them. We signed on odd bits of paper and on berdi reeds, cigarette packets picked up from the ground, notebooks, newspapers, family photographs, postcards of Warsaw and Budapest, pictures of Lenin, dollar notes and Iraqi dinars, matchboxes, wallets, passports. The atmosphere had varied between tragi-comic, desperate and hilarious.

As we had climbed up on deck to depart those who had nothing for us to write on got really wild. I had barely posed to improvise some words of farewell when a young and muscular Arab climbed aboard and tore open his shirt in front of me, wanting me to write on his bare chest. He was pulled away by the men around me only to be replaced by another young Arab, who clung to my arm and waved his bit of paper at me in the middle of my speech. I grabbed my pen to sign and get rid of him, but this made him even more desperate. He did not want my autograph at all; he had a bill from the resthouse for some beer one of the men had consumed after the final accounts had been settled. I grabbed in my pocket for money while I tried to give sense to my speech, which I discovered was being recorded by several microphones and echoed back to me from various translators. Someone whispered to me that the only man I had ignored when he wanted to salute me as I climbed aboard was the recent Minister of Information. In this utterly chaotic atmosphere the crumbling Tower of Babel could not have created greater linguistic confusion. We just had to get away. We would have time to adjust the jumble of cargo on board when we were alone and at peace on the long voyage down the calm river.

Only now, as we had anchored and fenced ourselves in with stakes far from the Garden of Eden, could we really relax and begin to enjoy the crazy comedy of the two weeks between launching and departure. Only now could I really lean back in comfort against the mast leg with a cup of hot Arab tea and take a closer look at the strange mixture of men I had assembled around me for this adventure.

There was – and he should not have been there at all – a Russian carpenter who had been busy lashing the last cross-pole to the steering bridge when we started. A good man. Nobody had protested. Now he could help us repair the bridge. Yuri, interpreting for him, said Dimitri was happy to travel as far as Basra, where he had to report back to work. He crawled to bed on the cabin roof in a sleeping bag originally intended for an Indian dhow-sailor.

As no one had to attend to the rudder oars, we were all eleven together for the first time, and one by one we crawled into the cosy cabin for a good night's sleep. There were men seated at our table whom I knew really well and others who were completely new acquaintances. Our ages ranged from twenty to sixty-three. Our nationalities and characters were no less diversified.

There was my old friend Norman Baker from the USA. Wiry and strong. All skin and muscle. He seemed small in a winter coat and big in swimming trunks. Our wakes had first crossed in Tahiti twenty years earlier; he came sailing from Hawaii and I with my expedition ship from Easter Island. Norman was now in his late forties; Commander in the US Navy Reserve, but a New York building contractor in private life, he had been my second in command on both the *Ra* expeditions. Norman alone was enough to make me feel like skipper Noah. He was as agile as a monkey, strong as a tiger, stubborn as a rhinoceros, had a canine appetite and in a storm could be heard like a trumpeting elephant.

At his side sat our robust Russian bear, Yuri Senkevitch, forty years old, built like a wrestler, as peaceful as a bishop, doctor to Soviet astronauts, who had also become a popular Moscow television announcer since we had last seen him. He had sailed with us as medical officer on both *Ra* expeditions and had later turned into a

bit of a globe trotter, introducing the weekly Sunday travel programmes for a hundred million Soviet television viewers. Yuri
could hardly open his mouth without laughing or cracking a joke.
He said he had acquired this habit when he flew to Cairo to join us
on *Ra I*: he had emptied half a bottle of Vodka on board the Soviet
aircraft as my letter of invitation to the President of the Soviet
Academy of Sciences had stressed that I wanted a Russian doctor,
but one with a sense of humour.

Carlo Mauri of Italy, in his late forties, had also been with us on
both reed-ship voyages across the Atlantic. Resembling Noah more
than I did, because of his impressive full beard, and being blonder
and more blue-eyed than any Nordic Viking, Carlo was one of
Italy's most noted mountaineers, a professional alpinist who had
climbed up and down the steepest and highest rock walls in all
continents and hung in more ropes and tied more and better knots
than any man I have known. Latin by temperament, Carlo could
turn from a domesticated lamb into a roaring lion, and the next
moment grab pen and paper to write poetic accounts of his experiences. Carlo could live without food and comfort, but not without a
rope in his hand. He was to take the expedition's still pictures. And
he was to twist his brain and improvise the most ingenious knots
and criss-cross lashings each time a cabin, a mast foot, or a leg of the
bridge began to wobble and dance.

Detlef Soitzek from Germany I had never known before.
Twenty-six years old, one of the youngest captains in the West
German merchant marine, he was also an enthusiastic sportsman
and a climbing instructor at Berchtesgaden. He was recommended
to me by German friends when I was looking for a good representative of post-Hitler Germany. Detlef was a naturalist and idealist.
Peace-lover, anti-war, anti-violence, anti-racist. He rarely spoke
without good reason, but was a keen listener and would chuckle
more than any at a good joke.

Gherman Carrasco, fifty-five, industrialist and amateur film
producer from Mexico, was our entertainer. There was no relation-
ship between his body and his soul. With his chubby build and
moustache he seemed likely to be most at home in a wide sombrero
under a cactus. But not at all. The motto of his private film
collection is: The World is my Playground. The urge within him
makes him leave his four rubber factories in Mexico City several
times a year to fly around the world. It was he who swam under the
Polar ice and filmed Ramon Bravo when a polar bear bit him in the
leg. A scar around his eye testifies to the day when he fell from a

jungle tree in Borneo while filming orang-outangs. He had been in trouble with sharks in Polynesia and the Red Sea, and he had filmed for his own pleasure in every nation on the map, with Red China as his favourite hunting-ground. Before I knew him he had asked in vain to join the *Ra*. But now I knew him. He had trampled with me in the burning sand of the Nubian desert, filming rock carvings of pre-dynastic ships, and we had waded together in the pouring jungle rain of the Mexican Gulf, filming Olmec and Maya pyramids and pre-Columbian statues of bearded men. Furthermore, he was two doors away from me when the Hotel Europa in Guatemala City collapsed and buried us in bricks and dust, while all those in the room between us and twenty thousand others lost their lives in the great 1976 earthquake.

I was never more surprised than when Gherman asked me to come to his office to see his museum. He pressed a button behind his desk, and a huge painting of three fat angels drifting like pink balloons on blue canvas swung aside and revealed an opening in the wall like the hatch in a submarine. Inside were four big rooms filled from floor to ceiling with shelves and glass cases stacked with properly labelled archaeological objects of Maya, Aztec, Toltec, Mixtec, and Olmec pottery of all shapes and colours, stone statues, ceramic images and figurines, reliefs, gold objects, a priceless fragment of a paper codex, and minute carvings in shell and bone. At that time this was all Gherman's property, but Mexican law has later confiscated all pre-Columbian art in the country for the nation's benefit. Four students and a professor spent many months cataloguing his collection of tens of thousands of items. But Gherman was appointed custodian of the collection and the museum remains where it was behind the angel painting.

And now Gherman sat with us on the boat. Beside him sat a completely new friend, Toru Suzuki, a Japanese underwater photographer in his middle forties. I knew very little about him yet,

except that he had spent several years filming marine life at the Great Barrier Reef and now ran a small Japanese restaurant somewhere in Australia. I had accepted him blindly on the recommendation of Japanese friends. In Japan, more than in most countries, national pride and self-control tends to reduce the danger of acquiring a problem-maker on an expedition. Toru's English was remarkably fluent; he was a well built athlete of few words, but always ready with a bright smile and a helping hand. I felt I had hand-picked him myself.

In this mixed company the two Scandinavian students seemed like twins. Both were chosen on the recommendation of the United World College headmasters, and were graduate students from Atlantic College in Wales. Asbjörn Damhus from Denmark, twenty-one, 'HP' Böhn from Norway, twenty-two, typical descendants of the Vikings. Men like Asbjörn were probably with the Danes when they invaded medieval England and carried off giggling girls, while HP might well have been waving in the mast top when Leif Eiriksson sighted Vinland. They were always up to something, working together to devise the most unexpected practical jokes. Full of resource they were technically minded and clever with their fingers. They were as much at home in turbulent water as in a bathtub, and looked forward to any form of adventure before returning to their university desks.

Youngest of all was Rashad Nazir Salim, twenty years old, an art student from Iraq. Slender but athletic, the young Rashad had a keen brain, always eager to listen and learn and yet not without his own strong opinions. He was an impassioned Arab patriot but full of good humour and far from aggressive. He had come to the

22–23. The archaeologist Geoffrey Bibby shows the author a survival from Bahrain's antiquity, the boat is made from *palm stalks* and Toru uses the same material to repair the bow of *Tigris. (p. 93)*
24–26. Prehistoric burial mounds on Bahrain are estimated to number one hundred thousand; a few are as big as pyramids and lined with dressed stones as shown by Geoffrey Bibby. *(p. 94)*
27–28. Under the sands of Bahrain Danish archaeologists have discovered a long-lost port city with a walled harbour basin. The quarried and beautifully fitted stones of a sacred well on Bahrain were of a kind unknown to this island and hence brought by prehistoric mariners of Sumerian times. *(p. 95)*
29–30. Prehistoric quarries on the prison-island of Jedda, from where the ancient masons of Bahrain had rafted their big blocks. *(opposite)*

Garden of Eden with his letter of recommendation and spoke modestly of himself in flawless English. He knew Europe since the days when his father was an Iraqi diplomat; now the diplomat had turned into one of Baghdad's most notable painters, and Rashad wanted to follow in his footsteps.

Half a head taller than all the others at the table, I could see the eleventh man: Norris Brock. Professional US cameraman, forty years old. Tall and thin but remarkably agile. I did not know him. I had not chosen him. He was, until we met him, an inevitable clause in the contract with the consortium that had lent me funds for the enterprise. Norris used his eyes more than his mouth. He seemed to be everpresent and always with his baby at his chest, a specially built and waterproofed sound camera with a long microphone on top that looked like a baby's bottle. He would nurse it at the top of the mast and even dive with it from the cabin roof while it was working. Until I got to know him and his abilities I thought the National Geographical Society had chosen him because of his height. He was supposed to film us even if we sank, and by the looks of it he would be able to keep on filming with his head above water while all the rest of us were disappearing under. Until they got used to his everpresent camera that had the privilege of recording anything we did or said, the men came to me to find out what kind of fellow this tall character was. I could not answer the question. All I knew was that during the previous year he had been sent in the same manner on an expedition with a fibre-glass double-canoe that sailed in the wake of the ancient Polynesians from Hawaii to Tahiti. Another photographer had filmed from an escort vessel, but tall Norris had sat in cramped quarters inside the canoe, filming all the way at close range. Nothing had happened on that journey except a violent psychological storm and a dramatic split between the Polynesian crew and the foreign leaders on board. Nobody had tried to hide that they expected he would have to film even worse mental storms among the mixed lot on board *Tigris*. The psychological drama and conflict story had been the theme of the recently released film of the canoe adventure, and as sails and waves can only be of interest to the viewers for a few minutes, our eleventh man obviously had his orders not to miss the moment when we began to punch each other's noses. This, I said – and I never tire of repeating it before every ocean expedition in small craft – this is what we know as 'expedition fever'. It is worse than any hurricane to men sharing cramped quarters for a long time at sea, and is as certain to come as any shark if one is not ready for it, to shut one's mouth the moment the urge to yell at one's neighbour is felt, because he has left his fishhook in your mattress or used the windward side of the raft for his toilet.

The men listened, and I began to suspect that Norris' ever-present baby could perhaps become the most effective little device any expedition leader could dream of to quench expedition fever at its beginning. Then I was not so sure. After all, Norris had carried the same baby on his previous trip, and had come back nevertheless with a record of squabbling on board. And Carlo had already begun to grumble because Norris was only free to shoot movies while he, Carlo, as still photographer had to cook spaghetti and store cargo.

When the rest were not listening, I had admitted to Norris that he was free to behave as a passenger; but for his own sake, I added, so as not to feel isolated as an outsider, he ought to take the same steering watches, kitchen cleanings and other routine duties shared by us all. He could be relieved any time there was something he wanted to film. Norris answered that he had in fact intended to ask for this. He wanted to become one of us. And he did.

We remained three days anchored beside the reeds, repairing and strengthening our steering-bridge with better lashings and more cross poles. Also, two men were ashore buying thick brown buffalo hides which we cut up and tied on to the reed bundles where the oar shafts and anchor ropes would tend to wear. We even began to build two tiny outboard toilets, one on either side aft, which we screened with coiled *chola* mats obtained from the Marsh Arabs. The sun-rise and the big southern moon were spectacular, but the north wind was biting cold at night and we pulled the canvas down on one side of the cane wall, which was so airy that we could see the stars between the wickerwork. The day temperature sank to 17°C (62°F).

On the afternoon of the third day we hoisted sail and continued the voyage down river. Norman had cut up one of our rowing oars and tried with the Russian carpenter to add something to the oval rudder-oar shafts at friction points to make them round. Failing this, they tried to plane off the thickest side. The oars still jammed. They remained as two monsters threatening to destroy the bridge whenever we turned and there was heavy pressure on the blades. Norman had to defend his system against growing criticism from Carlo, Yuri and me. The discussion died down when we saw Norris' head over the cabin roof and heard something that sounded like baby hiccups inside the forward cabin. That meant that his voice recorder was working. It was radio linked to a tiny mechanism in his back pocket. We quickly agreed that we had to pull the rudder oars ashore at the first place we could dock and reduce the steering colossi closer to their original size.

The sailing was good, with an estimated three knots, when the sun went down and we were seated around our kerosene lamps eating Rashad's Arab rice with raisins and onions. The two helmsmen on the steering bridge, with heads above the roof, shouted that ahead long flames were licking their way towards the river. We all climbed up on to the table and roofs. Norris was already in the mast top. In the darkness we saw three long horizontal flames from tall gas chimneys flickering over the river. We held close to the other bank with our reed-ship, which still rode high and very dry, and it was a spectacular and even dramatic moment as we sailed past the huge flames that seemed almost to reach us and lit up everything on board from sail to cabins and our own faces. They even lit up all the palms on the opposite side of the river. Shortly afterwards we sighted long, empty cement docks on the same right banks, and our pilot balam helped us by laying to between us and the cement as a wooden fender. Norman was again struck suddenly with a very high fever for a couple of days. Even Yuri now confided to me that he had severe pains in the chest.

At sunrise we began to see clearly the vast industrial complex to which the mole belonged. West German engineers came and with their crane helped us to lift the two gigantic rudder oars ashore. We cut off one third of each blade, and with his Russian adze Dimitri chopped down the sides of the oval shafts so that they became much lighter and fairly round. Friendly German and Swiss engineers invited all of us to lunch and to dinner. They were building a modern paper mill beside the large section already in operation. With the other mill under construction far up the river, these plants would suffice to make deforested Iraq self-supporting in paper manufactured from canes and reeds from the marshes. The kassab was especially suitable and was rafted to the mill as large gáré. An enormous field next to the plant was stocked with thousands of tons of cane, ready to be converted into paper pulp.

This was to be our next nightmare, never experienced by the Sumerians. We had observed that the Shatt-al-Arab was very polluted in this area, but not until we came to the pier in high spirits after a late party did we notice sheets of some white substance floating down the black water. With our flashlights we saw no water at all around our golden reed-ship; everything looked like whipped cream with streaks of yellow butter. In the chilly night wind we felt as if we had come to the Arctic. Large floes and flakes of ice, some capped with snow, appeared to come slowly drifting out of the night to build up like pack ice around *Tigris*.

Some of the men ran upstream and found the white foam coming in a solid flow down a canal from the big factory buildings. An engineer confirmed that the old mill was washed out at night. The modern one would not pollute the water in the same way when it was ready for operation. Meanwhile, here was our *Tigris*, afloat in the thick chemical spillage of a plant that converted cane into paper!

With the blades of our rowing oars we tried to scoop away the deep layers of white and golden foam, but it built up again against the reed bundles as fast as we got a moment's glimpse of black water. We wanted to escape but could not. Our heavy rudder oars were ashore in the dark, and the reshaped blades were so far only partly covered with new asphalt.

Next morning all the men swore that *Tigris* lay considerably deeper in the water. Norman was convinced the chemicals in the pollutants had penetrated and damaged the reeds. We hoisted the oars aboard and set sail as fast as we could. But it was lunchtime before we had everything ready for departure. As we sailed on we passed foam caps that sailed slower than ourselves, and all that day and all the next we had them with us.

It was a great relief to have round steering-oar shafts that now rotated against smooth buffalo leather. But the wind died down. Completely. The river still ran. Slowly. A beautiful, undisturbed landscape. Except for the white caps of foam. Date plantations. Water buffaloes. Geese, ducks and kassab canes. Small villages on the riverside with happy dancing children, some running with us, others climbing date palms. Barking dogs. Women in colourful dresses, but always covered by long black cloaks. Some with sheep, some with bottle gourds and some with aluminium containers on their heads. Pottery had gone out of use. We saw a couple of canoes with fishermen. Nothing else. Peace. The red sun set behind palms. The river stopped running. We anchored before it started flowing the other way.

Next day we again reached modern civilisation, as if with the wave of a magic wand. We passed Sindbad Island and had to take a tow between the colossal pillars of the new bridge which already spanned the other half of the river, beyond the island. I had been down before we started building and spoken to the German engineers who constructed this super-bridge with a frightening speed. In another month they would have the whole bridge completed, and so low that no masted reed-ship could ever again sail down to the sea. On the pillars of the bridge, and from now on, through all the hectic ports of Basra harbour, enormous crowds awaited us. Police boats escorted us and the balam that towed us. All the cargo ships blew their sirens and rang their bells. The Iraqi navy vessels had their officers and crews lined up in salute on deck and dipped their flags. Hooting and whistling were everywhere, so Yuri grabbed our own bronze fog-horn, jumped on to the roof and trumpeted right and left while the rest of us clung together to the wide mast ladder or waved and shouted from the steering bridge. I had to yell myself when Carlo pointed out a Norwegian ship from my own little home town of Larvik.

In the midst of the chaos our contact man from the BBC in London, Peter Clark, dropped out of the sky and boarded us from a motor vessel. He just had a bite to eat and returned to shore with Dimitri in a Russian motor launch.

It was hilarious. It was scaring. Supposing we got no further than down the river. Supposing the acids of whatever we had absorbed from the paper mill had damaged our reeds?

The noises faded behind us. We all climbed down on deck except the helmsmen. It had been fun. But we were hungry. Asbjörn served hot Danish stew: *hakkeböff*.

Our pilots had left us in Basra to refuel and we continued on our own, escorted by a couple of casual small boats southwards through another truly beautiful area. Here and there, in a fertile terrain of dense palm forests, lay attractive villas in splendid terrace gardens; now and then a primitive but most picturesque village of mud-plastered huts appeared half-hidden under the giant leaves of banana plants and date palms. But every now and then a new road reached the river at some newly built mole with industry under construction on the banks.

At about two in the afternoon we reached the border of Iran, and from here on Iraq's territory was restricted to the western half of the river. From now on all ships at anchor were on the Iranian side. Nobody reacted as we passed, and when the pilot balam finally caught up with us the men on board were in a panic that we might come too close to the midline of the river. When I had visited Iraq a few years earlier the two countries were enemies on the edge of war, now they were on friendly terms except that Iraq had just beaten Iran at football. I did not feel too certain about how we rated with the Shah after the violent protests from his Imperial Court and Embassy at our reference to the 'Arabian' instead of the 'Persian'

Gulf. Now we were heading for that gulf, so what were we to call it so as not to offend anyone? After all, Persia had changed its own name to Iran, and Mesopotamia had become Iraq, so why the problem with the gulf? We were travelling the Sumerian way, and the Sumerians too must have had a name for this gulf, before either the Arabs or the British Navy. So for our own internal use I suggested that we called it the Sumerian Gulf. The Sumerians themselves on their inscribed tablets refer to the 'Sea of the Rising Sun', the 'Lower Sea' and the 'Bitter Sea', but we cannot know if these descriptive names do not comprise the sea even beyond the gulf area.

But we had not yet reached that gulf. The pilots were so afraid of crossing the midline that they led us all the time as close as possible to Iraq's banks. So close that I occasionally suspected that the oars touched the bottom. I began to realise that our guides in the balam had never been further down than Basra harbour.

As night approached they insisted on stopping. They showed us where to spend the night and told us to throw out the anchor. We did, near a bend where a South Korean and a Monrovian ship were anchored near the Iranian bank. Detlef threw out the anchor, but the rope did not follow. When he pulled in he straightway had the anchor in his hand, dripping with mud. I tried to twist the starboard rudder-oar. It sat like a spoon in butter, as did the port-side rudder too. Rashad yelled to the pilots that we were in shallow water. They yelled back that just where we were was the deepest part of the river. We threw a line and asked them to tow us off. They tried, but failed. There was little more than a metre down to the loose mud that began to suck the whole bottom fast, like a quagmire. Even our punt poles sank into the loose bottom and were hard to pull up again.

We had to wait for high tide, said our pilots. But they admitted that they did not know when high tide would come until they saw it. It was never the same one day as the next, and tidal hours were different here from further up the river. We all poked our noses close to the dirty water. It did not move. It was high tide right now. In fact the water was slowly turning and beginning to run away into the gulf. We fought desperately, but either we were sinking ever deeper into the mud or else the mud was building up quickly around us. It was a frightening situation. Led by Carlo and Yuri, we managed to lift the rudder oars up and tied them on so that they did not reach deeper than the bottom of the vessel.

We sat there as the moon came up and had the horrible feeling of being sucked down into some bottomless liquid clay by invisible octopus tentacles. In the night big steamers, brilliantly lit, passed us going up river. At least we knew they had professional local pilots who certainly held them well away from our banks.

In the moonlight Detlef and Toru crossed the river by dinghy to the other side and asked the crew of the Monrovian ship if they would help pull us off with their winch. They politely refused, from fear of the Iranian police, as we were on the Iraqi side. Our two envoys then rowed over to the Koreans. They were willing to stretch their own rope to the midline of the river, but not into Iraqi waters. Anyway, we had to wait for the next high tide as they would otherwise pull our bundles to pieces. As both ships gave very contrary estimates of local tidal hours our two men came back from the Iranian side without result.

The situation got worse hour by hour. Surely our reeds would gradually be buried in running river silt.

Late at night the water stopped flowing out. The banks where we sat seemed almost dry in the moonlight. No further changes kept us awake, and one by one we dozed off with a night watch on the roof.

At 2.30 a.m. I woke as I heard water gurgling and swirling. I poked my head through the opening and the night watch showed me with his flashlight that the chocolate-coloured water was gushing up-river, splashing past the portside rudder blade under my nose. It was as if we ourselves were shooting downstream through the rapids of a muddy river, though in reality we were sitting where we sat last night. But the tide was rushing in from the gulf at a frightening speed. Now, either we would surely be buried or else torn loose. We launched the dinghy again and Detlef and Asbjörn rowed out and dropped both our anchors in the deepest part of the river; then we on board pulled on the anchor ropes to try to drag ourselves into deeper water.

For nine hours we had been stuck on the mud banks, when at last the undermining effect of the incoming tide, combined with our own struggles, began to have effect. At 3.30 a.m. the bow was slowly turning away from land. To help this movement we kept pulling in all rope-slack to the anchor, while we yelled to awaken our sleeping pilots. I would have preferred to pay them off, but somewhere around the next bend was Iran's large modern city of Abadan, and we would probably have to take a tow between oil tankers and refineries. By five o'clock we were able to lower the rudder oars into position and hoist our sail against a starlit sky. A huge bright halo surrounded the moon, which had been full two nights earlier.

We had hardly rounded the first great bends before we saw the silhouette of Abadan against a dawn sky. Tall smoke stacks, radio towers, a whole city of lofty oil tanks. A feeble wind turned against us and a faint current still ran against the bow, so we lowered the sail and let the balam tow us as fast as possible through the worst pollution we had ever seen. From a paradise of a kind our golden ship had suddenly found herself in a modern inferno. The surface between the big ships and the modern dock installations was neither sea nor river water, but a thick soup of black crude oil and floating refuse. In the cleanest spots it shone and reflected rainbow colours as the sun rose behind the industrial fog. Sumerians would have been horrified to see the environment modern man prefers. Even the lower half of all the green berdi lining the undeveloped banks on the Iraqi side was black from tar or oil, clearly showing the level of high tide. The air smelt of oil. We were ashamed of our proud vessel that now began to get dirty with oil and grease above the waterline from the wash of passing ships.

Rashad asked our balam to give full engine, even though it might be hard on our reeds. We had to get out into the open bay. To our surprise we found the water cleaner as soon as we had passed the big city, as if the mess was floating in and out at the same spot. But surely it was bound to catch up with us again if we did not reach the mouth of the river before the tide flowed back towards the sea. The water turned from black to brown. There were a very few date palms again in a naked landscape without beauty. It was indeed an area fit for industrial expansion. In the afternoon a town rose above the level wasteland on the Iraqi side. Fao. The last town at the river's mouth. In a sense the river continued; at low tide it wound its way like a shiny sea serpent through partly submerged bogs and empty tidal flats, until it sank where there was no more sign of land. We were longing to get there; out where brackish water would turn salt and seagulls waited to escort us into the freedom of the open sea.

CHAPTER 4

Problems Continue

THERE WAS no more river. No more land. The balam had been paid off in Fao and had returned upstream. The vast mud flats extending into the gulf from Iraq and Iran formed an indefinable coastline as we entered salt water and hoisted sail on what looked like a floating fruit basket. Bundled reeds and plaited canes bobbed merrily along with red tomatoes, green salad, yellow citrus fruits, carrots and potatoes topping hemp sacks and wickerwork containers. We were loaded with perishables for as long as they would keep on the sun-lit deck of an open raft-ship. In a few days all these fresh provisions would be consumed by eleven hungry men or else covered by mould just as fast as green seaweed would start growing on our submerged reeds.

We had docked at Fao at the transition from the sandy plains to the mud flats long enough to fill our vessel with these delicate garden products, and also carried aboard a good supply of onions, garlic, raisins and a variety of local nuts, seeds and grains that would keep at sea.

From the port of Fao a long and narrow channel had been dredged through the vast empty tidal flats to the open sea for big ships to be piloted in and out along numbered navigation buoys. *Tigris* had been towed through this channel by a professional Iraqi pilot tug. Mud, mud, nothing but mud. All formed by the never-ending deposits of fine river silt from Mount Ararat in Turkey, and desert dust from the twin river plains of Iraq. We passed lazily with the outgoing tide before sunrise.

We met the first slight swells from the open gulf as we passed the Khafka light-buoy and the pilot boat left us to ourselves. The sun rose red in morning mist over an open sea. We were filled with expectation.

By leaving the mouth of the outer channel I felt as if I were once more about to break a scientific taboo. Vessels like *Tigris* were not supposed to go any further. We were trespassing beyond the limits of what competent scholars had set for the range of a Mesopotamian vessel of berdi reeds. Not until the Sumerians invented wooden boats did they have access to the open gulf, according to what we had all been taught. Textbooks and teachers repeated what some long forgotten authority had assumed to be true: that people in Mesopotamia, like those in Egypt, began river navigation in ships built from bundles of reeds, but had no means to leave the outlets until they abandoned the early reed-ships and invented the first plank-built craft.

We were about to violate a well-established time barrier. Zero hour for marine history and cultural contact by sea were both tied to the change from compact bundle-craft to the hollow hull. So important was this transition that we were led to take it for granted that if there was an open stretch of water between them cultures and civilisations arose independently before that time.

I knew as we hoisted sail beyond the Shatt-al-Arab that to scientists in many fields this would seem as a vote of no confidence in long-accepted teachings in anthropology. Perhaps it was. But it was fair play. To those who really believed in the old doctrine we should now be about to prove that they were right, and I wrong. But with all respect for my own colleagues among the scholars, none of them had ever seen a berdi ship nor were they able to quote anyone who had. Nor was I.

And I did feel a bit uneasy myself after seeing the complete waterlogging of our two test bundles in the river Tigris. Less encouraging still: our own ship had begun to absorb. I was not exactly surprised when in Fao Yuri had pulled me aside for a quiet talk. The Russian reed-ship veteran had for once an expression as if he were about to summon me for a grave operation. He pointed to our slightly sunken waterline. Had I seen it?

I had. And I tended to agree with Norman that some acid or pulp-producing chemical from the paper-mill could have helped water penetrate the skin of the outer bundles. When the exterior reeds had swollen as much as the tight spiral rope permitted, they would probably block water from passing further in.

Yuri looked at me without comment. Then he said calmly:

'Carlo and I agree that we should unload everything not absolutely needed. Better to leave it with people in Fao than to throw it overboard in the gulf.'

I studied Yuri's expression. Was he becoming afraid? No more than I was. I knew what he was referring to. We had only the word of the Marsh Arabs to rely upon when it came to the water resistance of berdi cut in August. On *Ra II* water absorption began to reduce the carrying capacity of our thirty-foot papyrus ship as soon as we were past the Canary Islands. We had to dump all spare woodwork and extra food and water into the sea off the coast of Africa even before we started across the Atlantic, for fear of sinking due to overloading. Yuri's sombre expression was enough to remind me of the moments of horror we had experienced together on low-riding reed-bundles at sea. Moments when we sat waist deep in the salt Atlantic and sometimes had the waves washing over our heads. *Ra II* was not so bad as *Ra I*. We were waterlogged and had barnacles growing on deck, but at least we reached America without the loss of a single reed. *Ra I* had been the tough one. As the wave-beaten bamboo cabin had severed the bundle lashings the whole ship split lengthwise along the middle and we could see the bottomless ocean between our feet. Moments of horror mixed with moments of triumph and joy. Days and nights with our hearts in our mouths when we faced death opening and closing its jaws beneath us and we never knew what disaster the next second might produce.

'Yuri, you are right,' I said with a sudden impulse of support. 'Let us carry ashore everything we can do without.'

And at once Norman and Carlo were with us, making up a list. Of course, we did not need a compressor for Gherman's aqualungs. Nor any spare tanks for diving. No one needed to dive deeper than was necessary to inspect the bottom of our bundles, or to loosen a short-lined anchor if stuck in some rocks. This we could do without aqualungs. The boys who had never sailed a reed-ship before did not understand what had overtaken the four old-timers. Gherman protested. He had bought all this expensive equipment on my behalf, so that he and Toru could film sharks and other fish we knew would follow under a silent raft-ship.

His protests were of no avail. We had no need to film anything that required compressed air. In fact, we could do without artificial light as well, so ashore went Gherman's incredibly heavy underwater lamps and batteries, together with the compressor and a large pile of Mexican souvenirs and such personal belongings as he would never need on a raft voyage. All this was nailed up in cases and shipped back to his home. The others among us had to go through a similar stripping, although we allowed a certain tolerance and few had brought more than was needed. The greatest cut in cargo came when Yuri laid his hands on our spare timber. Here I felt as if he was stripping me personally to the skin. The four of us knew that the wooden parts were most vulnerable on a reed-ship. The rigging, the bridge, and particularly the shafts of the rudder-oars. While reeds flexed under strain, wood broke. The reed bundles were literally unbreakable. Like solid rubber. But not wood. Wood broke in combat with the elements.

As on *Ra II* I had brought beams, poles and hardwood pieces for splicing broken parts. To the amazement of the already puzzled newcomers Yuri now wanted all of it left behind. Our friends in the balam almost sank as they accepted this precious gift with open arms. I barely managed to save our long rowing oars, twelve in all, and a few hardwood odds and ends for emergency splicing. Better to re-establish mental peace right now if anyone already began to feel uneasy. In fact, was I not one of them myself?

Tigris was fully sixty feet long as compared with the thirty feet of *Ra II*. But we were eleven men on board instead of eight. And we carried far more water and provisions than for a two months' journey like that which confronted the *Ra I* and *II*. And again, we still had only the word of the Marsh Arabs to counter those of the scientists. All we had done apart from using berdi instead of papyrus, was to cut our reeds in August instead of in December.

We did not rise noticeably in the water by the few hundred kilos we had carried ashore. But we were in any case still floating incredibly high. So high that we could not bend over and grab passing flotsam or wash our hands, a fact which almost annoyed those of us who were used to raft-ships where we could do our morning toilet without the use of canvas buckets.

As the heavy yardarm was hoisted to the top of the straddle-mast our sail unfolded and the red morning sun seemed reflected in the big red sun on our sail. Ours rose behind a stepped pyramid. The real one rose freely above the misty sea.

Great expectations. What we had expected was a strong wind from Iraq behind us. In the open gulf beyond the mud flats there was nothing to give shelter. But the wind was just not there.

Strange. For this was the second day of December and during all the winter months a steady north wind was supposed to blow strongly from Iraq down the full length of the gulf. If we could, we wanted to call at the island of Bahrain, which would be almost on our way. Once they were able to import timber to build wooden ships, the Sumerians were thought to have had easy sailing away from their shores in the winter months, and could return with their cargo when the winds changed in the summer.

But the weather of recent years seemed to have forgotten earlier habits. It was as if the age of the sailing vessels was gone anyhow. Indeed, the winter rains had surprised us by arriving at the Garden of Eden over a month too early. In Fao, the Arabs had warned us that during the last two years the winds had been crazy. And now we had a feeling of being stuck in displaced doldrums.

We just had to accept the flimsy and feeble gusts we got and decide on a steering course. This was the place where we should have had some Sindbad-type with us. Until we had left the river I had hoped to find a dhow-sailor who would join us, just any old-timer from the gulf, but one who knew the tides and shallows close to the coasts. Such people were gone with the wind. The last vessels we saw in the river was a whole fleet of old gaily-coloured dhows anchored on the Iranian side, but every one of them had the mast sawn away and a motor installed instead.

The Sumerian world had changed at sea as much as ashore. Sand had conquered their fields and cities and filled their navigation canals. Silt had changed their coastline by adding mud-flats to their waterfront. And their open water was no longer a playground for many kinds of silent watercraft, gliding about propelled by oars and sails. Speedboats and supertankers had taken over. The clamour of motors and engines was everywhere. Independent of winds and reefs, Arabs and their neighbours and visitors coasted around oil installations and criss-crossed a network of modern shipping lanes. This once peaceful area beyond the Mesopotamian coast had turned into the worst possible place for novices to fool about in, experimenting with a reed-ship's steering.

Certainly, we had to keep out of the way of others. We had to sail away from this vast stretch of water separating the Arabian peninsula from the rest of Asia. The plan was to follow the Arabian side as close to shore as reefs and shallows would permit. There the streams of big ships would not interfere with our free movement.

Our two navigators had carefully studied the charts to find the best route beyond the traffic and clear of oil platforms and islets. Norman suggested steering 135°. Detlef proposed 149°. But a voice from the top of the mast recommended that we first take a good look ahead. We did. The morning haze was still quite thick. But in the binoculars we detected ships at anchor wherever we looked. From left to right I counted forty and then caught sight of something almost frightening: a colossal and lofty oil platform with a supertanker moored at its side. I forgot to count further. There were masts everywhere out there. Some were almost invisible in the mist, and there were doubtless others beyond them still.

No matter where we steered, we had to enter a chaotic forest of big ships, some anchored, a few slowly moving. As we approached, most of them turned out to be cargo ships. This was not only a filling area for tankers; it was the famous anchorage for all the ships waiting their turn to proceed up the river and deliver their cargoes to Iraqi and Iranian ports as far as Sindbad Island near Basra.

The tall bow of a Japanese tanker emerged from the haze. Its mighty hull passed close to our stern. I was at one rudder-oar, Yuri at the other. As the wind was too feeble to give us good steering control we had to struggle hard to avoid collision even with anchored ships.

Norman and Detlef agreed that the best we could do now was to keep a course close to Buoy 23, which I could barely make out in the mist as a red dot behind a cluster of ships.

We began to observe a system in the chaos. The tanker channel and the filling station were now at our starboard side. There were only cargo ships at anchor on our port side. The wind was so feeble that it took us two hours to pass the ships and get within a hundred yards of the large red buoy. The number 23 could be seen painted on it in black. Our plan now was to hang on to it until we got a better wind. We were close. But then we noted we had been still closer a moment earlier. There was no wind, but in fact we were drifting back the way we had come. The tide had changed. All the cargo ships began to swing round on their long anchor cables, turning their starboard sides towards us, their bows pointing away from the land.

The sail gave us absolutely no help. It was quite clear that the tide was sending us back towards the river's mouth. The Shatt-al-Arab was sucking up water now. At this moment we saw something orange-coloured being lowered from one of the anchored ships, about a mile away. It came straight towards us and turned out to be a strange kind of lifeboat. It approached us in complete silence. No motor, no sail, and no oars. Never had we seen a boat like this. It was packed with husky men, most with bare chests and with bands around their foreheads. They sat facing each other from either side in two rows, rhythmically turning one long common crank-handle. The handle obviously operated a propeller. No fuel. Rotating arms. Perfect team-work. Like some undulating, crawling sea creature, all arms revolving together. All the torsos swung up and down at the same time. Then a head rose up in front:

'Yuri! Yuri Alexandrovitch Senkevitch!'

They were Russians. They knew Yuri by his full name. Violent hand shakings. Carlo quickly threw them a long rope, which Yuri asked them to tie to the buoy which was now about 300 yards away. They immediately resumed their seats and started their combined cranking operation, now steering for the buoy. They could not tow us but they carried our mooring rope. The longer the rope in the water the more drag there was on their progress. Carlo worked desperately to tie on more and more sections of rope, until there was no more, and to his dismay the last end slipped away overboard. We waved to the Russians to come back with all our rope but they only waved back in salute. They continued to crank as fast as they could for the buoy. Dead tired, they dragged the hundreds of yards of heavy rope to it, making the end fast. Then they signalled anxiously to us to start pulling in free rope. We were already far from our own lost end, so far that we could no longer read the number on the buoy.

The Russians at first seemed quite bewildered. But not for long. We had at any rate no time to think of them or of the great length of lost rope. We had to prepare to run aground. To get ready our two small anchors. Our anchor ropes were not very long, but we hoped that one or both anchors would take hold before we became stuck on the mud-flats. The bottom where we were drifting was only loose silt from the river. The ships out at the good anchorage had long chains that could reach bottom anywhere in the gulf. Fortunately there were neither reefs nor rocks awaiting us.

'Look!' Norris on the cabin roof pointed to a big black cargo ship that had left its place among the others and came straight towards us guided by the orange-red lifeboat. They came right up to us.

Minutes later we had one end of the lost rope-line back again, with the other end now secured to the lifeboat, which in turn was towed away by the big ship. Straight back to the anchorage area.

A short, thickset sea-dog had led the whole operation from the bow of the lifeboat. With a broad smile he asked for permission to visit our vessel. He jumped on to the reeds like a short-legged kangaroo and introduced himself as Captain Igor Usakowsky. Ruddy, jovial, middle-aged and in shorts, our visitor was in command of the ship we hung on to, the 17,000-ton Soviet freighter *Slavsk* of Odessa. He was as excited as a boy on a rocking horse when he felt the supple bundles under his feet and tried the stability of the lashed-on straddle-mast and cane walls. He had to stand on our steering platform, lie down on our cabin floor, and sit on our sewn-together benches. And then, before we knew it, we were all sitting around two long tables in Captain Igor's own officers' mess on board *Slavsk*, eating Russian bortsch and admiring the appetising dishes piling up in front of us as well as the two shapely blondes who brought them in. There was vodka, wine and Russian champagne. There were pork chops, meat cakes, cabbage-and-carrot salad and cheese with butter and fresh Russian bread. Captain Igor rose to speak; I did, we all did. Our host was a great speaker, humorist and big eater. His glass seemed open at both ends. He was born in Georgia, the son of a Polish nobleman who had joined the revolution. At the beginning of the evening he called me 'Captain' but later he called me 'Father'.

'Cheers, you are my father,' he said each time he lifted his glass. I wondered if he could really take me for that much older than himself, until he added that he referred not to age but experience. 'Then you had better call me grandson', I retorted, well aware of my own status as a mere landlubber who enjoyed drifting about on prehistoric raft-ships, testing how long they would float. Captain Igor was a real sea-dog. He had begun at an early age, whaling in Arctic waters. Later he had mastered big ships on all seas. We ended with the popular captain among his forty officers and crew in the ship's spacious assembly-room. The night air was filled with loud metallic music from a Greek freighter anchored as nearest neighbour. Time was surely long for all the hundreds of idle people stuck in this floating steel village. Some ships had been anchored there for over a month. *Slavsk* expected to wait another week or two before her turn came to proceed up river to Basra.

In the night mist we had fun cranking ourselves back to *Tigris* with the hand-propelled lifeboat. We had seen the regular motor launch under repair on the deck.

We crept into our sleeping-bags in ample time before sunrise. It was still dark when I awoke, feeling a chilly draught on my face through the cane wall at my side. Wind. I woke the others. Wind! Sleepily we tumbled to the sail and the rudder-oars. Sleepily we discovered that for us this was the worst wind possible. Strong, but from the south-east. The very opposite to normal, and completely contrary to what we needed. We wanted to steer for the island of Bahrain, but that was exactly where this wind was coming from.

All odds were against us. Yet we wanted to make a try rather than hang on where we were. *Slavsk* pulled us clear of the anchorage area, and we hoisted again the only sail we had, the thin downwind canvas. The proper tacking sail, with cringles and reinforcements, still lay in pieces on the cabin roof. It could never be hoisted unless we first got to Bahrain where an experienced sailmaker could restore it. The problem was to get to Bahrain without it. But at least the thin downwind sail was bigger and gave us better speed.

Compared with previous experiences, we certainly rode very high on the waves three weeks after launching. And with the big, light sail we moved with great speed away from the anchorage and the tanker channel. The wind increased. It turned even more southerly. The best we could do with the rigging we had was to take the wind straight in athwart and hold a course of 240°–250°. If we tried to do better we quickly lost leeway.

The sea now ran white-capped everywhere and we saw neither land nor ships. With this course we were heading in the direction of Failaka Island, that lay in front of Kuwait. This was really good sailing and it did not yet take us totally off course. We had to get further west anyhow in order to avoid the busy shipping lanes in the central part of the gulf. We had to get closer to the shores of Saudi Arabia. Before we came too close, the wind ought to change and resume its normal course, and, with the fine speed we were making, a good north wind should take us to Bahrain in four days.

But the wind did not turn to its normal course. It blew ever stronger from the south and our speed westwards increased. The fresh southerly dug up a choppy sea with deep and narrow troughs coming at us athwart and making us roll heavily. We had to straddle to keep upright on the wooden bridge. Some of the men began to get drowsy. Asbjörn gave us a seasick smile and apologised as he crawled to bed in the forward cabin. The long mast legs began to jump and hammer. They were tightened to their shoes by Carlo.

We saw a few porpoises but no other sign of life and kept on the same steady course westwards in the direction of Kuwait. If this southerly wind were to last for some days we would have to drop anchor somewhere along the coast of Failaka Island and wait for a change of weather. Ancient navigators might have done the same. They lacked many modern conveniences, but were never short of time.

With this wind there would be shelter on the north coast of Failaka. But our navigation chart showed no harbour or anchorage in that area. In fact it showed no details at all in the broad belt between the island and the mainland north of it, for this area was marked as unnavigable, due to unbroken shallows. But with our flat bottom, compact bundles and modest draught we could venture where others ought not to try.

Norman searched for some information on that area in the *Persian Gulf Pilot*, published in London a year previously, and read aloud: '. . . this coast is rarely visited by Europeans. There are large tracts without villages or any settled inhabitants, and it is probably unsafe to wander away from the towns on the mainland without an armed escort.'

The wind blew stronger. Our speed increased. The day passed and the sun sank in the choppy sea ahead of us. It sank in the direction of Failaka Island and its shallows.

We had not intended to sail so far west as this. We wanted to aim for the island of Bahrain but here we found ourselves heading for Failaka Island. It seemed like a trick of fate, for Failaka was a competitor with Bahrain among students of Sumerian traditions about the legendary Dilmun, and I had no objection to getting at least a passing glimpse of the island before we left the gulf. We were now sailing in real Sumerian waters and no other island lay closer to former Sumerian ports. In the meeting at the Baghdad Museum the Iraqi scientists had mentioned Failaka several times. The great scholar Fuad Safar was inclined to believe that Failaka was the important place referred to as Dilmun on the Sumerian clay tablets. In his opinion Bahrain was too far away.

But as *Tigris* raced towards Failaka in the evening of our very first day, we asked ourselves: was not Failaka too close? Most scholars identified the legendary Dilmun with Bahrain, and that was a major reason why I wanted to go there.

One thing was certain. I could inform my men that in the darkness ahead of us lay a barren speck of land with a remarkable story to tell about man's activities at sea since the earliest days of navigation. Somewhere ahead was a low sandstone island seven miles long and three miles wide, full of vestiges left by prehistoric sailors.

Alexander the Great had personally named this island Ikaros when the Greek sailing vessels, built in the distant Indus Valley, came this way about 325 BC. The now barren island was still wooded then, and was conveniently placed for the Greeks when they conquered the gulf area. Although the mainland of the great country they named Mesopotamia lay nearby, they built a fort and a temple to their goddess Artemis on the tiny island. In a manner typical of Europeans even eighteen or twenty centuries later, the early Greeks considered themselves discoverers of any land they reached that had foreign gods and a different culture. Only modern excavations have been able to show us that, when Alexander the Great came to Failaka over two thousand years ago, the Sumerians had been there over two thousand years before him.

The Greeks who named the twin-river country and the gulf islands had never heard of Sumer. Sumer disappeared as a political entity with the destruction of Ur about 2050 BC. The Sumerians, their language and their culture were erased from the memory of man until their buried ruins with cuneiform records were uncovered and deciphered by archaeologists who thus brought the Sumerians back to life in the last century.

Yet it was not until our own generation that archaeologists dug on Failaka Island and found that the continental Sumerians had been there too. And not only the Sumerians; before them the Akkadians, and after them the Babylonians. Although the never-ending sedimentation of silt from the rivers has since Sumerian times brought Failaka closer to the growing mainland of Iraq by a good hundred miles, the island had never been beyond the reach of continental cultures since the days when civilisation was first established in Mesopotamia.

I had read about Failaka for the first time in a book by the British-born archaeologist, Geoffrey Bibby. He described how he and his Danish colleagues had dug up large numbers of potsherds and stamp seals from the many ruins and refuse mounds on the little island. The seals especially served as unmistakable fingerprints. They were incised with special symbols and motifs that linked them to specific areas and epochs of the outside world.

Most of these finds differed from those of nearby Mesopotamia. Nothing linked them to mainland Kuwait, which was extremely poor in archaeological remains. But an impressive number of the Failaka remains belonged to a lost civilisation that flourished on the island of Bahrain about four thousand years ago. There seemed to have been an intimate seaway contact between these two prehistoric island cultures 250 miles apart.

Bibby had been field director of a Danish archaeological expedition to Bahrain. They brought to light hitherto unknown harbour-cities and temples that rivalled those of Egypt and Sumer in antiquity. Their excavations had convinced Bibby and most other scientists that Bahrain, and not Failaka, had been the Dilmun of Sumerian records. But Bibby had begun to speculate as to whether Dilmun could not have been an extensive maritime empire that embraced the whole island area from Bahrain to Failaka.

The most remarkable piece which the Danes dug up on Failaka was a round stamp seal that could only have come from the distant Indus Valley.[1] It was thin and flat, with a high boss, and bore on its face an inscription in the still undeciphered Indus Valley script. Whoever brought it to this island had been in touch with people of the great civilisation that flourished on the banks of the Indus river and along the coasts of present Pakistan and adjacent India in Sumerian times. As suddenly as it had appeared in full bloom with the magnificent cities of Mohenjo-Daro and Harappa about 2500 BC, just as suddenly and inexplicably had this mighty civilisation bordering on the Indian Ocean disappeared completely about 1500 BC.

Although Alexander the Great had built his ships in the Indus Valley, he came too late to have brought the inscribed Indus seal to Failaka. The Indus script, like the entire cities of Mohenjo-Daro and Harappa, was completely buried and forgotten at the time of Alexander and only rediscovered by the archaeologist's shovel a few decades ago. Thus in the period between 2500 and 1500 BC, Failaka Island had contact not only with Bahrain but with a civilised nation outside the gulf area. People able to read and write in their own characters had ploughed these waters long before literacy spread from the Middle East to Greece and the rest of Europe.

The men on board *Tigris* knew that shortly before we set sail down the river I had gone to Kuwait hoping to visit Failaka. Their interest in what I had seen was clearly genuine now that the island seemed inescapable. In fact, on that visit to Kuwait I had never got as far as the island, although it was only a three-hour boat trip away from the capital. Failaka even had a small harbour with open entrance on the north-west coast, facing Kuwait, whereas all other coasts, including the east side which we were approaching, were blocked by reefs and shallows. But before I found the ferry I called on Kuwait's Director of Antiquities and Museums, Ibrahim Al Baghly, who led me to the local expert on Failaka archaeology, Imran Abdo, the Antiquity Superintendent. I got no further. What I had come for was no longer on the island. Abdo brought his keys and opened glass doors and cases.

Sure enough, here were those precious stamp seals with Sumerian and Babylonian mythical scenes incised upon them, used for sealing the cargo of merchant sailors frequenting Failaka in epochs lost from written history. Among them I caught sight of a motif more precious to me than that of any legendary encounter between Sumerian demi-gods and kings. A ship! A sickle-shaped ship with mast and with crosswise hatching along its curved body to illustrate rope lashings of a reed-boat just like ours.

Enthralled, I stood scrutinising the prehistoric seal in my hand. Abdo, a blue-eyed Palestinian with thirty years of local research behind him, looked at me in surprise. Did I not realise that Failaka had been a very early shipping centre? There was proof of very early contact not only with nearby Mesopotamia, Bahrain and the distant Indus Valley, but also with ancient Egypt.

He dug out of his archaeological treasure-chest a chunk of stone. Just an ordinary piece of rock, but clearly the fragment of something that had been worked, for one side was brightly polished.

'Egyptian granite,' Abdo said triumphantly. 'An American expedition from Johns Hopkins University dug it up on Failaka five years ago.'

We admired the piece together as an art object more precious than gold. Gold could have come to Failaka from anywhere. This particular kind of granite only from the remote Nile Valley. The Greeks did not quarry stones in Egypt to bring to Failaka.

I had hardly finished gazing at the lump of Egyptian granite when Mr Abdo began unwrapping fragments of worked alabaster. 'Look,' he said. 'Cream-coloured alabaster. As in Egypt. Not white as in Anatolia.'

Next he opened with great caution a little box with a thumb-sized sculpture of a beetle. A scarab! An unmistakable Egyptian scarab. The strange symbols incised on it were of local character, but whoever had carved it had somehow been under Egyptian influence.

A tall, slim jar of Egyptian style had also been excavated on Failaka. It bore a slight resemblance to those containing the Dead Sea scrolls, but was of a type not known in Mesopotamia.

Although all this was important evidence of long-range navigation, I had to return to the seals again. By the time we had gone through the whole collection Abdo and I had picked out a total of five Failaka seals that clearly depicted ships. All were sickle-shaped reed-ships with masts. One had a figure seated astern, hoisting or holding the halyard to a big, matted sail. His rope ran through the mast top. Another showed two figures standing, one on either side of the mast, each grasping the lower edge of a reefed sail above their heads. All five ships were engraved on stamp seals from about 2500 BC.

The *Tigris* crew listened to my story from the Kuwait Museum as we ourselves raced towards Failaka on a reed-ship. The night was black. Flickering lamp-light fell on attentive faces as we clung to a table that would have plunged overboard with all of us if not lashed to the deck bundles. Each of us had his own lifeline tied to mast-legs or stays, so as not to disappear in the night waves if an unexpected breaker should tumble in from the port side.

'Gherman,' I shouted across the table to the one who understood English least. 'Do you understand what I saw on those Failaka seals?'

'Ships.'

'But ships in every detail like those you and I saw incised on the canyon walls of Upper Egypt. Similar enough to match like fingerprints.'

I continued my story. I had told Abdo that we were ourselves building such a reed-ship in Iraq to test it in the gulf. He was not surprised. He replied that small boats of this prehistoric type had been used by Failaka fishermen until our own days. The last of them had just been abandoned. He had secured it for the museum. It was the same type as the reed-boats still used in Iraq a few years ago, but built from bundles of palm-leaf stems, because there were no reeds on Failaka Island.

'What about your fingerprints?' Asbjörn's smiling face appeared beside us on the doorsill of the forward cabin. I was slowly getting to my point.

The boats on the Failaka seals did not merely end in high points fore and aft, like our ship, or like those on ancient designs of reed-ships from the Mediterranean islands. On either side of the pointed bow was a long curved horn. I knew this peculiarity from reed-ship designs in Egypt and Mesopotamia. On the best representations they were really shown as animal horns. But on simplified designs the top of the bow just ended in three tips. I had already found this curious detail common to ships shown in numerous Egyptian petroglyphs and on Mesopotamian seals. This symbol even antedated the invention of man's first known script. The triple point on an upcurved bow was the symbol for the word 'ship' in the earliest Sumerian hieroglyphs, which scientists had found to be the same as the earliest Egyptian hieroglyphic sign for 'marine'. Moreover, the Sumerian word for a ship's 'bow' was the same as their word for 'horn'.

I began to see a meaning in all this when Mr Abdo showed me the same 'horns' on the Failaka seals and added that it had been a local custom to place the head of a gazelle in the bow of a ship, either a real gazelle's head or a carved one. And here we came to the fingerprint. Three of the five Failaka seals had the sickle-shaped ship depicted in a most peculiar manner. The deep curve of the deck was incised to coincide with the dorsal outline of a saddle-backed gazelle in such a way that ship and animal became an inseparable unit. The raised neck with the head and horns of the gazelle coincided with the high bow of the ship, while the curved-up rump and tail became the stern. The mast rose from the sagging back of the animal, and the ship's crew was thus at the same time both sailing and riding. In one case two persons flanked the mast while tending a reefed sail hoisted above their heads.

Common to these three seals was the idea of the men on deck navigating and yet at the same time riding a bouncing gazelle. This was perhaps a vivid symbolism of wave motion. But the combination of a sailing vessel's hull with the body of a beast of the field was as special and unlikely to be repeated as was a fingerprint. Repeated yes, but not without some communication between the boat designers. The design had impressed me and puzzled me when I first came across it, in a special publication on *Egyptian* petroglyphs. Ships were reported to be among the most common motifs incised in pre-dynastic time on the naked desert rocks between the Nile and the Red Sea. This had made me invite Gherman to join me in these dried-out canyons, looking for further information on prehistoric navigation. Where deep sand and fallen rocks stopped further progress by jeep, Gherman and I had continued on foot in the direction of the Red Sea. It would seem absurd to look for anything to do with boats in such surroundings: sand dunes, not a drop of water, not a green leaf. Not a place for living creatures other than vultures, flies and desert snakes. Through the barren Red Sea plateau desert wadis ran like empty tributaries into the Nile Valley. Their steep canyon walls and the waterworn boulders at their feet bore evidence that the wadis had been deep rivers in early prehistoric time. The barren plateau around them down to the Red Sea had formerly been amply watered by rain and covered by grass and forests. The change had taken place some five or ten thousand years ago, before the days of the Pharaohs. Whether the climate had changed because the forest disappeared or the forest disappeared because the climate had changed is still a matter of dispute among botanists and climatologists. It is a documented fact that Mesopotamia, too, had been wooded in early Sumerian times. The Sumerians describe their own hills as wooded.[2] The Pharaohs do not. But archaeology shows that there were people in Egypt long before the Pharaohs, and they have something to say as well.

It was well known that the desolate desert wadis we entered were filled with ancient petroglyphs illustrating forest animals and ships. And we found them everywhere. Many had never been reported before, but all repeated the same restricted repertoire: antelopes, waterbuck and other long-horned species, giraffes, lions, crocodiles, ostriches and, in addition to these animal pictures, hunters with dogs, and a great number of ships. Boats and ships of all sizes. Some were propelled by rows of oars, others by mast and sail. As could be expected from pre-dynastic art, all represented strongly sickle-shaped reed-boats. The size of some must have been quite formidable, since anything from twenty to forty oars were common, and a few were shown with a crew of fifty or more on deck. Many had two cabins, one on either side of the mast. A few carried horned cattle or other large animals on board, which were dwarfed in proportion to the big vessel transporting them.

As Gherman and I came out of Wadi Abu Subeira, the wide desert canyon between Aswan and the Red Sea, it was clearer to me than ever that water transport on a large scale had been of paramount importance in the Red Sea area long before man domesticated the horse and invented the wheel. Had the full story which this prehistoric desert art could tell us been properly appreciated? The fact that the Egyptian petroglyphs were surviving examples of the unsophisticated local art in pre-Pharaonic time seemed to have over-shadowed their deeper implications. To me their real value had ceased to be a matter of artistic quality, becoming instead the simple fact that they reflected what the artists had seen in a period leading up to known civilisation. Forest animals and watercraft. Besides the beasts around them the artists had cut into the solid rock their testimony of man's earliest achievements in architecture: huge watercraft for transport and security. Ships were built and depicted long before carts, forts and temples. On his huge raft-ships man was mobile and safe from wild beasts and enemies in days when the land was covered by dangerous forests without roads or walled cities. Beasts and big ships were all the artists thought of in the millennia preceeding Pharaonic time.

The dried-out wadis lead to the Nile. But the distance from the ship designs to the Red Sea is negligible today and was possibly shorter earlier. The wadis might even have been rivers running from the forests to the Red Sea; land lifting, linked with the Rift Valley movements, is fully possible here. The sea-going curves of these pre-dynastic boats were indisputable. With bow and stern elegantly swung high, as on our *Tigris* and often much more so, they bore clear evidence of a maritime background. While they may very well have been used on an inland waterway like the Nile as well, they were not primarily designed or developed as barges or rafts for floating cargo, beasts or people on a smooth river. They were designed for bouncing over ocean waves in gazelle fashion, as we ourselves were doing on *Tigris*, while skidding and jumping across the frothing wave crests. The most artistic composition I had seen from the hands of these early artists had been the combination of sailing ship and gazelle, giving motion to the vessel by letting it bounce like a horned beast over the ocean waves.

Someone knowing this ingenious composition had bounced from the Red Sea area to Failaka before us. The three seals incised with the Egyptian petroglyph motif of a sailing gazelle were indeed fingerprints left on that island. We would soon be there ourselves.

I had been steering most of the afternoon and proposed that we all left the table to get some early sleep. We were heading for troubled water and could expect a lively night. Asbjörn was now in good shape and he and HP took turns in climbing to the mast-top to watch for lights. The chart showed a tall lighthouse built to guide ships to Kuwait around the south-western end of Failaka and its shallows. The lighthouse was marked as seventy-five feet high and visible at sixteen miles. A good mile to the north of it a dangerous reef was marked, with a pile of rocks, which, according to the chart, had a light, but according to the pilot book had none.

I had barely turned in at 8 p.m. when Norman's head appeared in the cabin door informing me that from the mast they saw the light. It was where we had expected it. I still had time for a quick nap.

A moment later he was there again, now visibly worried. We were going fast, and we were heading straight for the lighthouse rocks. The helmsmen were unable to press us past the lighthouse on the left side and into the shipping channel for Kuwait. Too much leeway with this wind. There was no choice but to fall off from 250° to 290° and steer straight into the Failaka shallows to avoid collision with the lighthouse island and the rock pile to its starboard side.

The flashes from the lighthouse in the black night were soon visible even from the deck. Three short flashes in groups, followed by long intervals, while all around us the night was as black as a wall of tar. With our kerosene lamps we could see nothing but the yellow reeds and bamboo that surrounded us in a black universe. There was a glitter of phosphorescent plankton dancing wildly in the wakes of our two rudder-oars, nothing else. The lookout clinging to the swinging mast-top saw no light but the short blinks that approached us with good speed on the left. We were soon going to clear the light on the port side, but where was the reef with the rock pile? It was obviously not lit at all. No modern ship would come on this side of the lighthouse.

Asbjörn lay stretched out on the upcurved reed bundles forward, his head out beyond the bow, to scout for rocks without being blinded by the flickering lamps on our side of the sail and the cabins. The flashes from the invisible tower slid by at some distance on our port side. Were we far enough off to clear the unlit rocks as well?

While we all strained our eyes in vain from deck and mast, Carlo came crawling out of the cabin and said he heard surf. We all listened. Sure enough, we all heard surf rumbling against boulders as a growing rhythmic undertone in a hissing orchestra of seas on all sides. The growing rumble came out of the night somewhere ahead, and to the left, it seemed.

Land. Rocks. On the port side now. Very clearly. Were there more in front? For safety we turned even further to starboard, steering 320°, but saw nothing. The rumble of water on rocks slowly drowned again in the normal roar of breaking seas. Shortly afterwards even the noise of the sea became noticeably lighter. We also stopped rolling. We were in the shelter of something – probably the little island holding the lighthouse to our windward. This was surely the best place to anchor and wait for a favourable wind. Ahead of us were only the extensive shallows and sharp rocks of Failaka Island. Norman shouted from the mast that he saw several faint lights from Failaka along the horizon in front.

I ordered the ship turned all about and the sail down. Thus we could throw out the anchor from the bow where we had tied broad pieces of water-buffalo hide to the bundles to protect the reeds from being cut by the anchor rope. Detlef was in the bow with assistants ready with the anchor. Each man did his job perfectly with halyard, sheets and braces, and the sail was packed up around the yard-arm on deck before the little anchor was thrown overboard. Perfect team-work. It was 10.30 p.m.

Moments afterwards we heard shouts from the bow that drowned in the noises of the night. Norman on the cabin roof thought he had heard that we had lost the anchor. HP shouted the same words from somewhere up in front. At first I thought it was a bad joke. The reefs were just ahead of us. I refused to believe it until Detlef came fumbling out of the dark and reported that the anchor rope had snapped.

123

'Hurry! The other one!'

Fortunately we had a second, smaller anchor, Carlo and Detlef were already busy getting it ready. This time from the stern where it lay. We all checked and double-checked the knots. I repeated again and again: 'Make sure nothing fails. This is the last thing we have on board that can grab hold of the bottom!'

Our second anchor went overboard. We began to get outside the lee of the invisible isle now and with the high bow and stern catching wind we picked up speed. The anchor did not seem to take hold. Someone quietly suggested that perhaps we had lost that one too. Silence. Detlef tried the rope. Began to pull in the slack. An empty end came out of the black water. None of us uttered a word. Detlef remained speechless and motionless. The young German captain, used to weighing tons of anchor chain by pressing a button, just stood there with the short piece of rope dangling from his hand.

Once more we all searched our minds for just anything that could stop our drift if we threw it out on a rope. We had nothing.

We were picking up speed and drifting straight for the unbroken cliffs and rocks awaiting us, no matter how we turned our rudder-oars now. We could clearly see the many dim lights scattered along the invisible coast of Failaka. They soon spanned the entire horizon where the wind was carrying us, and we could not escape on any side, even if we hoisted our sail.

If only it had been daylight we would have seen land. We might have picked our course and steered towards some less rugged part of the shore. Perhaps there was some small opening in the barriers even though the pilot book said there was no landing on this side. But it was now just before midnight. We would strike the rocks before we saw them. We would be shipwrecked in complete darkness.

No reason for panic. A bundle-boat like ours was the safest craft we could wish to be on at a moment like this. Ropes and reed bundles might be torn to bits but would protect us from being beaten to death against the reefs. Thank heavens we were not heading for the surf on the coast in a plank-built boat; then we would all have been in the utmost peril. Now our proud ship was probably doomed, or at least would have to be rebuilt. And this even before it had been put to a fair test. Certainly both we and the ship would have been much safer if we could see something around us. We were like eleven blind men. How could we swim between reefs or jump on to rocks when we could not even see our own hands or feet without lamps?

There was obviously one thing to try: slow down as much as we could. Then we might delay our crash landing until dawn, and also hit the rocks less violently.

'Throw out the sea anchor!' It was hanging ready on the bridge floor under my feet.

The sea anchor was nothing but a semi-floating canvas bag to be trailed in our wake. It was a simple device used on sailing vessels to keep either the stern or the bow turned into the wind when sails had to be lowered in a storm: a long conical bag without a bottom.

The sea anchor went overboard and acted as a most effective brake. In fact it reduced our speed of drift so effectively that we began to notice that all the lights from the island remained in the same spot. And so did the blinking from the lighthouse behind us. Never had I seen a sea anchor function so marvellously as this. With our high stern turned to the wind and waves we just hung as if fixed in the sea. Little did we realise in the black night that the choppy sea around us was so shallow that the canvas bag had caught the bottom. Instead of floating just below the surface as it should, it had bobbed up and down in the wave troughs digging up mud until it was full and heavy and grabbed a good hold in the bottom mire.

The lights we saw here and there at long intervals all along the invisible coast were yellowish and feeble. They all seemed to be from kerosene lamps like ours. They certainly came from scattered houses or huts. We were close enough for our own lights to be seen from the shore, so I sent a few SOS signals towards land with flashlight. But no reply. To the left over Failaka the black night sky reflected the barely noticeable glow of a distant modern city: Kuwait, thirty miles away. We heard the sound of an aeroplane passing above the clouds.

On behalf of the consortium, the BBC had equipped us with a tiny radio transceiver for the dispatch of expedition news. Norman tried to contact Kuwait coastal station which had a twenty-four hour watch just on the other side of this island. There was no answer. Nobody could hear us anywhere.

We shared double watches for the remainder of the night and slept fully dressed with lifejackets as pillows. The sea anchor gave us some peace of mind. Some. Norman seemed to be constantly fiddling with another small transceiver which a radio amateur friend had given him because Norman did not trust the one the consortium had insisted on. He had rolled up his mattress and pulled up all his gear from the boxes he slept on. Since Norman and I slept feet to feet I had no room for my legs except on top of his pile.

I had scarcely fallen asleep after my midnight to 2 a.m. watch when I awoke and crawled out on deck again. This was too risky. At intervals we again sent light signals to the island. Three short flashes, three long, and three short again. sos. No response. Most of the lights ashore had been turned off anyhow, and I crawled to bed again and left the watch to Toru and Yuri.

I had been out and in again a couple of times to check the direction of the lighthouse and the wind when Yuri called me and said, 'They are answering!'

It was still pitch dark. 5 a.m. I crawled out and, yes! I too saw a strong light directed straight towards us in long, well-spaced flashes. It appeared just to the left of one of the few dim house lights still lit. A very strong electric signal light. From a boat? No, it had to be ashore. It was quite immobile.

For a long while I kept on morsing sos and each time they answered T, meaning 'received'. We rejoiced that someone would soon come to guide us through the reefs or tow us around the corner of the island. At last we had contact with other people.

Shortly before 6 a.m. the signals from ashore stopped. Now we began to detect the indistinct contours of Failaka. A long, treeless and very low island with its highest point some forty feet only. I kept on morsing, 'sos no anchor we need tugboat sos'. No further reply. Next we began to see the outlines of three small ships at anchor. Probably dhows. They lay together just where we had seen the signal light. There seemed to be a reef between them and us. One of the boats began to move, carefully, as if manoeuvring out through a difficult channel. With binoculars we could see that the men on board were watching us. But they only turned away in the opposite direction once they reached open water. Without sail they disappeared around the island and no other living soul could be seen anywhere.

Norman had gone to sleep at last after sending a blind radio message into the night, repeating several times the same few words we had sent with our flashlights. Detlef refused to sleep, as if it were his fault that we had lost the anchors. He was on the bridge, fiddling with the tiny radio box, when he heard a faint voice calling 'Tigris, Tigris, Tigris' followed by babbling in an unintelligible language. He called Rashad, who could only confirm that the voice did not speak in Arabic. Then the two plainly heard the word 'Slavsk'.

'Yuri! Yuri!'

In two jumps Yuri was on the bridge. *Slavsk* was gone. Nothing. Norman came and called 'Slavsk, Slavsk, Slavsk'. Suddenly a voice was there again: Captain Igor! Clearly now, so that we all heard him. Yuri's face beamed with pride and happiness in the first glow of the rising sun as he translated: *Slavsk* was already weighing anchor. Igor wanted our position. Yuri also learnt that a twenty-knot wind from the same southerly direction was expected in our part of the gulf that day.

As daylight broke we saw a few small huts far apart along the coast. No smoke. No people. The sea around us was not blue and clear like it was yesterday, but greyish-green, filled with sand or mud. Sloppy brown branches of loose seaweed were dancing everywhere in the choppy sea. We were deep inside the Failaka shallows, with hardly room for a flounder to swim under our ship. As we gathered in high spirits around the plank table to enjoy Carlo's Sunday morning oatmeal porridge we noted that all the seaweed suddenly began to swim. As new clusters came towards us from the bow, the others went away behind the stern. As we checked the nearest cluster, we found it just circling up and down with the waves in the same grainy water. It was *Tigris* that had suddenly picked up speed and resumed the drift with the south wind.

The tide had played us another trick. The incoming high tide had pulled in the same direction as the wind and at the same time lifted the sea anchor free from whatever it had caught. Full of muddy clay and heavy as a sack of cement it now dragged along the bottom and we would hit the reefs before *Slavsk* had time to reach us. The drift was straight for the island.

We had a choice. If we did nothing we would soon be wrecked against the limestone reefs and cliffs of Failaka. If we hoisted sail and used our rudder-oars we could take the wind athwart from starboard and sail parallel to the island coast. This would lead us directly into worse shallows, thirty miles wide, dotted with reefs, which separate Failaka from the mainland of Iraq. I favoured the second alternative, and everybody agreed. If we managed to sail safely between the reefs and across the shallows in that direction, we would be stranded on 'swampy low land' that would not damage our reed-ship. All the coastline from Iran to Kuwait would be more or less of the same type we had seen at the mouth of the Shatt-al-Arab. Our only problem there would be the lack of arms on board: we were now heading for the very coast described in the previous year's *Persian Gulf Pilot* as probably unsafe without armed escort.

The tidal current came in full force and the wind increased in strength. We left the sea anchor out so as not to be too far displaced from our given position before *Slavsk* found us. With our port side turned to the island we began to roll badly. Rashad and Asbjörn were seasick. Gherman crawled in and fell asleep. In the wave troughs the sea was so shallow that grey mud was whirled up in thick clouds from the bottom. No sign of any kind of life ashore, but in the binoculars we saw a few round elevations that probably were some of the prehistoric burial mounds. If the ancient navigators had frequented this side of the island they must have been truly expert mariners, unless the reefs and shallows were of more recent date.

It was not yet noon when we sighted *Slavsk* as a black spot on the eastern horizon. Half an hour later an orange-coloured motor launch was lowered and came dancing towards us across the green shallows. The big ship anchored in blue water three miles away, as close to the edge of the shallows as was prudent. We recognised Captain Igor in the bow of the bouncing launch. Water cascaded around him and his crew in orange lifejackets. With a broad smile and open arms he jumped on board our reed bundles and embraced us.

'My father!'

'My grandfather!'

Our sail was down. All was joy now. The men with the orange lifejackets in the orange lifeboat began pulling up our sea anchor. It was as heavy as lead, full to the brim with greyish mud. It had saved us from shipwreck in the dark.

No sooner was the sea anchor up and emptied before we and the lifeboat began to drift off together with great speed. The lifeboat crew started their engine and began to tow. But although they had a powerful motor the wind took a strong grip on the high curves of our reed-ship, with the result that *Tigris* began towing the Russians. This was not quite apparent at first, as the tidal current ran our way. Captain Igor and his heavily built second-in-command had a dinner of salt meat and peas with us around our crowded table, and we all rejoiced and wanted to make believe that all went well. But in the afternoon it became clear to all that the lifeboat was only drifting away with us along the coast and towards the much worse shallows east and north of Failaka. The houses moved even more to the left of their earlier bearing and the tall lighthouse on the little island began to sink. *Slavsk* became more and more indistinct, and in the afternoon it disappeared completely from view. The wind increased in force to twenty-four knots.

At this moment a primitive-looking dhow with a powerful engine and shallow draft suddenly appeared from nowhere. The former mast had been sawn down. It headed for us, but did not come close. It circled around us beyond hailing distance and although several men were on board they were clearly afraid of getting close. No wonder. A great many generations had passed since their ancestors had seen a vessel in these waters as strange as ours.

We inflated our tiny rubber dinghy, brought along for filming and barely big enough for three men. Asbjörn, as our dinghy captain, took Rashad with him and rowed close enough to the frightened fishermen for Rashad to shout to them in their own language. A moment later our two men came back and reported that the crew of the dhow wanted 300 Kuwaiti dinars, worth more than a thousand dollars if converted into American money, to tow us out to *Slavsk*.

I was ready to start bargaining, but Igor refused to listen and told Rashad to row back again on his behalf and offer six bottles of vodka and two cases of wine. The two boys went over a second time but came back with the message from the men in the dhow that they were Moslems and did not drink alcohol. The seamen in the dhow now realised our condition and ventured to sweep close up past our side before they took off with no further bargaining and seemed to resume fishing, barely visible on the eastern horizon.

The only solution left to us was to hoist sail on *Tigris* and abandon further attempts at a tow back to the long since invisible *Slavsk*. If we sailed as we had done before, continuing with our port side facing the coast, the Russians could try to pull our bow slightly into the wind and away from the coast. In this way we could possibly fight ourselves gradually out into deep water before we came into the much worse area awaiting us off the eastern end of Failaka.

Captain Igor had a walkie-talkie and spoke to his officer in charge on *Slavsk*, who reported that their ship, with a total of six metres draught, had ventured into four metres of water, but was now going out to follow us as close as possible along the outer edge of the shallows.

Very soon *Slavsk* reported on Igor's walkie-talkie that their radar showed us approaching the danger area with great speed. There was a terrible current. The bottom mud whirled up around us from ever deeper wave troughs.

Now a second dhow appeared from the direction of Kuwait. While the crew of the first had looked like fishermen, this lot looked like real bandits. They, too, kept at a good distance, and when Rashad went over and explained our trouble they doubled the price in Kuwaiti dinars and wanted the equivalent of two thousand dollars. They did not yield a penny and let us know in plain words that if we did not pay what they asked for they would get everything the moment we went on the reefs we were heading for. Again Igor was furious and refused to witness any deal with bandits who wanted ransom. His angry gesticulations left no need for translation, and uninterested in further bargaining with me the newcomers broke contact. The dhow picked up speed and left. The last Rashad heard was a cynical warning shouted back at us: without their help we would all be doomed.

We were all filled with contempt and anger. If Captain Igor could have waded with his head above water I am sure he would have jumped overboard and started to tow us himself. But for all their strength and willingness the Russians could not make the lifeboat pull more strongly, and we on *Tigris* had not yet discovered how ancient mariners in these waters had been able to tack into the wind.

Far ahead we saw the two dhows apparently anchored near the dangerous reefs, waiting like jackals for our disaster. We began to realise what awaited us next night if we ran aground. The pilot book was evidently not unjustified in its warning.

We had to take the sail down. The Russian lifeboat anchored in water barely six feet deep and we hung on together hoping for a better wind. No sign of *Slavsk*.

The sun was slowly on its way down towards the horizon when a third dhow appeared. The mast was cut down as on the other two, and with a minimum draft and strong engine the crew were clearly at home in the local shallows. Precisely the same process was repeated, for the third time. Even the ransom was the same as the last, as if this was a mere routine. But as the pilot book stressed that the low coast ahead of us was rarely visited by Europeans, I began to suspect that the three dhows probably had contact with each other by radio. By walkie-talkie like Igor's. They were certainly no fishermen.

'Look at that man on the pillows.' Detlef was at my side with binoculars. We all took a good look, and the men in the dhow seemed unimpressed by our long range attention; they seemed to have known before they came what our situation was.

What Detlef had pointed out was a fat, criminal-looking man with big turban and crossed legs who sat on pillows and scrutinised us with contempt and calculation. His fat hands had certainly never touched a fishing line and he was the archetype of a hardened crook. The others were a mixed lot, none of them to be trusted behind one's back. Some wore turbans and might be Pakistanis, a few less fierce-looking could be Arab seamen of a sort from Kuwait.

That these men asked ransom money was clear to all, and Captain Igor again vigorously opposed my entering into any kind of deal with them. It was equally clear that if we did not pay, they would just hang around with the other two gangs and harvest all we had in the black night if we were forced to jump on to some reef or the swampy land behind. We could not trust the Russians' anchor. In the vast shallows ahead no coastguard or customs officer would ever disturb these people in whatever business they were up to. It was certainly not fishing. Perhaps smuggling of drugs or dutiable goods to Failaka from the other side of the gulf. We had heard that even human labour was smuggled to rich Kuwait from Pakistan by organised gangsters. If people or goods could be brought ashore on this deserted side of Failaka Island, then the back door to Kuwait was open. Failaka was Kuwait territory and there was a regular ferry service from the other side of the island direct to the capital on the mainland.

In an hour or so the sun would set. This dhow was clearly the last chance to get out into deep water before the world around us again disappeared from sight. Captain Igor was still furious at the mere idea of dealing with gangsters. To me this had become a dilemma. I now felt a double responsibility. My own men had volunteered to confront the hazard of the experiment we were involved in. Our ship was of a type that could bring us safe up on rocks and reefs, so long as there were no vertical cliffs. But now we were dragging a lifeboat full of Russians with us into the shallows, their big ship circling around somewhere outside the reefs without captain.

'Captain Igor,' I said, 'now I accept that I am your father. Then I am in command. I will consent to pay the dhow.'

I could see how Igor fought to keep his mouth shut. He had no comment when Rashad passed my acceptance on to the man on the pillows. I crawled into the cabin and rolled up my mattress to look for my dwindling supply of cash. I had no Kuwaiti dinars. But the pirates had agreed to be paid off in Iraqi dinars, which was strong cash in these parts. It was lucky that I had enough in reserve for the gulf area.

131

The bandits were not willing to yield a penny and even insisted on keeping Rashad as a hostage until the ransom was paid when we reached the edge of the shallows. Without the slightest sign of fear, Rashad agreed.

The wind had somewhat abated and yet the dhow had difficulties getting us moving in the right direction when they threw a tow to the Russians who in turn had us at the end of their line. Slowly the procession of three small vessels began to work to windward without any sign of *Slavsk*.

The night fell over us again, as black as the last. Except for our own kerosene lamps and modest flashlights, there was nothing to be seen in any direction. No lights on the island. No lighthouse. No *Slavsk*. Whether the other two dhows were going away, or coming, was anybody's guess.

Captain Igor was still with us on the reed-ship and could report that *Slavsk* was about seven miles away when the towing began. We seemed to be going through an extremely shallow area, for the pitching and rolling was formidable. Our flashlights sometimes showed seaweed dancing in the waves of an almost milk-coloured sea. The rascals ahead of us in the pitch dark did a good job. It proved to be a long passage. Then, at last, Norris shouted from the mast ladder that he saw the lights of a ship on the starboard side forward. The lights grew bigger. It was *Slavsk*.

The sea ran really wild, with tall swells at the edge of the submerged reef when the pulling ceased. The dhow suddenly appeared close to our side, now with a kerosene lamp lit, like on *Tigris*. This time we feared that they might come too close, for as one boat rose up on a wave crest the other sank down in opposing rhythm. Their wooden gunwhales might rip the reed-rolls of *Tigris* to shreds. It was a truly violent dance, but we had to venture close to pay the ransom and get Rashad back. I stepped on to the side roll of *Tigris* with a grip in a stay and stretched out over the water as far as I could with a thick stack of dinars in my free hand. A dark-faced Arab sailor on the dhow stretched out as far as he dared to meet me, while the rest of our men worked with bamboo rods to push the two wild vessels apart if they should get entangled in the dark. The man grabbed the cash, and now we clung to the side of the dhow as the pack of paper money was brought to the chief on the pillows. He took his time counting the wad of notes one by one while someone held the lamp over his turban-covered head. Then he nodded and in a leap Rashad was with us. *Slavsk*, fully lit, was now coming to our side. The men on the dhow blew out their lamp and like Aladdin's genii they vanished into the darkness. There was no further sign of them.

We were dancing along with the men in the orange lifeboat and had to take great care not to be smashed against the steel hull of the big rolling ship that approached us. *Slavsk* was alternately pitch black and blood red as we were tossed up and down past its waterline. Both the lifeboat and *Tigris* were in danger, first from the suction of the big revolving propeller, next from the bottom platform of the staircase lowered from the lofty deck of the big ship. It rose and fell like a giant piston, one moment high above our heads and the next disappearing with a splash deep into the turbulent waves. It was difficult enough for the crew of the lifeboat to get on to the platform before it escaped over their heads or sank into the black sea. It was worse still for the four men who had to repeat this wild performance from *Tigris* to *Slavsk* by way of the riotous lifeboat. We all held our breath when Captain Igor and his mate jumped. Igor almost tumbled into the sea as the lifeboat shot skywards just when he jumped down. Yuri and Carlo followed in a fraction of a second. The four men were then lost in the darkness until we saw them all enter a light-beam as they hurried up the long stairways at the side of the *Slavsk*.

Yuri had whispered to me before they jumped that Carlo had a serious leg infection. He wanted to take this opportunity to clean and treat Carlo's leg carefully.

We pushed off with our long bamboos to avoid getting sucked in and cut to pieces by the propellers. Soon we hung on a long rope behind the empty lifeboat, now in the tow of its Russian owners. We felt violent jerks from the bow each time the ship, the lifeboat and *Tigris* rose and sank out of time. Igor had refused to let us loose before we were safely out of the reach of the jackals. He had promised to go as slow as his pistons could churn the propellers, for I had made it clear that nothing was harder on a reed-ship than to be towed in open sea. The risk was far greater that the short reeds would be pulled apart by a jerky tow than if they danced free in a hurricane. Norman and I, the only old-timers left on board, had our hands on our knives several times, ready to cut the tow-rope if we felt the bow might be ripped apart. But nothing seemed to happen, and we fell asleep, dividing the watches on the rudder-oars among the nine of us left.

How crazy it was to feel the breaking seas hammering against a reed-ship straight from the bow.

CHAPTER 5

To Dilmun, the Land of Noah

ADAM AND Noah have one thing in common: they are the only two men we all descend from, according to the beliefs brought by the Hebrew patriarchs from Ur. From the Garden of Eden we had come on *Tigris* to the waters where the story of Noah's Ark began. A thousand years before Abraham heard it in Ur, the Sumerians had told it to their children in the same city-port. In these waters, they said, a big ship had once been built by the progenitor of all peoples as he complied with the orders of a merciful god who wanted to save mankind from complete obliteration in a terrible flood.

While fields and homes were submerged the big ship resisted the fury of the raging elements by floating upon the waves. In the Hebrew version the builder of the ship afterwards thanked his God who set the rainbow in the sky as sign of his covenant with the survivors. In the Sumerian version he prostrated himself to the reappearing sun in gratitude.

The sea was moderately rough as the same Sumerian sun rose above the former Sumerian waters and filtered the first rays of the day through the cracks of the woven cane wall upon my closed eyes. With mixed feelings I awoke and gazed through the open doorway at the flowing disc that slowly rose with majestic dignity from its bath in the sea. Beautiful. Magnificent. Clean, virgin light was being lit for a new day. I welcomed the beautiful sight whole-heartedly, for I felt as if the sun had lit another hope in my sombre spirit. After all, we were safe and free, free to start all over again with a ship that was still in good shape.

It had been a restless, unpleasant night, with violent jerks from the tow-rope. When the snatches at the bow had been too brutal I had suffered in my sleep, as if it was I who was pulled by the hair and not *Tigris*. I had been out on deck to check that the bow was still with us. Each time it was with strangely mixed feelings of security, fear and disappointment that I observed the steady lights from a big ship right ahead. But of course it was the friendly *Slavsk*, with Captain Igor, Yuri and Carlo probably sound asleep aboard, not feeling, on their big steel ship, the nerve-racking snatches from the rope that held us together.

The strain on our ship was sometimes scaring. Terrible shrieks, cracking and gnawing noises came from ropes, lashed wood, bamboo and berdi. The huge shafts of the rudder-oars banged and hammered from side to side in their wooden forks so hard that they literally shook the ship and could be felt as veritable shocks through the wooden cases on which we slept. Detlef and I were out once and worked in the dark to lash the shafts into a tight position. This stopped the terrible hammering.

During the night Norman and I had also been out together to toss bright bits of berdi from the bow, which we followed with flashlight in the black water and timed as they passed the ten-metre mark on the side bundle. Thus we checked that the engineer of *Slavsk* lived up to Captain Igor's promise not to go faster than two knots, the speed we had been sailing towards Failaka. The lights of *Slavsk* ahead of us gave me continuously mixed feelings of relief and disappointment. Relief from the burden of responsibility I had felt for all the men some hours ago in the shallows of Failaka. Now the rascals and the reefs disappeared ever further behind us in the dark. But we were being towed away. We had failed to escape under our own sail. But for the human vultures somewhere out there in the night it would probably have been safer for our sickle-shaped reed-stack to have run its bow into the mud flats; it would have been safer than being dragged to windward, violently bumping every five seconds into the rising wall of a contrary wave.

Now the sun shone freely above the horizon. Someone whistled a merry tune in the open galley nook. I could discern the movements of the whistler through the thousand cracks in the cane wall and catch the pleasant odour of something reminiscent of pancakes. It must be HP. His sleeping bag was empty. The others, except the watch on the bridge, were still asleep, probably relaxed and happy to be towed along. Perhaps not Norman, for he was dead set on solving the sailing problem.

The purpose of our reed-ship experiment this time was not merely to float and drift, but to navigate. Therefore the beginning of the voyage had been a glamorous failure which we could only laugh at as we gathered at the breakfast table. The southerly wind was still dead against us, first of moderate strength, then increasing again in force. We even had to put on wind jackets to enjoy our meals at the unsheltered table.

'Are you sure the ancient people could have done better than us,' HP queried. 'Maybe they just hung on wherever they were until the wind blew more in their direction.'

'After all, we too could at least pick our course to span half the horizon,' Asbjörn added. We all agreed that we could steer successfully 90° to either side of a following wind.

Suddenly we observed that the violent hugging and lugging came at much shorter intervals. Water cascaded in front of the bow. We were going faster. We hurried up to the cabin roof and waved desperately for *Slavsk* to slow down. This was crazy going. But nobody on the ship ahead saw us or understood our signals. The lifeboat midway on the tow-line was empty, it bounced on the water worse than we did, being pulled in two directions with ropes fore and aft. Before I could make up my mind whether to cut ourselves free with a knife or try to contact *Slavsk* by radio, the tow-line broke in a last violent jerk. Our speed slowed down as suddenly as it had picked up. *Slavsk* went on alone with its empty lifeboat.

Asbjörn climbed over the bow and reported that the one-inch tow-line had broken close beside *Tigris*, but what was a real shock to all of us was that he had found a huge hole torn in our bow. Fragments of loose reeds were in fact floating behind us. In front of the hole the spiral ropes hung loose across a cavity as large as a dog-kennel. This was a frightening discovery. We forced our fingers under the lashings on deck up front, and pulled to feel the tautness where they passed the cavity. The spiral rope was still as tight as if glued to the reeds. The reed bundles had fortunately swollen so much that they squeezed the lashings fast between them. With a large hole in the bow we thus apparently faced no immediate catastrophe. But we had to find a way of filling the hole before it grew too big. The reeds around it would now loosen one by one and gradually cause the whole bundle-boat to fall apart.

Slavsk came back in a great circle and Captain Igor appeared with a megaphone. He refused to let us remain loose. He had contacted his shipping company in Odessa and they had approved his actions. He had even sent a message to the Ministry of the Merchant Marine in Moscow, and Minister Gujenko had personally authorised *Slavsk* to tow the reed-boat *Tigris* 'to an area of safety'.

'There is no such area short of Bahrain,' assured Captain Igor.

And his men began to throw us a new tow-line from their tall ship. No matter how they threw it, even with a life-ring on the end, it was sucked in by the colossal propeller, and so were we. The propeller had to churn around, otherwise the rolling ship would be as much out of control as we were without a sail. Even with the engine of *Slavsk* running, its tall iron wall was swaying over our dancing reed-ship while both vessels were tossed to and from each other. Our waving bipod mast and our elevated reed ends were close to being crushed from above whenever we tried to fish the rope end out of the whirls of the propeller. It was two hours before we managed to toss our own thin line up to the men on the rolling ship and pull it back with the thick tow-rope tied to it. Captain Igor shouted that the ship's engineer had unfortunately been tempted to raise the speed because going dead slow in these seas was harmful to the propeller shaft. But they would never again exceed two knots. We also made clear that we wanted to hoist our sail and continue on our own the moment the wind turned to normal and permitted us to set course for Bahrain.

We on *Tigris* were all curious to see our position in relation to Failaka and Bahrain, and Norman rolled out a very illustrative map sent us by the National Geographic Society. It was a sort of historical map entitled *Lands of the Bible Today*, with archaeological annotations such as Abraham's route from Ur and other pertinent data taken from the Bible as well as from archaeological finds. The 'Persian Gulf' showed up beautifully in blue with yellow islands. Norman put his finger at our approximate position. He then read aloud the text that happened to be printed beside his finger: 'Earliest Sumerian records refer to shipwrights and seafaring people. Some of man's earliest ventures on the sea occurred in the Persian Gulf.'

This fired the curiosity of everyone. What were these records about? Had I read them? Certainly not all. But probably all that dealt with seafaring provided that they had been translated from cuneiform script into European languages. Perhaps I had under-estimated the interest of my companions in the actual background of our adventure. We could hardly expect a better opportunity for a quiet get-together than now as *Slavsk* towed us at sailing speed past ships and oil platforms. I crawled into the cabin and came back with a bag full of pocket notebooks replete with scribblings from my own researches. Notes from museum exhibits and store rooms and quotations from scientific books and learned journals, like those I had studied in the Baghdad Museum Library. I opened them one by one and looked for hand-written extracts underlined in red.

There was a quotation from an essay entitled 'The Seafaring Merchants of Ur' published in a scientific journal[1] by a noted authority on Sumerian culture, A. L. Oppenheim. He was of the opinion that the most interesting information contained in some of the inscribed tablets from Ur

has to do with the role of the town of Ur as the 'port of entry' for copper into Mesopotamia at the time of the Dynasty of Larsa. The copper was imported by boat from Telmun [i.e. Dilmun], today the island of Bahrain, in the Persian Gulf. This 'Telmun-trade' was in the hands of a group of seafaring merchants – called alik Telmun – who worked hand in hand with enterprising capitalists in Ur to take garments to the island in order to buy large quantities of copper there. Since the island hardly yielded any ore – not to speak of the fuel needed for smelting – we are faced here with a situation which is typical for international trade on a primitive level: Telmun served as 'market place', a neutral territory, in which the parties coming from various regions of the coastal area of the gulf exchange or sell the products of their countries. . . .

Since Telmun was only a market place, two possibilities have to be envisaged: the ivory obtained there by the traders of Ur could have come either from Egypt – through some unknown commercial channel – or from India brought across the Indian Ocean on boats sailing with the monsoon. In favor of the second alternative speak the well established links between southern Mesopotamia – especially Ur itself – and the civilization of the Indus Valley. The discovery of Indian seals . . . and of specially treated carnelian beads . . . in Mesopotamian excavations has proven beyond any doubt the existence of such trade relations. We now may very well add ivory to the list as an item based exclusively in Mesopotamian sources on philological evidence, while we have from Mohenjo-Daro actual ivory combs . . .

Thanks to the deciphering of the inscribed tablets, scholars like Oppenheim can give us a good idea of what life had been like in the gulf ports of Mesopotamia in Sumerian times. Shipbuilding, navigation and maritime commerce was the second largest occupation in ancient Ur, surpassed only by agriculture. Maritime activities were extremely well organised and formed the basis on which Ur founded its economy. From Ur river boats carried the gulf trade up the two rivers to other peoples as far north as present-day Turkey,

Syria and Lebanon. The harbours as well as the network of inland canals were dredged and well maintained under the supervision of high officials directly responsible to the king. The harbour authorities imposed taxes on importing ships, and the captains carried sealed documents concerning vessel and cargo. Oppenheim quoted part of a legal document in the form of a clay tablet from Ur, in which the captain's responsibility is spelt out in cuneiform characters: '. . . the well-preserved ship and its fittings he will return to its owner in the harbor of Ur intact . . .'

Ships were so much part of the daily life of the Sumerians that they even entered into their proverbs: 'The ship bent on honest pursuits sails off with the wind, Utu [the Sun-god] finds honest ports for it. The ship bent on evil sails off with the wind, he will run it aground on the beaches.'[2]

Gherman jokingly commented that with this proverb true we were damned close to getting ourselves a bad reputation by our manoeuvres in the river and off Failaka. Detlef wanted to know more about Sumerian ships and their cargo. I looked for another notebook: one with extracts from the writings of Armas Salonen. Nobody had done more research into these topics than he had. I admired his intellect although we were probably going to prove his verdict on the buoyancy of berdi wrong. But I knew nobody who could present his findings in a way less likely to become a bestseller. In a most academic and almost unintelligible treatise intended only for his fellow readers of *Studia Orientalia Edidit Societas Orientalis Fennica* this erudite Finnish scholar presents more than two hundred pages in a mixture of German, Greek, Hebraic, Latin, Arabic, French, English, Sumerian, Babylonian and Akkadian languages, amassing all that the learned world has recorded of fragmentary references to ancient Mesopotamian ships and cargo since the days of Alexander the Great. Most of his sources were precious extracts from cuneiform Sumerian and Akkadian clay tablets.

Salonen stressed that the first ships in the twin river country were reed-ships, and that, with them as models, the earliest wooden ships were built later. He begins his treatise by pointing out that the evolution of shipbuilding in Mesopotamia is the same as that of ancient Egypt, where reed-ships also formed the prototypes for later wooden boats. He shows that until historic time all the original types of Mesopotamian watercraft, in one form or the other, continued to survive side by side: the reed-boat, the goat-skin pontoon-raft, the basket-boat of coracle type, and the plank-built wooden ship.

139

First of all he assembles references to the *ma-gur*, which he refers to as 'the sea-going ship', 'the god-ship', 'the ship with high bow and stern'. This, he says, was the type depicted in the oldest ideographs for 'ship' before the cuneiform script was invented. It is also the traditional vessel incised on the earliest Sumerian cylinder seals. It was the type of ship used by the demi-gods and divine ancestors before Ur was settled, originally built from reeds, not from wood.

Salonen specifically refers to the Egyptian reed-ships as built of papyrus, but he has no comment on the kind of reeds used for the original sea-going ma-gur of Mesopotamia. He translates the Babylonian word for reed-ship, *elep urbati*, into German as *Papyrus-boot*, but there is no evidence that papyrus ever grew in Mesopotamia. For botanical reasons we cannot escape the conclusion that the sea-going Sumerian reed-ships were built from the same berdi as that which dominates the local marshes today. It was now up to us to find out whether berdi cut in August might not float just as well as papyrus.

The national heroes of the Sumerians, the important ancestor-god Enki and his contemporaries, sailed to Ur from distant Dilmun in ma-gurs of reeds. But in subsequent Sumerian times ma-gur also remained the term for the largest of the ships used in the gulf for merchant adventures, even those that followed the original reed-ships in form although built from split timber. Timber became, together with copper, one of the principal cargoes freighted to Mesopotamia in the largest of local sailing ships.

Salonen shows that the Sumerians had names also for four other types of wooden ships, two for mere river traffic, one for normal sailing both on river or sea, and one a simpler freighter or cargo barge. However, in the functions of the temple priests and other religious performances, it was the 'god-ship', the original ma-gur, that was invariably represented.

A common measure given for a ma-gur was 120 *gur*. A gur was unfortunately a measure of varying value, sometimes given as the equivalent of 80 gallons and sometimes as 121 litres. In either case it would be ships roughly within the ranges of the modern dhows. Oppenheim, subsequent to Salonen's study, came across references to ships from Ur recorded as 300 gur. He referred to them as exceedingly large, and they were indeed: they would have made the two *Ras* and *Tigris* seem small.

Salonen quotes excerpts from tablet texts referring to passenger-ships, ferry-boats, fishing vessels, battle-ships, troop-transport ships, and privately-owned as well as chartered merchant vessels. 'Life-saving boats' are also mentioned, indicating that larger ships carried lifeboats in case of emergency. The ships even had personal names, like ours today, after towns, countries, kings and heroes. Some even had more romantic names, like 'The Morning', 'The Life-protectress', or 'Heart's Delight'. Many carried names referring to their special cargo. To judge from these names the transport was by no means restricted to timber, copper, ivory and textiles, but included such trade goods as wool, canes, reed-mats, shoe-leather, bricks, quarried stone, asphalt, cattle, small live-stock, hay, grain, flour, bread, dates, milk-products, onions, herbs, flax, malt, fish, fish-oil, vegetable oil and wine. One tablet refers to sixteen men needing two days to pull their ship into port and unload it on the docks and another day to move all the unloaded cargo into the warehouse.[3]

Our conclusion from Salonen's analysis was that *Tigris* clearly would have to be classed as a ma-gur, a 'god-ship' of early model. This made sense to everybody on board, especially as I had stressed that we were in search of the very beginning, and Sumerian history began with navigating gods, not with merchant seamen.

And I was not joking. It was sometimes easy for us, who have inherited Abraham's monotheistic religion from Ur, to forget that the term 'god' had a different meaning for the ancestor-worshippers who had another religion. Semitic tribes blended at an early date with Sumerian intruders in what is today southern Iraq, and there must have been a blending of old religions as well. Like Abraham, the Sumerians traced their list of kings back to the boat-builder who saved mankind from the flood, but whereas the Hebrews believed that kings and men alike descended from Adam, the Sumerians made a clear distinction between commoners and kings. Like the Egyptians and the early culture peoples in Mexico and Peru, they believed that their royal families were the divine descendants of the sun. They were sun-worshippers combined with ancestor-worshippers. The King was venerated as a human god even while still alive, and his rank among the deities was higher the further he was counted back in the royal genealogies. The sacred kings who first came to the Sumerian coast, and whose descendants founded the First Dynasty of Ur, would necessarily be classified as gods by the scribes who recorded ancestral events on tablets in the centuries before Abraham's departure. If we dismiss the Sumerian gods as mere mythical creatures, we should have also to dismiss all their royal families from first to last. The real problem is to disentangle the transition between Sumerian history and Sumerian myth way back where terrestrial god-men are blended with bird-men and celestial bodies.

I had on board also a little book called *The Sumerians*, written by another noted authority on Middle Eastern archaeology, Professor C. L. Woolley. In his first chapter, termed *The Beginnings*, he goes straight to the point:

Sumerian legends which explain the beginnings of civilization in Mesopotamia seem to imply an influx of people from the sea, which people can scarcely be other than the Sumerians themselves, and the fact that the historic Sumerians are at home in the south country and that Eridu, the city reputed by them to be the oldest in the land, is the southernmost of all, supports that implication.

The same scholar ends his book by stating how difficult it is to estimate the debt which the modern world owes to the Sumerians, a branch of mankind so recently rescued from complete oblivion. The Sumerians, he says, merit a very honourable place for their attainment, and a still higher rank for their effect on human history. Their civilisation lit up a world still plunged in primitive barbarism. We have outgrown the phase when all the arts were traced to Greece and the Olympian Zeus, he says, and continues:

. . . we have learnt how that flower of genius drew its sap from Lydians and Hittites, from Phoenicia and Crete, from Babylon and Egypt. But the roots go farther back: behind all these lies Sumer. The military conquest of the Sumerians, the arts and crafts which they raised to so high a level, their social organiza-tion and their conceptions of morality, even of religion, are not an isolated phenomenon, an archaeological curiosity; it is as part of our own substance that they claim our study, and in so far as they win our admiration we praise our spiritual forebears.[4]

When we praise the Sumerians as our spiritual forebears we praise a people with no known beginnings, a people who came by ma-gur from the sea.

Summarising present knowledge gained from excavations, Woolley points out that Sumerian civilisation made no progress in the period from the First Dynasty of Ur to the last. On the contrary, all archaeological evidence shows that Sumerian civilisation had attained its zenith already before the First Dynasty was founded: 'By the First Dynasty of Ur if there is any change it is in the nature of decadence, and from later ages we have nothing to parallel the treasures of the prehistoric tombs.'

The incredible art treasures left by these, our spiritual forebears, in pre-dynastic time, have more to tell us than a story of unsurpassed craftsmanship and a most refined taste. The mere materials chosen for use by the artists may tell us something about their former homeland or routine range of foreign contacts. Contacts of no superficial or casual nature. For in the land they came to settle, southern Iraq, they would neither find nor learn about metals or precious stones. There was not even common rock to quarry. Their only riches were fertile soil and navigable waters: vast expanses of alluvial plains for growing their crops and grazing their herds, and a perfect location for commercial activities by sea and river. Indeed, the Sumer they found had a richer vegetation, as appears from a tablet describing the arrival at Ur of the first god-king from Dilmun: 'To Ur he came, Enki king of the abyss, decrees the fate: "O city, well supplied, washed by much water, . . . green like the mountain, *Hashur*-forest, wide shade, . . ." '[5]

This was the Sumerian description of a place today without a drop of water, buried in sand, deprived of every green leaf and without shade. But the beauty of the landscape did not yield what a goldsmith or a jeweller needed for planning and producing the magnificent treasures interred with the earliest royal families. With this in mind, Woolley's vivid descriptions of such tombs are thought-provoking:

> At Ur has been found a cemetery of which the earlier graves would seem to date to about 3500 BC and the latest to come down to the beginning of the First Dynasty of Ur; amongst them are the tombs of local kings not recorded in the king-lists. . . . It is astonishing to find that at this early period the Sumerians were acquainted with and commonly employed not only the column but the arch, the vault, and . . . the dome, architectural forms which were not to find their way into the western world for thousands of years.
>
> That the general level of civilization accorded with the high development of architecture is shown by the richness of the graves. Objects of gold and silver are abundant, not only personal ornaments but vessels, weapons, and even tools being made of the precious metals; copper is the metal of everyday use. Stone

vases are numerous, white calcite (alabaster) being most favoured, but soapstone, diorite, and limestone also common, while as rarities we find cups or bowls of obsidian and lapis lazuli; lapis and carnelian are the stones ordinarily used by the jeweller. The inlay technique that was illustrated by the Kish wall-decoration, carried out in shell, mother-of-pearl, and lapis lazuli, occurs freely in the graves at Ur.

A description of the contents of the grave of a prince, Mes-kalam-dug, belonging to the latter part of the cemetery period, will show the wealth of this civilization. The grave was an ordinary one, a plain earth shaft, at the bottom of which was a wooden coffin containing the body with a space alongside it wherein the offerings were placed. The prince wore a complete head-dress or helmet of beaten gold in the form of a wig, the hair rendered by engraved lines and the fillet which bound it by a twisted band also engraved; the helmet came down to the nape of the neck and covered the cheeks, the ears being represented in the round and the side-whiskers in relief; it is just such a head-covering as is represented on Eannatum's Stela of the Vultures. With the body were two plain bowls and a shell-shaped lamp of gold, each inscribed with the name of the prince; a dagger with gold blade and gold-studded hilt hung from his silver belt and two axes of electrum lay by his side; his personal ornaments included a bracelet of triangular beads of gold and lapis lazuli, hundreds of other beads in the same materials, ear-rings and bracelets of gold and silver, a gold bull amulet in the form of a seated calf, two silver lamps shaped as shells, a gold pin with lapis head. Outside the coffin the offerings were far more numerous. The finest of them was a gold bowl, fluted and engraved, with small handles of lapis lazuli; by this lay a silver libation-jug and a patten; there were some fifty cups and bowls of silver and copper and a great number of weapons, a gold-mounted spear, daggers with hilts decorated with silver and gold, copper spears, axes and adzes, and a set of arrows with triangular flint heads.

The royal graves with masonry tomb chambers had been even richer, and these presented a feature to which there was no parallel in the plain shaft graves. The burial of the kings were accompanied by human sacrifice on a lavish scale, the bottom of the grave pit being crowded with the bodies of men and women who seemed to have been brought down here and butchered where they stood. In one grave the soldiers of the guard, wearing copper helmets and carrying spears, lie at the foot of the sloped *dromos* which lead down into the grave; against the end of the tomb chamber are nine ladies of the court with elaborate golden head-dresses; in front of the entrance are drawn up two heavy four-wheeled carts with three bullocks harnessed to each, and the driver's bones lie in the carts and the grooms are by the heads of the animals; in another grave, that of Queen Shub-ad, the court ladies are in two parallel rows, at the end of which is the harpist with a harp of inlay work decorated with a calf's head in lapis and gold, and the player's arm-bones were found lying across the wreckage of the instrument; even inside the tomb chamber two bodies were found crouched, one at the head and the other at the foot of the wooden bier on which the queen lay. In no known text is there anything that hints at human sacrifice of this sort, nor had archaeology discovered any trace of such a custom or any survival of it in a later age; if, as I have suggested above, it is to be explained by the deification of the early kings, we can say that in the historic period even the greater gods demanded no such rite . . .[6]

One point that should not be passed over lightly is what the same scholar emphasised; the wealth of Lower Mesopotamia is purely agricultural: 'there is no metal here and no stone, and not the least interesting point about the treasures recovered from the site of Ur is that the raw material of nearly all of them is imported from abroad'. How can the first civilisation known to us, antedating the first known dynasties in both Sumer and Egypt, be based entirely on imported materials? Extensive unrecorded travels must somehow have antedated the known beginnings.

Unless we recognise that the god-men described and depicted on the earliest Sumerian tablets and seals were people like those buried in the earliest royal tombs of Ur, we shall never have an explanation of the riches in those tombs. Enki, the 'god' who came from Dilmun and found Ur still washed by water and shaded by *hashur*-forest, reflects the memory of one of these mighty kings whose fleet of ma-gurs must already have carried capable craftsmen and merchants to many distant lands. How else could they have been acquainted with the great variety of precious metals and stones they needed to create the royal treasures? Nowhere in the vicinity of their own kingdom would they find gold, silver, electrum, copper, lapis lazuli, carnelian, alabaster, diorite, soapstone, or flint. Before coming to Ur experienced members of the king's party must already have thoroughly explored many distant lands to acquire an expert knowledge of all those foreign materials, where to locate them and how to work them. The shipwrights and seamen of the earliest god-kings buried at Ur must therefore have been of the same high standard as the goldsmiths and the jewellers.

No wonder that people of today who have heard of spacecraft but never of ma-gur could be fooled into believing that these god-men of sudden appearance had come to Mesopotamia from outer space. We ourselves had every reason to feel humble hanging on a rope behind *Slavsk*.

The Russians continued to haul us at sailing speed towards Bahrain. The abnormal winter wind did not relax its grip on the gulf. During the day it maintained its southerly direction in varying strength. In the early evening Norman took his radio equipment out from under his mattress. He had a radio appointment to transmit our whereabouts to the BBC with the set they had provided, but was unable to reach any coastal station. He again rigged up his own set and immediately had the eager voices of radio amateurs raining down upon him from all directions. Our station call-sign, LI2B, was the same as formerly used on *Kon-Tiki* and *Ra* and a sort of collector's item to amateurs able to establish two-way contact. So whenever Norman called blindly his earphones sounded like a disturbed wasps' nest, with the humming of numerous distant voices who wanted contact. But as soon as he had answered one of them, all the others got off the air and waited for the next chance to come in. As an enthusiastic radio ham Norman spent all the evening accommodating eagerly calling amateurs from east and west, north and south, with the standard exchange of international codes for identity, locality, strength and clarity of contact.

On one occasion Norman was about to answer a call when another station broke in and said: 'Don't answer this one, or you will get a lot of trouble!' The first station then came back, saying: 'This is a hobby, not politics!' Norman was immediately on the air again and said that he agreed: radio amateurs had the right to run their hobby in any country; we were for international friendship irrespective of frontiers; this was a hobby, not politics. He then began to contact the first station again, which proved to be in Israel, and this time all the stations on the air held their peace until the exchange of customary phrases had been completed. Norman had won a peaceful victory on the air. Rashad looked at him in silence as they sat down together at the table. Norman was the son of a Jewish family. Rashad was an Arab from Iraq, the nation most hostile to the state of Israel. Perhaps their ancestors had been in boats together

146

before. Arabs and Jews both count their pedigrees back to Shem, the ancestor of Abraham from Ur.

This was 5 December; the sky was clear and the evening stars just began to sparkle in the firmament when I climbed up on the bridge platform to relieve the steering watch. Rashad and Asbjörn were up there and jokingly asked if I had seen how the new moon looked like our ship. At that very moment it certainly did. As always in southern latitudes, the new moon hung like a hammock in the sky instead of standing on end as in northern countries, but just then it rested with its bottom on the black waters of the horizon, precisely like a golden, sickle-shaped reed-ship. We were really looking at a true god-ship sailing parallel to us on the horizon. The similarity to our own ma-gur was stunning. We went on gazing at it until our shiny companion lifted itself free of the waters and began sailing among the sparkling astral plankton of the black sky.

The sight had made on me a deep impression. For years I had followed this as a basic motif in prehistoric art. I was back in the days when the great reed-ship builders of Sumer, pre-Inca Peru and lonely Easter Island shared the tradition that the new moon was a god-ship, on which the sun-god and the primeval ancestor-kings travelled across the night sky. The ancient Sumerians and Peruvians expressed this belief both in words and in art. The Easter Islanders of our own days had forgotten the original symbolism, but the traditional badge of sovereignty, hanging at the chest of all their divine kings, was a sickle-shaped wooden pectoral which was known by two names: *rei-miro*, meaning 'ship-pectoral', and *rei-marama*, meaning 'moon-pectoral'.

Later in the night we saw black clouds with distant flashes of lightning over Iran. Next day we had rain showers, but the tenacious south-east wind kept blowing and Captain Igor refused to let us loose. Without anchors and with a big hole in the bow we must either have a fair wind or else reach a safe place for repairs. During the day the wind again turned due south and increased in violence until all wave-crests broke and sent the spray over us as we hit them with our bow. The only station Norman could reach with his official set was that of *Slavsk* at the other end of our tow-line, and Captain Igor advised us to keep steering in their wind-break. But *Slavsk* was too far ahead to give any shelter to our battered bow.

In the early afternoon of the fourth day we reached Bahrain. That is, we came to the buoy marking the entrance to the navigable channel through the limestone shallows surrounding the island. The *Slavsk* received radio orders from Bahrain to stop right there.

Captain Igor's request to continue into port was turned down. Someone would come out for *Tigris*. *Slavsk* anchored, and we hung on.

The legendary island was still hidden below the horizon and with our binoculars we could barely see indistinct clusters of tall chimneys of oil installations.

Before long a modern coastguard vessel came racing into sight, and shortly afterwards a helicopter too. We were filmed from the air, but the officials of the coastguard vessel only waved and continuously circled at a distance, almost like the dhows off Failaka. The helicopter left, but the coastguard never came closer. *Slavsk* lowered its lifeboat and Yuri and Carlo came back to *Tigris* in top spirits with Captain Igor. They had evidently had a grand time. The coastguard just went on circling round while we hung on to *Slavsk* on our rope so as not to drift off into the shallows.

The afternoon passed and we scouted the horizon, but nobody else came out. The coastguard patiently kept on circling *Slavsk* and us. As the sun sank low, Captain Igor finally greeted us for the last time. He returned to his own ship followed by our last shouts of thanks and good wishes. It was like losing a fine companion, a former stranger who had shown so much courage and unselfish humanity.

The sun began to set, and we were now prepared to hang on to the anchored *Slavsk* overnight. Then to our surprise the coastguard vessel came close to our side and asked if we would not accept a tow to Bahrain. We accepted with thanks, but I asked if we were not waiting for someone else. No, we learnt. It was they who had come to fetch us. But they had waited for the Russians to let go the rope; we had seemed to them to be held as some sort of captives.

On the contrary, I explained. They had helped us. They had saved us from the reefs and towed us to Bahrain. This explanation made no difference. No Russian ship was admitted to the Emirate of Bahrain. The friendly coastguard officers immediately grabbed our tow rope as soon as we shouted to the Russians to let it go. There was nothing we could do. With heavy heart we waved to Captain Igor, who was not allowed into harbour, and while we were towed towards the lights of a modern city port that slowly rose into sight, *Slavsk* weighed anchor and began its return voyage to the other ships that waited patiently at the mouth of the distant Shatt-al-Arab.

We were heading for a tiny independent nation booming in modern development and with seemingly limitless wealth. The great wealth of Bahrain lies not so much in its own now rapidly

148

dwindling oil supply as in its geographical location as a convenient terminal for pipelines from Saudi Arabia, where tankers from all continents can dock in deep sheltered harbours. The very location of this island has made it a crossroad for travellers and merchants in all epochs. Today its airport has become a junction for airliners from all directions. Even Concorde calls at Bahrain.

We were blinded by floodlights and illuminated modern installations when late at night we were towed past anchored tankers and between concrete breakwaters to an enormous mole not yet officially opened. Popularly known as ASRY, the Arab Shipbuilding and Repair Yard, it was the largest drydock in the world, just ready to accommodate supertankers of up to 450,000 tons. It so happened that tiny *Tigris*, with its topmast hardly visible above the lowest platform of the mole, was the first ship to enter and dock, two days before the official opening. We tossed our reed fenders outside the side bundles to save them from friction against the concrete wall when the tide sent us up or down, and climbed a long iron ladder to a crowd of official and unofficial spectators admitted through the police gates to see the reed-ship.

First to stand out snow-white in the floodlight was the long Arab attire of some cordial and straightforward dignitary who welcomed us and wanted to know our verdict on the reed-boat. His Excellency Tariq Al-Moayyed was the Minister of Information. I could tell him that we were all exceedingly happy with the body of the vessel. It was still strong and sturdy in spite of all that had happened since the day of the crash launching, and floated very high. But we had failed to solve the problems of the sail. We now had come in the hope of finding a sail-maker here in Bahrain, and an Arab or Indian dhow-sailor to join us from here on.

'Khalifa will help you with that,' said the Minister and introduced us to a young man dressed like himself. 'But what would you like to see while you are here?'

I looked up at the outlines of colossal structures of steel and concrete that rose like obelisks and pyramids against the night sky. We had hoped to see the remains of the earliest seafarers that had come to Bahrain. By this remark I had rubbed Aladdin's Lamp. The Minister turned his head to someone who now appeared out of the dark with a broad smile and saluted us by removing the huge curved pipe that hung from his teeth. I recognised the familiar face of the famous archaeologist, Geoffrey Bibby. It was this British-born scientist and his collaborators who had shaken former beliefs about the beginning of civilisation when they dug up temples and tombs buried in the sand of this island, testifying to merchant activity and maritime trade with remote countries more than five thousand years ago.

Bibby had flown down from Denmark when he learnt that we were heading for Bahrain. He wanted to get a personal impression of a ship built after the earliest type known in the gulf area. He reminded me laughingly that I had followed his advice. Long ago, in a review of my book on the *Ra* expeditions for the *New York Times*, he had challenged me to test a Mesopotamian reed-ship next. He reminded us of Captain Igor by the cheerful way he stepped on to our deck and enjoyed the unusual craft. Bibby's interest was deeply rooted in his own work. His discoveries on Bahrain proved that seafaring had been a basic element in human society since the very beginning of civilisation. He had done more than anyone else to demonstrate that Bahrain was identical with the distant trading centre of Dilmun, recorded in ancient Mesopotamian inscriptions. He repaid his visit to our ma-gur by lecturing to the *Tigris* crew on the island's prehistory. And during the next couple of weeks he took us to the sites of his main excavations.

When Geoffrey Bibby in his picturesque turban came to fetch us the following day, and we drove in two cars away from the Manama city area and the ASRY docks, we plunged with space-craft velocity back through five thousand years of human history. The world's biggest in ships and aeroplanes and most expensive in luxury hotels were quickly exchanged for a few modest Arab dwellings of braided palm-leaves and sun-baked bricks, all ready to vanish with the bulldozer's shovel. Then we passed through a similarly doomed plantation of beheaded date palms, resembling a graveyard of telephone posts. With their majestic crowns removed and their roots dug free, the dying palm-trunks seemed to have a message for the hurrying passer-by; the oil pumps may draw their sap from the ground for several decades, but for thousands of years, since the days of Dilmun, it had been the roots of the date palms that drew the sap that had fed the nation. With oil and industry a better bargain than agriculture, it seemed as if no one on Bahrain cared to harvest dates or till the land. With the money that poured in from development areas one could buy the dates, fruits and vegetables. The city markets and shop windows excelled in fresh and colourful garden products. Flown in from three continents, however. And as modest dwellings and tall palms fell before the mechanical shovel, modern housing area and suburban industry moved across the green fields ever closer to the edge of the barren desert that today dominates the island. We quickly reached this open wasteland.

'This is Dilmun,' said Bibby and pointed with his pipe towards a landscape of giant pimples which stretched like a choppy sea of fossil wave-tops to the horizon and beyond. 'You can see why Peter Glob and I were tempted to come here and start digging.'

Prehistoric tombs. Burial mounds. According to the estimates there were supposed to be about one hundred thousand such man-made mounds on Bahrain. This was the largest prehistoric cemetery in the world. On this island there had been more to cope with than Bibby alone could handle, so over eighty archaeologists of half a dozen nationalities, but most of them Danes, and several hundred workmen from almost every Arab land had worked with him. Their first effort had been distressingly fruitless: not one of the numerous tombs opened had been spared by ancient grave-robbers. Apparently every one had been dug and plundered, an indication that their stone-lined burial chambers had contained more than withering human bones. All that had been left for the archaeologists were bones, the shell of an ostrich egg, potsherds, a couple of copper spearheads and fragments of a copper mirror. These tombs had evidently belonged to people who believed in a life after death and therefore left personal treasures and other funerary gifts in the grave for the deceased to use in his after-life.

The tombs varied greatly in magnitude. We first came to an area named Ali, where a large cluster of them exceeded the pyramids of Egypt in number and compared favourably in size with a medium-large Mesopotamian pyramid. Arab houses of one, two or even three floors, were built between them and were completely dwarfed. The modern residents of Ali had made a regular industry out of quarrying these huge man-made hills, using the limestone they extracted for burning lime. The result was that between the work of the ancient grave-robbers and the modern lime-burners the Ali mounds looked like gaping volcanoes that rose above the landscape. By climbing them one had a magnificent view over the endless stretches of smaller tombs that lay there like innumerable spawn left in a breeding-place behind the giant turtles of Ali. The colossi of Ali were amply spaced and majestically located closer to the sea, whereas the adjacent cemetery of smaller, dome-shaped hills continued inland and across the naked landscape, so closely packed that there was barely room to walk between them. I could not help feeling that the colossi, with all the space between them, antedated the closely packed fry. The big ones seemed to have been built while there was still room to spare

151

in this locality, and the multitude of smaller mausolea was packed close to them in a desire to be their neighbours.

It is usually taken for granted that things begin small and afterwards grow into more impressive proportions. But not always so with civilisations. There may be two reasons. Cultural growth ends in most known cases with stagnation and cultural decadence. The reasons for this might be anything from over-affluence to war, pestilence or natural catastrophe. But in addition, at the peak of evolution most civilisations tend to possess ships and be involved in some kind of seafaring. At this advanced stage they may suddenly escape invaders or travel in search of a better land. Families or entire organised colonies may settle with an advanced cultural level in areas previously uninhabited or occupied by some primitive society. We should not be surprised then to find that most ancient civilisations seem to appear without local background and often to disappear again without a trace. We dig in search of the roots, and expect every civilisation to have grown like a tree in the place we find it. But civilisations spread like seeds with the wind and the current once the tree is grown and in bloom. It would therefore be wrong to suspect that only primitive savages could have settled at Bahrain and that the giant tombs of Ali represent the local evolution from the countless small ones, grown large through experience. The pyramids of Egypt did not grow with time: the biggest were built by the first Pharaohs; later they got smaller. The same happened in Mesopotamia. And in Peru. Everything in Egypt started big with the first dynasties, as in Mesopotamia and Peru. Subsequent changes do not testify to cultural growth but to imitation or even decadence. In Peru the famous Inca culture never attained the height of its Tiahuanaco or Mochica predecessors, either in art or in magnitude of architecture. And now one was left with a similar impression on visiting the Ali cemetery.

I had seen groups of burial mounds and prehistoric cemeteries in many parts of the world, but nothing like this. There was just nothing like it. And from now on Bibby did not have to argue to convince me that Bahrain was Dilmun.

One of his premises had been that Dilmun was to the Sumerians a holy land, a land blessed by the gods who gave it to mankind after the flood. Dilmun was the place where man, in the story of Ziusudra, the Hebrew Noah, was given eternal life. The symbolic meaning is probably that Ziusudra's 'seeds' were given eternal life while all other men drowned. In the oldest of all known epics, King Gilgamesh of Uruk, the Biblical Erek, sailed to Dilmun in an effort to seek the flower of eternal life in the sacred home of his fore-fathers. A Sumerian poem says:

> The land of Dilmun is holy, the land of Dilmun is pure,
> the land of Dilmun is clean, the land of Dilmun is holy.[7]

For ancestor-worshippers with this belief it would seem tempting to bring, or even to ship, deceased persons of some importance to Dilmun in a funeral party. In Dilmun the spirit of the dead person would join the ancestral gods. Many, like Bibby, had found it difficult to see why the little island of Bahrain should house the world's largest cemetery, dateable to Sumerian times, unless the local soil at that period had a very special importance to people even outside the island itself.

The same Sumerian poem has another useful reference to Dilmun. The sea-faring god Enki had asked the supreme god of the heaven to bless Dilmun with fresh water:

> Let Utu (the sun god) stationed in heaven
> bring you sweet water from the earth, from the water-sources of
>   the earth;
> let him bring up the water into your large reservoirs (?);
> let him make your city drink from them the water of abundance;
> let him make Dilmun drink from them the water of abundance;
> let your wells of bitter water become wells of sweet water;
> let your furrowed fields and acres yield you their grain;
> let your city become the 'dock-yard'-house of the land.[8]

Water was a divine gift to any people in the gulf area. Deserts and dry wasteland dominate coasts and islands. Bahrain is a striking exception. On Bahrain water rises from the dry ground in springs and fountains, and flows endlessly into the sea. Around the coast there are even underwater springs, where divers can swim down to drink sweet water and refill their air-filled jars.

The amazing way in which nature has brought an abundance of fresh water to this low limestone island surrounded by the salt sea is almost enough to make anyone who drinks from these springs believe in miracles. On our way to the tombs Bibby took a side-road to a true oasis of palms and green grass. Inside was a deep and divinely beautiful pond of crystal clear water retained within the ancient stone walls of a circular reservoir. Young Arabs were diving and swimming, and one of them was sitting soaping himself in the water while three women washed clothing. Yet the water was constantly renewed and clear as morning dew. Every one of the smooth stones on the bottom was seen as clearly as if the pool were empty. In the centre the water was welling up to the surface like a fountain, and a constant overflow sent the soap-suds away down a fast-running drainage ditch formerly used for date palm irrigation.

Bibby could tell us of several such pools and springs on the island. No wonder that their origin had been ascribed to divine interference. This water came from the distant mountains of the Arabian peninsula, where the rain sank into the naked rocks and was lost for the mainland. Through a freak of nature it filtered into subterranean cracks and fissures, some of which carried the fresh water under the bottom of the gulf to reappear as springs on the island of Bahrain.

From the slopes of the island's central hills prehistoric engineers had constructed hidden water channels deep under the desert sand. They were walled and roofed with slabs and ran for miles, often twenty feet or more below the surface, to end in formerly cultivated fields. About every fifty yards circular stone shafts rose like buried chimneys from these subterranean channels up to the surface. Perhaps they had served as gutters for maintenance. Without them there was nothing to disclose the existence and route of these prehistoric pipelines. To Bibby and his colleagues these masterly examples of engineering remained a puzzle. Had the stone-lined aqueducts been built on the ground and the chimneys gradually extended upwards as wind-borne sand accumulated over them? Or had they been dug as deep underground passages from the very beginning? This would remain a riddle to me, too, until a few weeks later chance gave me and my reed-ship companions a clear hint when visiting another legendary Sumerian land.

By the time we had seen 'Enki's' wells and pools and the galaxy of burial mounds we realised that nothing else in the gulf could better aspire to be identified as Dilmun. But Bibby still had his trump cards: Dilmun was more than a playground for the gods and the lesser god-kings of Noah's dimensions. He drove us from the cemetery to a buried city where common people had once made their living by such profane activities as trade and shipping. We were told to bear in mind that Dilmun from the days of the Flood continued as a very real place, not only to the maritime Sumerians, but also to their cultural heirs of later Babylonian and Assyrian times.

On a stele and a clay tablet from about 2450 BC, King Ur-Nanshe, who founded the powerful dynasty in Lagash, recorded that the ships from Dilmun brought him timber. After him, the mighty Semitic ruler Sargon the Great, who lived about 2300 BC and subdued all nations from the gulf area to the Mediterranean Sea, erected memorial stele and statues at Nippur on which he boasted that ships from Dilmun, Makan and Meluhha docked together in the ports of his capital Akkad. Dilmun continued to figure as a place-name in Akkadian documents, and the Assyrian King Tukulti-Ninurta used in his title the epithet 'King of Dilmun and Meluhha'. King Sargon II of Assyria received tribute from a king in Dilmun named Uperi, and in the days of King Sennacherib soldiers were sent from Dilmun to help raze the rebellious city of Babylon to the ground.

The Danish scientists, under the leadership of the noted archaeologist P. V. Glob, have been responsible for all the amazing discoveries on Bahrain, and for over twenty years Bibby had been the field director. When these archaeologists laid eyes on Bahrain a few decades ago, there were no prehistoric houses to account for the presence of the hundred thousand tombs. The island seemed to have served only as a funerary centre for the gulf area. No ruins other than Arab mosques and Portuguese fortifications were seen above the ground. Failing to find much but sherds of a hitherto unknown type of pottery in the plundered burial mounds, Glob, Bibby and their companions had begun to search the island for flint chips, potsherds and irregularities in the naked terrain that might reveal former habitation. Thus they located a buried temple and next an unknown city, which he now wanted to show us.

There is always something intriguing about a buried city. Stepping down from a present ground level higher than the former roofs is like climbing through a trap-door down into the unknown past, following streets untrodden perhaps for thousands of years, as in the town discovered on Bahrain. On the north coast we were to climb from the sand-dunes at the foot of a sixteenth-century Portuguese fort down into a sunken city that had been teeming with life in Sumerian times. The fort on top, now nothing but picturesque ruins on a prominent bluff overlooking the sea, had been built by the Portuguese shortly after they conquered Bahrain from the Arabs in 1521. They had rebuilt a fort originally constructed by early Arabs who came to this island immediately after the days of Mohammed the Prophet. The Arabs, in turn, had used stones, some of which had been taken from even older buildings of unknown origin, possibly found emerging from the local sand.

155

Next to this fort Bibby's party had also been tempted to dig. Here was something hidden under a large sandy mound overlooking the sea, and as the white sand was removed a complete buried city appeared, and under it another, and still another. The streets and buildings were always carefully laid out east–west and north–south. The site witnessed of prosperity as well as disaster. About 1200 BC the city had been burned. Underneath were the walls of buildings dating from 2300 BC, on which the subsequent town had stood before the fire. This earlier city was contemporary with the mound burials and Sumerian voyages to Dilmun. They dated from what Bibby termed the Dilmun period. We walked together down into the oldest city. We stopped at a huge and solid city wall with a gate directly facing the sea. Here we found ourselves surrounded by tall stone walls, in an open square where a main avenue flanked by stone buildings led to this big gate and the sea. From the open plaza inside the city wall other streets took off at right angles.

Bibby pointed through the lofty gate. Huge sand dunes had blown up in front to bar the view, but when the wall was built this gate had led directly to the water, which was still right down below.

'Here ships docked four to five thousand years ago, in the Dilmun period, to load or unload their cargo,' said Bibby. He turned from the gate and pointed to the ground in the open square we stood on: 'And here we found the evidence of oversea trade. Here the cargo was unloaded. It was here and in the streets of the

31. Under sail in the gulf. *(opposite)*

32–33. Mountain-climber Carlo preparing extra ropes for the bow as the flexible reed-ship sails along the dangerous cliffs of Oman, aiming for the narrow exit from the gulf. *(opposite & p. 158)*

34. Turning into shelter outside the traffic-filled Hormuz Strait, before sailing along the Arabian peninsula. *(p. 158)*

35. A terraced temple mound unlike any Arab structure but closely resembling a Mesopotamian ziggurat with ceremonial ramp had just been discovered in Oman by mining prospectors. *(p. 159)*

36. The temple was in the midst of vast prehistoric copper mines; in one of them an entire mountain had been transformed into a valley and the only piece remaining is a monumental gateway left where the long-forgotten miners had first entered the rock. *(p. 159)*

37. Probably the output from the mines had been brought out to waiting reed-ships from Bahrain and Mesopotamia in vessels like these still in local use. *(p. 160)*

38. Sightseeing in Oman, a nation closed to tourists, the expedition group visits Al Hamra. *(p. 160)*

town that we found numerous scraps of unworked copper. Also copper fishhooks, bits of ivory, steatite seals and a carnelian bead. All represented materials foreign to Bahrain.'

He added that the nearest source for copper would be Oman. And ivory could only have come from India or Africa. The carnelian bead, and also a very special type of polished flint weight found among the ruins, must definitely have come from the now extinct Indus valley civilisation. The five flint weights found provided a surprising disclosure. It showed that the Bahrainians of the Dilmun period adhered to the weight system of the Indus valley people rather than to that of the Sumerians.

As we spoke a fine dusting of sand blew in over the town wall from the dunes along the shore. This was how the ancient port had become a buried city. But this ancient port was on the north coast, and it dawned upon me that the wind had changed. At long last it blew from Iraq.

Bibby adjusted his turban and laughed. 'You have had bad luck,' he said. 'This is the way the winter wind always blows. From the north. How much time do you think you would have needed to come here with this wind?'

'Three-and-a-half to four days,' I said. 'We were towed at the speed we sailed to Failaka. But we sailed at right angles to the wind. Probably we could have come faster if we had come straight to Bahrain with the wind at our back.'

'That makes sense. One Sumerian record speaks of Dilmun as thirty double-hours away. They reckoned distance in time of travel.'

39–40. Rowing a reed-ship out of Muscat harbour was a heavy job for eight men. *(p. 161)*

41. *Thor I.* The Norwegian merchantman *Thor I*, a recent replacement of the *Thor I* that in 1947 brought the *Kon-Tiki* raft back from Tahiti. *(p. 162)*

42. In the shipping lane. *Tigris* sailed under the United Nations flag. *(p. 162)*

43–44. We sail north to Pakistan, reach the snake island of Astola and follow the limestone cliffs of the Makran coast towards the former realm of the Indus Valley civilisation. *(p. 163)*

45. Ashore on the desert sands of Ormara bay. Homes are made of plaited mats. *(opposite)*

46. Inhabitants of Ormara bay. The man sleeps in a vaulted hut and the woman uses a scale, both of the types used from Mesopotamia to the Indus Valley four or five thousand years ago. *(opposite)*

'They would most certainly have done better than novices like us,' I admitted. 'Failaka must to them have been almost a home port and in a race from there to here a professional Sumerian crew would undoubtedly have beaten us by a few hours. We might have needed thirty-five double-hours with a good north wind.'

'The bottom shape of your ship with two bundles and a shallow draft is interesting,' commented Bibby, and took all of us up on the wall to see the shallow tide flats that reached right up to the sand dune in front of the maritime gate. At high tide the water must have flowed right up to the wall of the city.

'I can see how a flat-bottomed reed-ship could come all the way in at high tide,' he continued. 'With its twin body it could settle on the limestone bottom without capsizing even when the water went out. While beached at the gate it would be perfect for loading and unloading cargo.'

We had noted that the limestone bed all around Bahrain, and not least off this port, would only permit a manoeuvre just like the one Bibby described: sailing in at high tide and beaching as the tide returned. And Bibby had now seen with his own eyes that a berdi-ship would be able to carry a twenty-ton burden, such as had been the weight of some Dilmun cargo according to the early tablets.[9]

'But a merchant vessel must be able to return to the port it came from,' insisted Bibby. 'Do you think they waited half a year for the seasonal wind to change?'

Maybe. Maybe they even made a point of coming shortly before the wind was to change, to shorten their waiting time. But I did not believe so. I was afraid the fault was ours, who had not yet been able to imitate the old skills.

The seasonal winds would at any rate not seem to favour visits to the Indus valley the way it would with the Sumerian ports. What Bibby had first pointed out years ago was therefore a real puzzle: why had Bahrain used the standard weights of the Indus valley? The Sumerians and Babylonians used a completely different system. Not only were the weights different, but they worked in a different ratio, in thirds and tenths and sixteenths. There could only be one of two explanations, according to Bibby: either the first commercial impulses must have reached Dilmun not from Mesopotamia but from India, or else India was a far more important commercial connection with Dilmun than was Mesopotamia.

One thing was clear to both of us. The reason for Bahrain's importance on the trade route was its convenient location as a unique watering point. Nowhere else in all the length of the gulf could ancient sailors obtain fresh water in unrestricted quantities.

The scraps of unworked copper had their own story to tell and filled a gap in a jig-saw puzzle. Copper was perhaps the most important of all raw materials imported to Mesopotamia in Dilmun times. As Bibby had pointed out in his book on the quest for Dilmun,[10] writing was an extremely important art in ancient Sumer, and the clay tablets found in private houses and shops range from school exercise books to the account books of the money-lenders. There was also the regular business correspondence of a copper broker living in Ur. He was referred to as a Dilmun trader and yet it appears from one of the written tablets found in his house that Dilmun was not the place where the copper ore was quarried. Dilmun was only a trading centre where copper was bought and sold. The weight of some of the ship-loads of copper moving in the gulf in Sumerian time was by no means inconsiderable, to quote Bibby's expression. He figured out that in one case the shipment acquired in Dilmun was no less than eighteen and a half metric tons, 'which at present prices would fetch something like twenty thousand dollars'.

The correspondence of the copper broker also included a tablet sent him by a discontented customer: 'When you came, you said, "I will give good ingots to Gimil-Sin". That is what you said, but you have not done so; you offered bad ingots to my messenger, saying, "Take it or leave it". Who am I that you should treat me so contemptuously? Are we not both gentlemen? . . . Who is there among the Dilmun traders who has acted against me in this way?'

Dilmun was indeed a reality to our spiritual forebears. And Sumerian merchant vessels must certainly have been among those docking in front of the gate of the now buried harbour city of Bahrain, because it had the peak of its activity in Sumerian time. This, perhaps, was Enki's 'dock-yard house of the land' referred to in the Dilmun poem, unless another port of the same magnitude lies buried elsewhere on Bahrain. That seemed unlikely. This was a major trading port for such a small island. Ma-gurs from Ur must have been amongst those that came here to barter copper for Mesopotamian wool and garments, as the records and accounts on the tablets show. It required a considerable trade to keep prosperous such an island city as this. The vast complex of ruins extended inland from the sea wall, with streets and palaces.

It was clear at a glance that the city had been rebuilt. It was equally clear that the best stone masons had lived in the earlier period, that is, the original Dilmun period, contemporary with the burial mounds. Superimposed on the older city blocks were younger walls, some of which were ascribed to Assyrian times. There was a majestic interior gate known among the archaeologists as the 'Assyrian doorway', built from perfectly fitted quarried blocks with a single threshold stone bigger than a double bed and with a round indentation in one corner, to hold a door-post.

The Assyrians are famous for their stone work. But on Bahrain they had been excelled by their Dilmun predecessors. Could this mean that the Dilmun people had come from a stone-cutting area with a higher development or longer tradition in the stone-shaping art than the Assyrians? There were, indeed, such people in the ancient world. But they were not terribly many. Archaeologists are experts on pottery, and can with remarkable accuracy recognise cultural ties by identifying ceramic ware, even when found in sherds. But I doubt if any archaeologist has ever tried to shape a stone block like those left on Bahrain from the early Dilmun period. If they had, they would have discovered, like me, that they were unable to do it. Not even with iron tools. And the Dilmun people had no iron. In other words, the walls left by the first city builders on Bahrain had something to tell us. The founders of this port included masons from one of the very few areas where the secrets of this utterly unmanageable art were known and commonplace.

Reed-boats and carved stones seem to be separate topics and a reed-boat sailor should not care about stone walls. Not so in ancient times. Life-long research had shown me that reed-boat builders very often have had something to do with stone walls of this very kind. Usually they were the people who had made them.

Bibby looked rather surprised when I kneeled down to examine the perfectly plane and smooth surfaces of his Dilmun blocks and the way they were fitted together. The stones were carved with right angles but no two of them were alike, and some had inturned corners, but all were made to fit adjacent blocks with such precision that no crack or hole was left between them. My friends from *Tigris* looked at me like some kind of Sherlock Holmes trying to find fingerprints or tool-marks that might lead us on the track of those who did it. The beautifully dressed stones were shaped and joined together in a special manner I began to know all too well by now. I had to tell my puzzled companions why these stone walls had any bearing upon our voyage and upon the voyage to this same island by the people who had once made them.

It was clear to all that this intricate and specialised masonry technique had to represent some aesthetic or perhaps magico-religious tradition and was not dictated by any practical need. Never, in any period, had walls been built in this manner in Europe or in the Far East. The distribution nevertheless spanned two oceans, but in doing so followed a clear pattern. In a most conspicuous manner it followed the distribution of the peoples who had built reed-boats.

I had come close to such walls for the first time among the reed-boat builders on the world's loneliest inhabited speck of land: Easter Island. There unknown master-masons had used this technique in some of the oldest megalithic temple terraces supporting the large statues.[11] I found them again among the reed-ship builders of South America, among the same people who had helped us build the *Ra II* and *Tigris*. Here the technique appeared as the characteristic note distinguishing the megalithic temple walls of pre-Inca and Inca Peru, the area from where we ourselves sailed the *Kon-Tiki* down-wind past Easter Island to Polynesia proper.

The next encounter was unexpected. I stumbled upon the same technique in the titanic temple walls at Lixus on the Atlantic coast of Morocco. I had come to Lixus to see the local reed-boats before we sailed away from that coast to America with the *Ra*, but had never heard of these impressive ruins. They were assumed to have been left by Phoenician colonists who settled the Atlantic seaboard on voyages from Carthage and Asia Minor. If built by Phoenicians, no wonder the founders of Lixus had known this technique. They would have learnt it in the Middle East, the only area where it had been commonplace outside Inca territory and Easter Island. The finest megalithic masonry in the temple walls behind the great pyramids of Egypt had been made in this way. Yet the real centre and acme of the art seemed to have been in Hittite territory. The Hittites, the extinct, forgotten and recently rediscovered predecessors of the Phoenicians, once inhabited the entire area forming a bridge between Upper Mesopotamia and the Mediterranean Sea.

Now came the problem. The Hittites had inherited their customs and beliefs, and nearly all their arts and crafts, from the Sumerians. But the Sumerians did not build stone walls, none is left in the territory we know as Sumer.

How then did this art reach ancient Bahrain? Did the early masons of this one island in the gulf have anything to do with either the Hittites or the Egyptians? The total lack of any kind of stone wall in Sumer, where the Dilmun contact took place, seemed to create a conspicuous blank in the otherwise coherent distribution pattern. But for this there was a good reason. There was simply no stone in Lower Mesopotamia, only fertile river silt and clay to bake into building bricks. But to judge from the clay tablets, Ur and other Sumerian ports had as much trade and contact up river as across the sea, and as soon as rock appeared along the upper reaches of the two rivers, the Mesopotamians carved it, dressed it, and, in the earliest periods of Hittite-Sumerian contacts, jointed the blocks together with the peculiar technique that reappeared below the soil on Bahrain.

The absence of stone in most of Mesopotamia caused the Sumerians and their followers to build their ziggurats, or stepped pyramids, from millions of sun-baked bricks. A brick pyramid is a strange exception in Egypt, where nearly all the pyramids are built from quarried stone, even the oldest of them all, the Sakara pyramid, which was stepped just like the ziggurats of Mesopotamia. To many this has seemed a fundamental difference excluding a common origin. But this would be a hasty deduction. Never had I seen better dressed stone than the truly giant slabs from Nineveh, with beautiful reed-ship flotillas carved in relief. Nearby was the biblical site of Nimrud, with a colossal man-made hillock, representing the eroded remains of a former pyramid now covered by rubble from Assyrian bricks. No one would have suspected that quarried stone had been used in this structure. Nevertheless the recent removal of part of the Assyrian rubble has exposed a large section of the original wall, built from big quarried stones. Archaeologists estimate that this sun-oriented pyramid, now about 140 feet high, was probably 60 feet higher originally. Inside was found a vaulted chamber 100 feet long, 6 feet wide and 12 feet high, empty. The river Tigris had originally run along the western base of the pyramid, which rose from a twenty feet high quay of carefully dressed and fitted limestone blocks.[12]

The Nimrud pyramid had indeed originally been built from stone blocks, like those in Egypt. But there was a difference. The pyramids of Egypt were built from blocks all quarried to the same size to simplify work. Not so the Nimrud pyramid. And in addition, while checking the blocks, I found the jointing to be once more the one I was in search of, the one that now struck my eyes as soon as we descended between the excavated Dilmun walls with Bibby.

When Bibby noted my unexpected interest in stone masonry, he took us back again to one of the colossal Ali mounds, perhaps the largest of them all, not inferior in size, it seemed, to the Nimrud pyramid. Really a lofty hillock to ascend. He took us around to the back where a portion of the limestone rubble cover had been carefully removed, as on the pyramid of Nimrud. Inside emerged the section of a solid wall of quarried stones. The stones seemed to be of equal size, as in Egypt. These big mounds, said Bibby, had been a sort of round, stepped pyramid.

Nobody had ever doubted that the Ali mounds must have been built as some sort of mausoleum for defunct kings. These giant tombs were of such preponderant size that they had required organised mass labour and thus undoubtedly represented the resting place of extremely powerful monarchs. Their numbers were sufficiently restricted to represent successive generations of sovereigns, while the vast adjacent cemetery could have been reserved for lesser chiefs and anyone worthy of entombment in the vicinity of such important personages.

Gazing over the Arab roofs down below and across the scorched landscape, with tombs everywhere except in the direction of the decapitated date palms, I began to look upon these miserable surroundings with other eyes. This had to be Dilmun. But, admitting this, it was the land recorded by the Sumerians as the one-time abode of their early ancestors, the home of Ziusudra, the venerated priest-king praised as the one who by his ship had given eternal life to mankind – the culture hero later borrowed as their own by the Babylonians, Assyrians and possibly the Hittites with the name of Utu-nipishtim. The same important personage who finally found his way into the teachings of Hebrews, Christians and Moslems as Noah.

What an amazing thought: here I was, probably sitting on top of a mountainous burial mound overlooking the land of Noah. Perhaps this very mound was the burial mound of Ziusudra. It seemed to be the biggest. According to Sumerian texts Ziusudra never left Dilmun, *alias* Bahrain. It was his descendants that finally came to Sumer. If he ever existed, he would probably be buried in one of these giant mounds. Who could tell. Perhaps I was really sitting on the tomb of Noah.

The idea was not all that crazy. Noah, no. For Noah was only a dressed-up version of Ziusudra. No one landed here with a floating zoo. But Ziusudra was a very real person to the Sumerians in their time. He brought only domesticated cattle and sheep on board, and their bones are found from that early period both on this island and in Ur. Ziusudra never returned to his birthplace at Shuruppak on the Euphrates, his tomb logically had to be one of those left in Dilmun.

It would be foolish of us today to underestimate the early Sumerians just because they lived five thousand years before our time, when the world at large was peopled by savages. They were not illiterate. From them we learnt to write. They were not stupid. From them we got the wheel, the art of forging metals, of building arches, of weaving cloth, of hoisting sail, of sowing our fields and baking our bread. They gave us our domesticated animals. They invented units for weight, length, area, volume, and instruments to measure it all. They initiated real mathematics, made exact astronomical observations, kept track of time, devised a calendar system and recorded genealogies. When they spoke of Dilmun, Makan and Meluhha they knew where these places were; they were well at home in geography. How else could they know where to travel to locate copper, gold, lapis lazuli, carnelian, alabaster and the great many other materials precious to them, unknown locally, and yet firmly imbedded in their material culture? History was to them a major subject. They worshipped their ancestors and their minds were focused on the events and heroes of the past. Royal lines back through the ages were sacred lessons impressed on growing generations by the priesthood and the learned men. Noah is a legend to us, but Ziusudra was history to them. Dilmun may seem a castle in the air to us, but to them it was a trading centre thirty double-hours away.

I was brought back from Dilmun to Bahrain by the shrill sound of a claxon from one of the waiting cars on the plains below. Time to return. Scrambling down the steep rubble hill that concealed the fine stone walls, I drank in the view of the exclusive group of majestic man-made hills around me. This island contained the tombs of the Sumerian forebears. In a wider sense, the tombs of our own spiritual forebears. Would this bring us a step closer to our own lost beginnings?

It was almost with a feeling of awe that I descended to the ground and walked back to the cars across a terrain that must have seen strange processions. Here mighty leaders had been carried to rest. This plain had probably witnessed the funeral ceremony of some mighty seafaring hierarch whose maritime adventures, in ever more embellished versions, were to survive the ages and be retold even in my own childhood classroom. The vessel that became the 'Ark' had probably been a ma-gur beached on this coast. The fabulous procession of disembarking animals had probably been a few cows, a bull and a little flock of sheep. Lowing and bleating after days on board, they must have waded across the shallows from the broad and sturdy reed-ship, searching for the nearest waterhole. Man and beast survived here because the god of the venerated priest-king had let sweet cold water from the distant mainland mountains well out of the ground.

I looked with horror at the limestone burners digging away at some of the biggest of the giant mounds. While all the Arabs on this island emirate enter their mosques to pray, while they read about their progenitor Noah in their holy Koran, these Moslems were burning lime by assaulting with pick and shovel a notoriously ancient mausoleum that in fact might once have been venerated by Ham, Shem and Japhet as their father's tomb.

The early Bahrainians had been a religious people. From the nameless seaport Bibby brought us westwards along the coast to a locality known as Barbar. Here his team had made their first major discovery: a temple. And it was a very special temple. They had found it by digging a test pit into the lowest of a whole string of sandy mounds larger than even the biggest of the gravel mounds at Ali. Their interest in this particular mound had been fired when Bibby's Danish companion, Professor Glob, noted two colossal blocks of shaped limestone protruding from its slopes. The blocks proved to weigh over three tons each and stood on a paving of limestone slabs. Two square depressions cut in the top suggested they might have served as pedestals for big statues. Trenching the wide mound inwards along the elevated stone paving they struck a wall forming a step up to another terrace where the stone flooring slabs continued until they struck another wall forming a step up to a terrace higher still, where the stone paving led again to yet another wall enclosing the tallest central part of the structure. What amazed the archaeologists was that there had never been an open space or temple enclosure inside these walls. The building had been a compact, solidly filled elevation with right-angled corners, rising in steps above the terrain like a Sumerian temple-pyramid. The facing slabs of each of the superimposed terraces were blocks of fine, close-grained limestone, laid in three courses and carefully cut to fit together without mortar. The stones on the top platform were different; they were perfectly shaped like the tapering ice-blocks of an Eskimo igloo, to form a circular enclosure only six feet approximately in diameter.

Excavations revealed that four thousand years ago the original ground surface on which the structure stood must have been eight or ten feet lower. The central temple must then have been much more imposing, standing on its platform above sheer terrace walls oriented to the movements of the sun. As excavations revealed the staircases and ramps that led up to the summit temple, Bibby realised they had hit upon a religious structure that began to qualify as a ziggurat, the terraced temple mound of Mesopotamia.

With the Sumerian deluge story in mind, Bibby found it highly significant that they had discovered a temple on Bahrain with Sumerian affinities.[13] Nowhere in the gulf area outside Mesopotamia had structures resembling a stepped ziggurat ever been found. And the excavations of the temple uncovered quantities of potsherds, lapis lazuli beads, alabaster vases, copper bands and sheet copper, a copper figure of a bird, a cast bull's head with originally inlaid eyes, and, the final proof of Sumerian contact, a little copper statuette of a naked man with large round eyes and shaven head. He stood in the special attitude of supplication typical of Mesopotamia between 2500 and 1800 BC. One of the alabaster vases was of a shape used in Mesopotamia in the final centuries of the third millennium BC.

It was fascinating to visit this temple-mound with Bibby and hear how he linked it to long-range Sumerian sailing and Dilmun trade. This had indeed not been an isolated island civilisation. And much more was to be found under the ground. In fact, stairs led down the south wall of the sun-oriented structure to a deep excavation at its foot. Down there, below ground level, was a basin enclosed by a cellar-like wall, and we were only halfway down into it when I caught myself exclaiming: 'The fingerprint!'

There it was again, and in an even finer version, the masonry I had been looking for. Stone blocks of uneven size squared as if cut with a laser beam. Some intentionally shouldered, all dressed and polished almost to a shine and fitted together without cement so exactly that hardly a knife blade could be inserted between them. Again the temple walls of Vinapu on Easter Island stood clearly before me, and the whole series from the pre-Inca walls in Vinaque and Tiahuanaco to Lixus and the Hittite walls of Bogazköy. This was deep below the ground, built as a catchment basin to hold what had probably served as sacred water for ceremonial functions conducted on the structure above. Since the temple architects could hardly have found this spring merely by digging at the feet of the temple stairs, it must be assumed that the temple had been placed there because of the presence of the spring. This would again mean that the splendidly fitted wall of the basin dated from the first building phase of the temple. This observation was supported by the fact that the temple itself showed evidence of more than one building phase, and, as in the buried port city, some of the finest carved stones in the rebuilt structure appeared to be reused from an earlier period. But more was to be deduced from these dressed stones. The people who quarried them must have been sailors. The masons had benefited from the choice of the most suitable kind of stone and the knowledge of where to find it.

'A really fine limestone they used,' I remarked to Bibby.

'Right,' he said, 'and there is no such stone on Bahrain. They probably went to Jidda to fetch it.'

'Jidda?' I sensed evidence of marine activity.

'Jidda is a small island some five kilometres from the north-west point of this island.'

'Have you looked for quarries there?' I asked.

'No. There are supposed to be traces of that sort, but no one can go there as Jidda is the prison colony of the Emirate.'

My interest in that island was fired and I felt an irresistible urge to go there.

I went straight to my new acquaintance, the friendly Minister of Information, and decided to attack this bastion from every possible angle. In the meantime there were more urgent problems to solve.

We had come to Bahrain with a huge hole in our bow. Little was visible above waterline, but Gherman came up horrified from a dive and told us there was room for himself inside the cavity. The ropes were not broken but hung in loose loops and the inner bundles were exposed. We had to repair the hole, otherwise there was nothing to stop the continuous loss of reeds until the whole ship came apart.

We carried a modest quantity of spare reeds with us for minor repairs, and we could also use some of the berdi fenders to fill the hole. But it would still not be enough. What could we use? If we stuffed wood or other hard materials into the hole they would gnaw on the reeds and wriggle loose in violent seas. We had to look for something suitable in the meagre vegetation still growing on the island.

I had an idea and started skimming through a copy of Bibby's book which was part of the *Tigris* library. In it he had a line drawing of a bundle-boat with sail and double rudder-oars astern, resembling those of Failaka. It was the caption I was looking for: 'These boats, about 15 feet long and made of bundles of reeds, are used by fishermen of Bahrain. They are buoyant but not, of course, watertight (and are therefore technically rafts). Similar boats of papyrus reeds were in use in Egypt over 4000 years ago.'

I ran to Bibby. Nobody else I talked to had seen anything but tankers, yachts and motor vessels on Bahrain. Bibby clearly recalled having seen the craft he sketched and mentioned in his book. Together we drove across the island, past cemeteries with tens of thousands of mounds I had not seen before, and reached the tiny fishing village of Malakiya near the south-west coast. Women and children with gold teeth, jingling with jewellery, came out of small cement houses in colourful draperies, pointing out the trail to the sea. Bibby recalled having seen nicer and healthier houses of woven date-palm stalks when he was there a few years earlier. Plastic and modern refuse tossed about reflected sudden wealth, as did the half-eaten food thrown away to the benefit of buzzing clouds of flies. Our Arab driver assured us that these people had now so much money that their main problem was to find ways of spending it. We left our car at the fine motor road where no horses or donkeys had been seen, and passed through a shady palm forest with evidence of former intensive irrigation and cultivation, till we stood in the baking sun on a long and beautiful sandy beach. Outside were anchored a few small dhows without masts. Drawn up on the white sand lay a small raft-boat of the type I knew so well. It had just been pulled up from the water's edge and was still wet. We caught sight of an old man with white turban and mustard-coloured cloak about to escape in between the palm trunks with a bundle of shiny fish. We called him back and he came willingly. Willing also to answer all our questions, for this was his fishing-boat and he had built it himself. It was his *farteh*, and only four men on the island still knew how to make one.

This was a professional job. Beautiful symmetry and exact in every detail. The material was not reed, but the slender mid-stem of date-palm leaves, just as on Failaka, where reeds were equally lacking, at least today. Apart from the usual lashing, each stalk had been sewn neatly to its neighbour with a result strikingly similar to the cane-boats I had seen in use among the Seris Indians of Mexico. Apparently canes and palm-stalks were too hard, not spongy enough, to be lashed together with outside loops only. Nor would they probably maintain their buoyancy as long as reeds. I asked the old man. He did not know. They used to sail these farteh to Saudi Arabia in two days in former times, but after use they were always dried ashore. He doubted they would float more than a week. Probably the palm stems would not even survive more than a week in sea water. He now propelled his farteh with oars, but drew the shape of the former sail in the sand. He referred to it as a *sherá*. It was precisely like the trapezoidal sail of the former Iraqi dhows, now surviving only as a symbol on the date boxes. In all essentials it was nothing but an ancient Egyptian sail set at a slant. In theory, nothing more should be needed than to tilt a square sail to make a bundle-boat tack to windward.

We walked along the beach and found two more boats of precisely the same kind pulled up among the palms. One was quite new and a masterpiece of workmanship. These raft-boats obviously had the one advantage that they could come right in across the limestone shallows and be pulled ashore while other boats were anchored far out.

I found a single palm-leaf stem tossed up by the waves on the beach, which I first picked up for its beauty. It looked like a splendid white flower, as the thin end was densely overgrown with a colony of chalk-coloured, conical molluscs. I was showing it to Bibby as a curiosity when it dawned upon me that here was silent testimony that these palm stalks did not dissolve quickly in sea water. This one was still as complete and tough as new, and it must have been months in the sea for all these molluscs to have grown to such a great size.

We ran after the old fisherman. He promised to bring two hundred such palm stems to the ASRY docks next Friday if we sent a car for him. And he would personally repair our big farteh.

Friday came, and so did our driver with the big load of palm stems. But no fisherman. His wife had told the driver that some years ago another foreigner had come and asked the fisherman to build him a farteh. The foreigner had filmed the fisherman as long as he was building it, but when the boat was ready he had departed, saying, now they could keep the boat. So, no more boat-building for foreigners.

But the fisherman had delivered the two hundred stalks against cash. And that was, after all, what we really needed. Carlo worked above water and Toru and Gherman below. All the Iraqi berdi reeds we could spare were stuffed into the gaping wound. Then the long and tough but slightly pliant palm stalks were stuck down under the loops and sewn on, Dilmun fashion, side by side like a breastplate. To look at, the palm stems were amazingly similar to papyrus reeds. But they were heavier and harder. When the slack spiral loops were so full and tight that not a single further stalk would enter, a criss-cross net of string was tied over for extra security. Then *Tigris* looked as trim as when we raced towards Failaka.

We had another problem to solve. Khalifa, the fine young Arab the Minister had chosen to help us on arrival, came back the next morning with a discouraging report. With the bearing of a true gentleman and always in spotless white but for his shiny black shoe-tips, Khalifa looked like an Arab film star; he had also played as such in some Walt Disney production. His delightful old father had been one of the last pearl divers on Bahrain, a local profession which Bibby had traced back to Dilmun times. He could tell us at once that there was not a single soul left on the island today who could sew a sail, and the last dhow-sailors were so old that they had left the sea to the younger generation. Perhaps we would have better luck if we went to Pakistan.

We could not continue with our main sail in separate pieces. Besides, we had discovered that we ought to experiment with a much larger sail to get the speed we needed for tacking. The moment Khalifa brought his discouraging news, I had a crazy plan ready. I dug into the box under my mattress and counted my dwindling supply of cash. Barely sufficient to risk sending Detlef back to Germany with the dissected sail. The Hamburg sailmaker could put it together again. It was he who had made it. And he could also make us the big dhow-sail I had hoped to have made in Iraq. Norman, our sailing expert, begged to have it as big as the Southampton University test had suggested. Only then could we do justice to our ship. Taking off to look with Bibby at Dilmun archaeology, I gave Norman carte blanche to design the sail. With him were Detlef and two old pearl divers, former dhow-sailors, brought by Khalifa to give us advice.

Waiting for Detlef to come back we remained for more than three weeks on Bahrain. *Tigris* rose and sank with the tide, up and down the lofty wall of the concrete mole. We even swung madly to and fro as the tanker harbour was not protected. The Emir and dignitaries from all Arab nations inaugurated the world's largest drydock with a supertanker entering a lock a few yards from our side. Modern mariners gazed down upon us in wonder from *Texaco Japan*, a vessel of 325,000 tons. They shook their heads at the idea of going into the Indian Ocean in a haystack like ours, but I felt dizzy and unsafe as we clambered up the endless gangway hanging down the monstrous iron wall of their island-size tanker.

We got the same friendly treatment in Bahrain as we had experienced in Iraq, although the leaders of these two Arab nations were not on speaking terms. They represented opposite political systems. From the day I heard of the prison island I never missed a chance of trying to get there. Gherman commented that usually the difficulty was not to get in but to get out of a place like that. He had a whole assortment of criminal suggestions of how to get inside, all according to how long I would like to remain there.

In the meantime Sheikh Abdulaziz Al-Khalifa, the Minister of Education, gave an unforgettable Arab dinner for the expedition. And the son of the Emir and heir to the Emirate, Sheikh Hammad, invited Bibby and me to the palace and gave me an ebony walking stick with handle of pure gold, seen by Bibby as a peaceful modern substitute for the sword once donated by Arab rulers. And on the last day before Detlef's return I got the great news. The Commander of the prison colony, Major Smith, would personally fetch me with the police boat on the pier of Budayia village on the northwest coast by sunrise next morning. Khalifa would take me there. Bibby had unfortunately already flown back to Europe. I was permitted to bring two of my men on condition that their nationality was approved: I was not permitted to bring the Russian expedition member. I asked to bring Norris, my cameraman from the USA, and Carlo, my photographer from Italy, and this was agreed.

Major Smith, a husky police officer and former professional British soldier of long service in the colonial army, was punctually at the spot to receive us. His police launch was moored at a little pier between modern plastic boats and old dhows, some complete wrecks. The friendly Englishman apologised that he had to get us up so early, but we had to reach Jidda island before the tide went out.

No land was seen for the first quarter of an hour, until a most beautiful little palm-clad island rose into sight with a single but rather impressive house. Umm-al-Saban island, said the major. There was a well on it, like those on Bahrain, and the whole island belonged to Sheikh Hammad, the Emir's son.

Then Jidda island appeared on the horizon. High cliffs. White like Dover, with a single small house visible; the major's. The water was certainly shallow long before we docked at the end of a long, crude stone pier. To the right of the landing area was the high land. To the left were nothing but date palms and in front of them a number of huge rocks that seemed to be the remainder of a bluff blasted away by man.

I walked over to examine the surfaces of the fractures. This was indeed the work of man. Old. But not old enough to date from bronze age quarrying. On a well-sheltered overhang I detected an Arab inscription cut into the rock face. I called Khalifa and he translated: 'In the month of *shah'ban* the year 978 rocks are being cut by the honourable Mahmud Sar Ali to renew the towers at Bahrain fort.'

Translated into the Christian calendar this would become AD 1556, thirty-five years after the Portuguese had conquered Bahrain. This honourable Arab had apparently worked for the conquerors to rebuild the Arab fort above the buried Dilmun city. Fortunately neither he nor his commissioners had suspected that a whole city of quarried stone lay under their own feet at the building site.

There was apparently nothing more of interest among the blasted blocks, but Major Smith said he would show us something else. And he did. I sincerely regretted that Bibby was not with us. Walking inland with the major as our guide along a much trodden trail, we were greeted by calm and polite prisoners. They apparently walked about with no other guards than the sharks swimming around the island outside the wide shallows. Major Smith led us past his house, which seemed empty, and up on the elevated limestone plateau that was by far the largest part of the island, scarcely a mile long and much less in width. And it seemed as if a fair part of it had been taken away by ancient stone workers.

We could hardly believe the evidence we saw of the extent of former quarrying. In a few places were obvious traces of the early Portuguese, or of the Arabs working for them, as we had seen at the landing place. But these quarries were easy to distinguish from those that dominated the whole island. The quarries from Portuguese times had large flat surfaces which, after more than four hundred years, still lit up the mountainside with a yellow-grey colour and occasionally showed the marks of the drill holes used for powder. But these quarries were superimposed upon and surrounded by other quarries that filled almost every part of the island hills and the coastal cliffs. It was difficult to locate an area not cut into terraces, escarpments, niches and steps in times so long ago that all surfaces had so darkened as to be indistinguishable from the natural rock face, and so eroded as to lose the sharpness of all edges and corners. In these by far the most predominant quarries the stone worker, ignorant of explosives, had removed his blocks by rubbing deep grooves behind them. And no two blocks had been cut the same size. Most were no larger than four men could carry to the shore on poles, but some must have been truly gigantic to judge by the gaps where they had been removed, and by some unfinished blocks still in place. In some areas bizarre formations remained for no apparent reason, resembling petrified mud houses or cubist monuments. On the north-west plateau all rock had been removed and had thereby created a convenient site for the cluster of small fenced-in prison barracks. Extensive screes of eroded quarry rubble filled the area, sometimes in large heaps resembling burial mounds. In the midst of all this old gravel a square outcrop rose like a lonely building. It seemed as if intentionally left there to give an impressive idea of the quantity of rock that had been removed from all around it.

I was quite familiar with prehistoric quarries. I had lived with them for months on Easter Island and had studied those left by the pre-Inca master-builders of Peru and Bolivia. Also those that yielded the largest one-piece blocks for Egyptian, Phoenician and Hittite megalith builders. The bone-hard limestone cliffs of Jidda island had not been worked by amateurs, but by a people belonging to the great old clan of true stone experts. Everywhere were vestiges of an incredibly skilled activity, but no sign of buildings: large portions of the island had literally been carried away, not only from the quarries, but from the coast. So much rock had been removed that it far surpassed the sum total of quarried blocks in the structures so far excavated on Bahrain. It would therefore be tempting to prophesy that more buildings are yet to be discovered beneath the Dilmun sand. There was another reason for this suspicion: no columns had as yet been discovered in the Dilmun palaces or temples, whereas colonnades of round stone pillars were common in antiquity from Egypt to Mesopotamia. Among the niches in the Jidda quarries were some circular cavities, marks of the removal of cylindrical blocks the size a man could barely encircle with his arms. Stones of this shape do not reappear in the known buildings on Bahrain. They can hardly have been used for anything but segments for a column.

Most of the prisoners we saw went to and fro at a distance as if they did not notice us. A few came close and looked at us with big eyes and an expression as if they were pleased to have visitors on the island. Some even ventured a broad smile. The major explained that many of them were dangerous fanatics against the government. Perhaps these were not among the men who walked around loose. As we sat down among the niches to have our picnic lunch a friendly prisoner came up to us with a jug of milky tea. The only domesticated animal we saw was a white mule. But an incredible number of cats lurked everywhere, and the major estimated there were about four hundred of them running wild among the rocks. And never had I seen so many cormorants. Packed together into regular cloud formations they could blacken the sun as they sailed past and out over the sea.

The southern part of Jidda was so remarkably different from all the rest that I began to wonder whether it was due to the work of ancient man. The limestone rocks here suddenly fell away to slightly above beach level, and were no longer naked but covered by fertile black soil. Date palms and a few ornamental trees grew here, and between their trunks were truly luxuriant vegetable gardens. They were so unusually tidy and well kept that it could almost be suspected that the rulers of Bahrain hated gardeners and had sent all the best to this penitentiary. The whole area surrounded a basin with a large natural well from which ice cold, crystal clear water welled up from somewhere below the bottom of the sea in such a quantity that we had to jump aside when the proud major somehow forced it into a garden hose.

No wonder that the Sumerians who had been to Dilmun thought there were two seas, one of fresh water below the salt one. Bahrain also literally means 'Two Seas'. Jidda island and tiny Umm-al-Saban were in every respect typical satellites of Bahrain, barely seen from the highest cliffs. Paradise to Noah and the early Sumerians was Hell to the modern Bahrainians who were squeezed together on this same piece of quarried rock. If I had been the Emir I would have picked Jidda island for myself and sent the prisoners back to the oil fields on Bahrain. Carlo and Norris seemed to agree.

Major Smith was visibly flattered at our enthusiasm for what he could show us, but none of us envied him his little green corner when he confessed to a little loneliness. For many long years his life had been that of his own prisoners, except for his fine uniform, and except for his loneliness. They after all had each other for company. He seemed to apologise as much to himself as to us when time forced him to rush us back to the landing place. The tide was in again. The police boat could come right to the tip of the pier and take us to the free world before the salt sea withdrew from Jidda once more.

Major Smith stood alone like a statue at the tip of the pier as long as we could see him. The emptiness almost seemed measurable both in time and space. At his back was the rock with the written message from an Arab who left four centuries ago. All around him were the empty niches abandoned by Dilmun workmen four thousand years before.

The Dilmun transporters had certainly loaded their tons of burden from Jidda island on to strong, sturdy and shallow vessels, whether they called them farteh, like the Arabs today, or ma-gur like the Sumerians in the days when the quarries were worked, and the big blocks of stone were surely not unloaded on the nearest part of the Bahrain coast where we stepped ashore from the police boat. There was no need to drag them overland to the building sites when the floating vessel with sail or oars and punting poles could bring the tons of cargo straight to that part of the coast where the stone was needed. The tidal changes over the shallows seemed a gift from the ocean god just for this purpose. The vessel came easily in to Jidda island on the tide and was beached as the water withdrew. Sitting sturdy on the rock bottom the broad reed-ship would be as steady to load as a four-wheeled cart ashore. It would be ready to float seawards with its cargo when the tide next came in, and might even reach Bahrain to come in with the same flow, ready for unloading as soon as the water went away. No wonder these people built temples in gratitude to the gods of nature.

No sooner were we back from Jidda when Detlef was back from Hamburg. We rolled out the two sails on the ASRY mole for Rashad to add the emblem of the sun rising behind a stepped ziggurat. This symbol seemed all the more appropriate now that we had seen that such a structure had been built by the maritime sun-worshippers of Dilmun too. It was a relief to see our original mainsail back in one piece again, small and easy to handle. But I took an instinctive dislike to the new colossus when we all gazed at its size. Even Norman and Detlef scratched their heads while Rashad painted on it a huge pyramid.

It was Christmas in a Moslem world and irrespective of beliefs or disbeliefs the eleven of us celebrated the occasion in a dignified manner together ashore. Very early on the morning of 26 December we reloaded our ship for departure from Dilmun and, we hoped, from the Sumerian Gulf. Norman and I had barely managed to solve another major problem before the big new sail arrived: we had to prepare a very long yardarm to hoist it on. Khalifa had shown us that we could choose from literally hundreds of abandoned *dhow* masts and yardarms in the docks and lumber yards of Bahrain. But as the weeks passed and we had seen them all they were either too short, too crooked, or too worm-eaten and rotten. At the eleventh hour Khalifa found us an old carpenter who helped us splice two fairly healthy booms together to form the forty-foot yardarm our new sail would require.

Rashad declared his paint dry; we tied the new colossus to the new yardarm and would be ready to depart as soon as we got this important renovation on board. When the canvas was folded to the wood, it took all eleven of us to lift it off the ground. My scepticism grew into clear disapproval. It seemed as heavy as an elephant. It began to dawn upon me that we had acquired a white elephant.

'This is crazy,' I shouted to Norman. 'It will break our mast.'

Norman wiped his perspiring forehead. Poor man, he had been struck again with an inexplicable fever and was ill for the fifth day in a row. He admitted that the sail was too heavy. He had told Detlef to order the thickest cotton canvas the Hamburg sailmakers had, but had never realised they had anything *this* thick. We managed to drag ourselves in procession with our burden to the edge of the mole alongside *Tigris*, which was down below.

'How shall we get it on board?'

'We'll have to hoist it across with the halyard.'

'But how can we get the darn thing past all the backstays?'

Norman, to my amazement, only scratched his blond hair. I needed no more to realise that we really had a white elephant. And a big one. Norman, always so meticulously exact and foresighted in any planning, had planned a sail that would make Southampton University applaud in approval. But neither he nor I could see how to get it on board. While we all stood there in a line, suggestions rained from left and right and even from the spectators behind us. We could pass behind the stays if we pulled one end in from the stern. No, not that way, but from the bow. No, we could loosen all port side stays and let the yardarm under while someone held the masts. We could leave it alongside, outside all stays. We could, in fact, do many things. But we would be wedded to an unmanageable monster out at sea.

I felt that I was really losing my temper. The situation was ridiculous. I was just on the edge of letting my anger show above all the confused shouting when everybody down the line put on a Sunday-go-to-church expression and lowered their voices. It was not the Emir who arrived, but Norris with his baby. He had slipped his part of our common burden into the hands of Khalifa and came sneaking up with his sound-camera.

I took a brutal decision on the spur of the moment:

'Let this white elephant down. We are going to sail without it. We might get it on board here in port, but how the devil are we going to handle it in a storm at sea. The mast will rip off before we find a place for it on deck. All aboard! We're off!'

CHAPTER 6

We Gain Control of Tigris

WE UNFOLDED our good sail in the mythical waters of Dilmun and felt as if we had opened wings and taken off with the freedom of the air. The bow was once more as solid as a bird's breast, with every feather in place. The sail was perfectly set to catch the wind and gave us an uplift we could feel, a thrilling sensation known only to the winged species, gliders and sailsportsmen. The days of being towed had been like bumping in a truck off the road on punctured tyres. Now we barely seemed to touch the soft waves, ready for take-off in the manner of the moon-ship we had seen lifting from the sea Sumerian style.

This was exciting. Real fun. Norman ignored his fever and beamed with joy. We stood with a tiller each on either side of the steering platform, twisting the long slanting logs that ended in oar blades, and confirmed with satisfaction that for the moment our vessel responded marvellously to our manoeuvres.

We were eleven men as free as man can be. Free as the seagulls that accompanied us. Neither they nor we had any preconceived itinerary. Nobody expected us anywhere. We had no fixed port of call, no cargo to deliver. Free, except for one little snag. Unlike the seagulls, who know no boundaries, our freedom ended where land began. We had to get out of this gulf to secure the unrestricted freedom of the boundless ocean. But the outlet of the gulf was a needle's eye. Would we be able to hit it?

The winter wind had long since returned to its normal course and blew in full force from the north. A perfect wind to sail a reed-ship from Iraq to Bahrain. But after Bahrain the whole gulf curves at a right angle and we had to steer west-north-west to hit the needle's eye, out into the open ocean. We could hardly expect a following wind to turn with us up to the outlet of the gulf. Today, as in

Sumerian times, this was the one leg that really demanded navigation.

The gulf, when seen in correct proportions on a globe and not distorted on a flat world map, is the size of England and Scotland combined, and shaped like a stomach, with a single entrance and a single exit. We had entered by the throat, coming down the Shatt-al-Arab at one end, and were now heading for the tube-like exit, the Hormuz Strait at the other. Just there the Arabian peninsula stretches out a long dagger that points from south to north, which would have struck the bulging belly of Asia across the straits but for the coast of Iran, which withdraws in a deep inflexion before the dagger's tip, thus creating the extremely curved and tricky passage between the land-locked gulf and the free ocean outside.

There was more to it, as we rushed merrily ahead over the waves, than defying the main wind direction and getting out through a narrow neck, flanked by rocks and dotted with islands. The Hormuz Strait is renowned for its incredibly dense shipping traffic, with tankers and merchant vessels from all over the world rushing through in both directions, making it an extremely hazardous playground for small sailing vessels. Rightly or wrongly, we had been warned that this narrow passage represented the busiest shipping lane in the world, and our pilot chart showed it as a marine autostrada with one lane reserved for incoming and the other for outgoing traffic. All seemed well organised for superships chasing through at full speed with radar and automatic steering, but apparently the security for smaller vessels of wood or reeds was not up to the same standard. The captain of a Norwegian supertanker told me that, patrolling the ship one early morning, upon entering this strait, his watchman had discovered a dhow-sail hanging from the bow. Nobody had seen the dhow itself, nor did they ever hear a thing about its crew.

With Khalifa as interpreter, Norman and I had repeatedly visited the dock for small ships at Manama, speaking to the owners of the many motor-dhows. There were some small boats coming from Oman to fetch bottled water and other cargo from Bahrain, and one of the dhow captains said there was a narrow passage, sheltered from the Hormuz Strait shipping lane by some rocks, where they used to pass to avoid the busy thoroughfare of the fast big ships. He agreed to pilot us through if we followed in his wake. There was no need to be reckless when there was every reason for precaution, and as we set sail outside the tanker anchorage of Bahrain, an old and rather weather-beaten dhow without mast was running in our company at shouting distance. Its black captain, Said Abdulla, looked more African than Arab, although he was from the Sultanate of Oman. Three of his crew had abandoned the dhow at the moment of departure; they must have had a hunch of what awaited them. But four men remained on board, wanting to return to Oman. One was an African Swahili, two were from North Yemen, and one was a compatriot of the captain from Oman. Rashad happily joined this mixed lot to serve as interpreter and liaison between the two ships, roughly of the same size.

Said had a compass, but no map. He did not set course straight for our destination, the Hormuz Strait. What he did might reflect an old tradition, as he took a route apparently followed by all the dhows. As soon as we had passed a low, white, sand island on our port side and Bahrain with all its ships had sunk behind us, he set course for the tip of the Qatar peninsula, another long but blunt dagger jutting northwards into the gulf from the Arabian side. We had been slow getting away from the ASRY harbour, and Rashad yelled that Captain Said insisted we must hurry in order to pass Qatar before night. Norman consulted the Persian Gulf sailing directions, and we could well understand Said's desire to get away. We read: 'All the villages on the north-western coast of Al Qatar were in 1951 deserted and in ruins, having been sacked in recent years; a few fishermen sometimes camp temporarily among the ruins.'

We reached this north-western coast of Al Qatar just as the sun set and night fell upon us. It did not seem deserted. We saw several lights along the shore on our starboard side and our experience of Failaka was fresh in mind. Captain Said made a speed of four knots, but the best we could do on our own on an eastward course with a north wind was two knots, so he insisted on towing us. But luckily his towline broke and when he came back to tie us up again I refused. The danger of being towed by this reckless captain was greater to us than any fear of Qatar, which was very far away. Said was furious, and Rashad had the greatest difficulty in conveying the bad-tempered messages back and forth. In the midst of it all a huge sea-bird with a long beak, hooked at the tip, landed amongst us and created havoc until it calmed down in Carlo's arms; later it took off into the night and landed on the sea. Shortly afterwards a big falcon sailed into the light of our kerosene lamp astern and kept on sailing so low over the heads of the helmsmen that we could see the nostrils in the hooked beak. It, too, slid away. The moon was full, the sky was clear; it was great to be sailing on our own. But the north wind was biting cold and it felt fine to creep under a blanket when the dhow picked up speed and left us at peace with the wind and the natural rhythm of the waves.

We had been told by Said to steer from Ras Rakan, the northern tip of the Qatar peninsula, towards Halul island, which was supposed to be high and to have a light tower. I was sound asleep when the steering watches woke me up. The lights from Qatar were gone but the silhouette of the dhow was back with us and Rashad shouted that Captain Said now insisted that we must be towed. I was surprised and refused blankly, as the steering watches confirmed that we were making precisely the course Said had determined. We now learnt that the proposed tow was not for our sake. The sea was too rough, and the old dhow could break to pieces because it had sprung a leak and was being pumped out continuously. Its motor could not go as slow as we were sailing, and if it went in circles the big waves would hammer the hull to pieces. If it had us in a tow, we would steady the dhow and keep the motor working at a proper speed. Captain Said refused to continue unless he had us in tow.

This was a bizarre situation. Now, when we were at last able to sail as we wanted, our pilot was upset because we did not let him tow us. To gain time I asked for an hour to make up my mind. Said reluctantly agreed and they disappeared out of view. The last we heard from Rashad was his suggestion that the dhow be sent back to Bahrain whilst we should sail on alone. Many on board took the same view, but Carlo and Yuri agreed with me that until we were outside this confusing gulf with all its obstacles and traffic, its changing winds and tides, we had better have the dhow as a safety precaution.

An hour later the dhow was back and Captain Said was now all goodwill. He was here to serve us; anything we said he would obey. But could we please give him our position and estimated time of arrival at Halul island, as he had lost all bearings due to circling around. In the moonlight we could see the dhow rolling badly, and Rashad confirmed that *Tigris* looked like a sturdy mole compared with the frail cradle he was on. Their pump was running continuously. Their lifeboat seemed even less secure than the old dhow itself, but Rashad insisted on fulfilling his liaison mission. Clearly, Said would feel safer with our sturdy raft-ship lashed on astern, but fearing to be sent back to Bahrain alone he agreed to honour the original arrangement. We sailed on in triumph, independently.

Except for some wooden crates and other flotsam the sea appeared surprisingly clean; I had expected it much worse in the gulf. In the early afternoon the dhow had not been seen for hours

188

when HP shouted from the steering platform that he saw a strange white box floating right in front of the bow. Next moment we felt a violent jerk throughout the ship. HP completely lost steering control and *Tigris* swung around with the sail and the long loose loops of the guide-ropes hammering and snapping at all and everything. In a few chaotic seconds our ship had come to a dead stop; it was as if we had been caught by a giant fish-net in the middle of the sea. And that was just what had happened. We saw a thick, red nylon rope encircling us like a sea snake as the white box began dancing and then vanished beneath the waves. Others could now be seen bobbing up and down on the waves. We never saw the net they held. Like a Tarzan, Detlef leapt to the side and swung himself overboard, hanging in the stays, and with his sheath-knife he cut the thick rope that held us captive and threatened to rip the reeds apart. We had no other choice to save our vessel and its rudder oars. The red snake lost its tremendous grip; dead and powerless it floated up on the wave crests beside us and let *Tigris* pass as HP regained steerage. Hanging alongside with goggles Detlef could detect no damage to the bundles.

A few hours later *Tigris* was surrounded by real sea snakes, some red as the nylon rope. For a couple of days they dominated the sea around us. Horrible-looking creatures embellished with the most gorgeous colours. Most of them floated sleepily over and between the waves at our side, like severed bits of rope about the length of a common adder. Others undulated independently of the waves; they swam on the surface with all the attributes of real snakes. And they *are* real snakes. Twenty such marine species are known in the area where we now found ourselves and in the Gulf of Oman immediately outside the Hormuz Strait. Nineteen of these species have a deadly bite. But they are drowsy and hardly ever attack. The different kinds are distinguished by their great variety of colour and design.

I was prepared for their presence, but the first one I saw happened to be wriggling on the surface right beside our reeds just as Detlef threw a rope with a canvas bucket over the side and lifted it above his head for a refreshing shower. The snake was not in his bucket but Detlef thought it was when I shouted a warning, and he almost fell overboard before seeing the snake twisting where he had just filled his bucket. Brown on its back, yellow under, with black zig-zag designs on its sides. The next we saw was yellow with black spots, and some were bright red. For days no one dared to take a bath or a bucket shower before checking the water for the infinity of snakes that ruled the surface throughout this area.

That night, when I crawled out of the cabin to take over my midnight steering watch, I found *Tigris* surrounded on all sides by lights, as if in port. On one side was a lighthouse flashing on and off, and behind it in three different places the night sky was coloured deep red from the reflection of some very large invisible flares. On our other side, and very close, was an oil flare brightly burning, illuminating our sail and the starboard walls of our cabins. A fully-lit ship closely crossed our wake, and far ahead we could still see the mast-light of our dhow. It was suddenly much warmer. Few stars, but the moon, still almost full, danced madly about above the sail on either side of the mast head. We had reached Halul island and were now manoeuvring through a network of oilfields. Detlef estimated that we had been sailing at well over three knots.

We barely escaped severe collision with the dhow next morning when two hissing seas threw both boats down into the same trough. The dhow had come back to inform us that they now had trouble with the pump and wanted to alter course slightly more northerly to put in at Sirri island for repair. We located this island on our map and found to our delight that, in spite of leeway, we ought to be able to make it. But at our present speed we could never get to Sirri by sunset as Captain Said had estimated. The earliest possible arrival would be in the late afternoon of the following day. There was complete disagreement between Captain Said and our own navigators on this point, but we kept on more or less in company, the dhow sometimes disappearing far ahead of us and sometimes far behind.

Soon Said was once more dancing dangerously close at our side to let Rashad shout that we on *Tigris* were steering much too far north into the wind. We had a wrong course for Sirri island, and we could fall off comfortably since the island was much more to the east than we were steering. This offended the professional pride of both Norman and Detlef, who agreed on our hard course. They said that Said had obviously lost all sense of direction since he had now been turning around and away from his usual course too often.

We had reached no island by sunset, nor when the sun rose next morning, and Said gave up squabbling and ran peacefully in our company, now mostly far behind. This night we had again sailed between some oil platforms, just as our navigators had predicted. The rolling throughout the night had been terrific and we had to let down canvas in front of the port side cabin opening because the spray was whipping in over the side bundles. Rashad shouted, as the dhow ventured up to us after sunrise, that they now had a problem even with the rudder. We wanted him to return to us, but he was keen on sticking it out.

As the sun passed the zenith at noon, Norman and Detlef predicted that Sirri island would soon be within sight. Shortly after that Norman climbed the mast ladder and after a moment we heard his triumphant yell from the swinging mast-head: Sirri was to starboard of our bow! There was plenty of room for leeway.

Slowly a rather low but hilly island rose into view. It was a triumph for our navigators and for the reed-ship that had passed its first severe test in navigation. In the afternoon the island grew big. We saw off-shore rocks rising like castle ruins stormed by frothing white seas. We even saw some low land with big trees, and, less attractive for us, some enormous buildings and oil installations ashore.

By this time the dhow had come up alongside and informed us that there was the island! It began to rain. We prepared our new anchor and agreed to sail into the harbour facing us and wait for the dhow to be repaired. Said again took the lead, speeding up on a long cruise ahead of us. Then he came full speed back, with Rashad shouting: 'This is not Sir island! This is the Persian island of Surri! Said told you all the time that you were heading too far north!'

Complete confusion. Further map-reading on *Tigris* in the pouring rain. Spelling shouted back and forth in Arab, English and German. Surri island? There was no such island. This was certainly Sirri island, and it did belong to Iran. But with Said's and Rashad's new pronunciation and controlled spelling it appeared that Captain Said had wanted to go to Sir island, not Sirri. And there was an island mapped as Sir Abu Nu'air in a completely different direction, quite near the coast of Oman. Said recognised that full name, and stressed that that was where he had wanted to go.

No harm done. Where we were was for us a more favourable position; we could sail with better wind straight for the Hormuz Strait. We could go in to Sirri together and have the damage to the dhow repaired there. But no thank you. The mere thought of being in Iranian national territory made Captain Said desperate. He had no Persian flag. He had no documents permitting him to sail in Persian waters.

We had a Persian flag, he could borrow it, we said. But no use arguing; Said did not have time even to listen. He now confessed that there was something wrong with his engine too, and his water-tank was leaking. We were not in the Arabian part of the gulf; this was a serious matter. And before we had a chance of reaching any kind of agreement the dhow, with the defective pump running, water splashing in the hull and the rudder damaged, put up full speed downwind in a direction far south of the Hormuz Strait. We shouted that we had now to head for the Hormuz Strait and no other destination. Through the roar of the sea we thought we heard Rashad's voice instructing us that we had to make a rendezvous somewhere on the Arabian mainland north of Dubai. The name of the place, whether a bay or a coastal islet, was totally lost, and all we had to guide us was that, as the little dhow left, never to come back, Said had steered with a course south-east.

There was no need for *Tigris* to call at Sirri; this island had lost its charm for us anyhow, once we saw all its oil installations and hangar-like buildings. But neither did we see any reason to escape like Said. We had no visa, but nobody would suspect that smugglers or spies would sail the gulf in a reed-ship, so we ventured into the sheltered water on the south side of the island and sailed very close along the shore. The sudden silence of the sea, and the abrupt end to the violent pitching and rolling, left us with a comfortable feeling of theatrical unreality, and we lowered our voices as we seated ourselves on the benches along the plank table when Carlo called us to dinner. Fried fish-roe with biscuits and hot soup of dried fruits were enjoyed seated in oil-skin jackets while the last drizzle died away. What scenery! Calm sea, rising clouds, a long row of lamp lights very close to our port side, as if we were sailing slowly down a river, and on the other side violent red flares everywhere, as if the world was on fire along the horizon, particularly in the direction we would be heading. Occasionally, we thought we could still see the mast-top lights of the dhow, but in the early morning it was gone. With Rashad on board.

The peaceful hours in the shelter of Sirri were few. The wind came around the mountains with redoubled force, and the swells rose bigger and wilder, as if to take revenge for our attempt to escape their grip. We began to feel the steering bridge wobble as the ropes stretched and the many pieces of wood became loosened in their joints, where bound together or lashed to the reeds. The orchestration of crashing seas and grinding superstructure again became deafening. The thick hardwood block serving as a fork for the port-side rudder-oar split; it first began to gape and snap at our bare feet, and then threatened to come all apart and let loose the thick oar-shaft that would hammer the whole stern to pieces. Carlo was straightaway on the spot like a cowboy with his lasso, and with Yuri's help the gaping block was noosed and trapped motionless inside a network of rope. All the wriggling parts of the rebellious framework on the stern gradually fell quiet like passive prisoners. In our battle with the seas during recent days neither reeds nor palm-stalks had been damaged or lost, but two broad hardwood boards we had tried to lower at our side to reduce leeway broke at mid-length like chocolate bars. Our growing agony was the disappearance of Rashad and the dhow. The compact bundles of *Tigris* gave us a feeling of complete security in any sea. But how would the old dhow take this weather? The night fell on us again.

Our own immediate concern was the jungle of red flares blocking the road ahead and growing in threatening magnitude as we approached, as if we were heading for a battlefield. As the black night sky began to fade towards dawn, so did the intensity of the flares, but we knew they were there and the chart showed a vast area closed to all traffic due to oil installations. We had to steer north of it into a narrow passage between another oilfield and an island surrounded by reefs. North of all this again was the main shipping lane, which we had to avoid at all costs.

We entered the problem area shortly after midnight, when we first saw flashes from several lighthouses and then the contours of an island with a big burning flare that came up close to our starboard side. We were making terrific speed, and at that time could still see a small top lantern and red running light in front of us which might well have been from the dhow. With a defective engine it could not possibly go very much faster than ourselves, with a strong wind that turned increasingly in our favour. As I was called to take over my night watch at 2 a.m. I was surprised to find Norman's berth empty; it was not his watch yet, but I heard his voice on the bridge. Detlef was sound asleep, convinced like me that steering involved no problem with this good wind and the lighthouse coming up far ahead of us seen early enough for the helmsmen to avoid collision course. Outside the cabin I immediately realised that something was wrong either with our position or with the chart. The sky was pitch black on our port side, with no light-flashes as there should have been. All fixed lights and flares were on our starboard side, but there were several ship lights to be seen on our port side. Could a current have pulled us north of the whole barricade of oilfields and too close to the shipping lane?

On the bridge I found Norman with three of the other men eagerly discussing the chart. Something was obviously wrong, or at least different from what we had expected.

'What has happened, Norman?'

'I had to alter course, we were going by the wrong light!'

Norman had been lying restless on his mattress inside the airy cane wall when he heard agitated voices from the bridge. It was Toru, Asbjörn and Norris in a confused discussion from which it sounded as if they saw more lights than was to be expected from the chart. Norman had rushed up on the bridge to find us headed for the reef. The island lighthouse was out, and the one we were steering by was flashing the wrong signal.

'The safest thing we could do was to shear off to the north,' said Norman. 'We squeaked by the reef with about a mile to spare!'

Of all the oil flares we saw, only one appeared to be in the right position, the others were not as plotted on the chart, or not shown on the chart at all. We soon passed them one by one, and although none of us quite seemed to understand just what had happened, it was clear enough that Norman's alertness and quick action had saved us from the reefs. We agreed that we seemed somehow to have bypassed all the obstacles on our starboard side; in some inexplicable way the reefs and all had been marvellously out-manoeuvred, so we permitted ourselves to turn further to starboard and set our course at 80°, straight for the Hormuz Strait.

Just before sunrise the wind went mad. It turned more westerly, with violent gusts, and the sea was as chaotic as one would normally expect it to be only where there is interference from reefs or currents. We had just discovered some strange formations on the port side far in front of us, and we strained our eyes to understand what they could be. Through the binoculars in the twilight they looked like crazy castles from Arabian legend, with white foam from an angry sea shooting up along ramparts and towers. We also got a glimpse of a tiny speck that came and went on the horizon in that same area, probably our dhow with Rashad on board. Then we had to throw aside our binoculars and turn all attention to ourselves and our own ship. A sudden treacherous gust of wind, helped by a twisting wave, unexpectedly threw us side on to the weather and before the sail could be adjusted or the unfortunate helmsman could get us back on course, all the devils in the universe seemed to thunder down upon us. From a course of 80° the bow danced around, past 0° to 340°, and within seconds all hands were on deck in a terrible fight to regain control of our vessel. The wind threw itself upon the rigging and bamboo walls with a violence not yet experienced on this voyage. The thick sail battered with a force that would lift any man off the deck, and loops and rope-ends from sheets, braces and leach-lines whipped left and right and struck at everything on board. Like savages we clung to the canvas and ropes, and in the mad fight that followed the wooden block that held the port side topping lift split asunder and the yardarm with the sail sagged to port. The flapping and slashing sail had to come down quickly before all the rigging broke. But the loops of the halyard, normally easy to loosen were under such pressure where they were wound around the bridge railing that no combination of men were able to unfasten it, and the canvas, flapping in fury on a slanting yardarm, had to remain up at the mercy of the storm. I gave thanks that this was our smaller sail and not the giant we had rejected.

I stole some seconds to look for the strange structures ahead of us while the battle raged on deck, bridge and cabin roofs. There they were. But the dhow was gone and we saw it no more. The ghostly castles in the blue haze were much closer now and caught the first light of a breaking day. In one place was a huge platform resting on round pillars, tall and thick like towers, like a castle upside down. Two other formations were quite different; they really looked like mosques or oriental castles, filled with masts and spires, even houses. We did not know if they stood on reefs or shallows or were anchored and afloat, but the sea that ran densely white-capped everywhere was really wild, trying to climb the columns and ramparts of these solid impediments. During our desperate fight we could see that we would drift clear to the south, avoiding collision, even if we failed to straighten out our sail and regain control of the steering. But these man-made obstacles were clearly quite new, for there was no sign of them on our chart; therefore we did not know what more might lurk beyond. At this time two gigantic tankers crossed in front of us, and a third passed along our side. Too big to roll like us, they split the swelling seas into white geysers that rose high up their bows. We had no such geysers striking our bundles, otherwise cabins and all would have been washed overboard. But we danced about like a duck, preventing the seas from getting any sort of grip on us. Our only dangers were land or ships.

It was Carlo's mountain-climbing fingers that eventually loosened the jammed rope while half a dozen of the men hung with all their bodyweight on the halyard to reduce its drag on the knot. Deep imprints of the twisted rope were left in the wood of the bridge rail as the halyard was untied and the sail came down. Norman replaced the broken block, and, with HP lashed on to the waving mast-top, the sail came up again at an adjusted angle, enabling the two helmsmen to turn *Tigris* on to course. Breakfast porridge was now consumed standing, as the choppy sea sent heavy spray into cups and pots left unguarded. A single wanton cross-wave managed once to chase across from side to side between the two cabins, sweeping everything off the table and leaving us all drenched to chest-level.

If the little spot we had last seen ahead at sunrise was our dhow, it was not heading for the Hormuz Strait, but for some part of the Arabian shore further south. With this westerly wind we were now in a perfect position to sail for the Hormuz Strait but we could not abandon Rashad penniless among unknown sailors. We turned more south-easterly, in the approximate direction taken by the dhow.

This was a desperate situation; the sea was now so rough that the two helmsmen had to pay the utmost attention to every wave and deviating wind-gust in order not to lose steering control once more. We trailed our red buoy aft in case anyone should be washed overboard, but each of us had a personal lifeline tied around the waist, with strict orders to lash the loose end to any part of the rig or superstructure except when in a safe location inside the basketry walls of either of the two cabins. A six-foot shark came for a while to play seemingly in a friendly manner with the dancing buoy. It was the first big fish we had seen in this area apart from some tall, sharp fins that on a few occasions had emerged from the waves around us, resembling those of swordfish.

By midday we found ourselves for the first time in a terribly polluted area. Small clots and large slices of solidified black oil or asphalt floated closely packed everywhere in a manner that clearly testified to recent tanker washings. But the black tar soup was all mixed with bobbing cans, bottles and other refuse, and an incredible quantity of solid, usable wood: logs, planks, boards, cases, grids and large sheets of plywood. One such sheet carried a deadly yellow snake as passenger. All the wood was smeared and clotted with oil from the seas that tossed it about. None of us had ever seen pollution this thick out of sight of land. The contrast to the rest of the gulf was so marked that it made us fear that some marine disaster had occurred. No local dhow would willingly dispose of all this precious timber, nor would a dhow probably carry such quantities of crude oil on board. Only with difficulty did we avoid collision with some of the heavy logs and beams that rose like torpedoes on the waves. The smaller ones we could not dodge. Never had we raced this fast with a raft-ship. Detlef measured our speed as more than four knots.

Time and again we lost steering control but were able to return on to course. The moment the sail threatened to flap the helmsman on the leeward side of the steering platform had to manage the tiller with a single hand and use the other to pull sheet and brace until the sail turned back into the wind. Palms and fingers were scored and blistered by rope. Carlo and Yuri could hardly open their fists; they were invariably called upon when rope-fighting was at its worst.

Attentive to the rigging, and searching for the dhow, we were racing ahead at full speed when we heard Asbjörn's calm voice from the steering bridge: 'Look, what is that? Is it a cloud?'

The sky was blue above us, but there were white cloud-banks along the entire horizon ahead. Cloud-banks, but what the devil did we see above the clouds? I grabbed the binoculars and what Asbjörn had asked about jumped clearly into view. For a moment I could hardly believe my eyes. Above the cloud-banks, raised above the earth, was land, like another indistinct world of its own. Solid rock was sailing up there, still so far away that the lower parts seemed transparent and did not even reach down to the clouds; the upper ridge seen against the clear sky was of a different shade of blue. What we all were staring at seemed far too high up to be real. Were we heading for the Himalayas? Was this an optical distortion, a Fata Morgana?

Our navigation chart had given us no warning of what we were to see. It showed nothing behind the coastlines. Land masses were all equally white. We were so tuned in to the low profiles of the Iraqi plains and the mud-flats and limestone shelves we had so far seen in the gulf area, that we were not mentally prepared for a spectacular sight like this. We dug out of our boxes a land-map of Oman. It showed that this Arabian dagger, with the Hormuz Strait at its tip, rose steeply to an elevation of 6,400 feet above the gulf. This was what we saw ahead of us. The whole peninsula was a lofty mountain chain with rock walls dropping almost perpendicularly into the sea on the gulf side that we were now approaching.

Detlef had just measured a record speed of almost five knots. But when we saw what kind of land we had before us we instantly threw the rudder-oars over to try to turn away from the coast while there still was time. We had clearly come much further south than we ought to have done while trying to keep up with the dhow. They had an engine, and had possibly turned still further south to find a suitable port. Calculations made by our two navigators on the basis of wind and leeway convinced them that we were already so far down from the tip of the peninsula that we had to go in to the coast somewhere; we would never be able now to make our way straight for the Hormuz Strait. But where could we sail in? It became an ever greater puzzle to guess where the dhow could have sought shelter. With an engine it could get into any tricky inlet. But our charts showed no harbour on this coast, no settlement, not even a single lighthouse. There was no kind of beach or landing place, and nowhere even to anchor, for the tall cliffs fell straight into a deep, turbulent sea.

The map showed only a single slight indentation where Said might have taken Rashad and the other men in among the vertical cliffs to get the shelter needed for making their repairs; Ras al Shaikh. It would seem to be a most inhospitable cove between rock walls, to judge from the skyline we now saw. We were soon to find out. We had to clear Ras al Shaikh on our way up to the final cape marking the entrance to the Hormuz Strait.

The mountains were still far away, although near enough already to take the form of real rocks, first rising out of the sea and then piercing right through the cloud belt. We no longer approached this coast voluntarily. We ignored the few remaining pieces of drift wood and had no time to look for pollution; we had to try to save our own skins. A raft-ship like *Tigris* could better than any other vessel surf-ride on to beaches or banks, even be tossed safely up on a reef or a rocky shore, but no craft could tackle vertical walls.

It was time for Norman's scheduled radio contact with coastal stations. We now had to report the disappearance of Rashad and the dhow, and give our own estimated position alongside the empty rock walls of north-western Oman. Bahrain Radio was calling us, but too many strong voices were on the frequency for anyone anywhere to pick up Norman's call. Our whereabouts remained as unknown to the rest of the world as that of the dhow to us and, presumably, as that of *Tigris* to the dhow.

We were fighting our way up the coast, taking the weather in athwart, but we were also getting ever closer to the cliffs we wanted to avoid on the starboard side. A more forbidding land I had never seen: sky-piercing in every sense, and yet not even scrub or a green tuft of grass to brighten up the sterile ascents. A petrified desert tilted on end. The stormy weather struck these walls head on, and assaulting seas were violently rebuffed and rebounded with full force for many miles, stirring up a chaotic turmoil of tossing and leaping waves of a treacherous kind never encountered in the free ocean spaces. How utterly illusory it was for armchair anthropologists to believe and teach that pre-European voyages were possible only so long as the navigator could hug a mainland coast, and that ocean crossings were impossible before the days of the Spanish caravels. Nowhere is the sea worse and the problems more acute than where rocks are or where waves and currents encounter shores and shallows. To hug a coast can be the most demanding task for any primitive voyager. Ancient seafarers must have felt like us in similar situations, unless they were far better prepared. Never have I or my companions been tormented by more problems when travelling on primitive craft at sea than when we have struggled to clear the last mainland capes to get into the open ocean, or when upon an ocean crossing we have approached land on the other side. To hug this Arabian peninsula gave us not the slightest feeling of security. On the contrary, it was quite a nightmare, from which we would have wished to wake and find ourselves safe in the middle of the Indian Ocean.

The modest leeway we made confirmed the sinister prediction of our navigators from the moment we saw land: we would confront the cliff walls before we were able to clear them all the way back up to the latitude of the Hormuz Strait. The wind was too northerly again. In the agitated backwash from the cliffs we were repeatedly lifted up and turned around 40°–50° on a conical surge before sliding down into the next trough, completely off course, with all hands on deck and roofs in another desperate battle with canvas and ropes. There is nothing like common danger to weld men of all ages and ways of life together for common survival. Confronting perils together one rarely thinks of nationality or differences of upbringing. None of us believed he could benefit at the cost of the rest or impress others with reckless bravery; it is successful team-work that counts in achieving a victorious outcome of a struggle. Anyone acting otherwise becomes like a drummer trying to play a symphony without the conductor and the rest of the orchestra.

As the rocks drew nearer the spirit and determination of all on board was exemplified by Norman when he shouted in triumph. Hurrah, he cried, we are defying the wind! His observation was borne out by the red buoy towed astern. It revealed the degree of leeway and showed the direction of our true progress through the water. Clearly we would have done better with bigger oar-blades or more lee-boards, but even so we advanced a few most important degrees into the wind that now filled the sail slightly from forward of athwart. This triumph was enough to maintain the fighting spirit. Yet we all could see that land was still coming our way. We could not even get it away from the bow, unless we turned completely about and headed for the Arab emirates. The cliffs we had all the way along our starboard side ended in a cape that barely projected beyond our dancing bow. If we turned further into the wind to try to clear the headland our sail would flap and we would lose all steerage way. My only hope was that these conditions would change when we came still closer to land. The elements themselves would be forced to change course the moment they hit the lofty cliffs. The current would be turned parallel to the coast instead of against it, and be compressed to gain in speed, and so would the wind when striking the rocks at sea level. The only opening in the compact wall was the Hormuz Strait, way up at the tip of the peninsula. If nature was forced to follow such an escape route, we would be dragged along too.

We continued our ill-fated course, confident that we could improve upon it and turn to safety closer to land. We were close enough now to see the foot of the precipices where they fell into the frothing surf, yet there was no sign of the dhow between us and land. At one place, two small white houses appeared as if painted on the rock wall near sea level. They seemed deserted, if ever meant for people. No sign of any kind of life between us and the rock. I wrote in my diary:

The coast is scaringly close now. The two helmsmen have difficulties riding the huge waves on steady course. From my cabin corner the view through the bamboo-framed door opening is no longer one of burning flares, ships and chasing seas, but one big, continuous mountain side. The slanting late-afternoon light on the cliffs brings out vertical folds and furrows closely matching in cheerless shades the grey and sombre evening sea with veils of froth on the combs. It seems to me that we are travelling across the sterile screes of a naked highland plateau, the barren rocks rising vertically above us to still higher summits. Yet we are now desperately trying with two oars to force ourselves to 52° to avoid wreckage against Ras Shaikh, the first of the capes blocking our way up to the Hormuz Strait. With good luck, with westerly winds and probably northbound current along this coast, we have a chance to make it – barely – *insh-Allah!*

Soon after, at 4.45 p.m. shipboard time, I made a note that the sun had just set. We had by now sailed eastward into another time zone and were almost ready to set our watches an hour on. We began to see several ships' lights outside us. The sombre cliffs turned into ever darker shades, with a single star twinkling above them. Night fell on us as a full blackout, just as we coasted northwards excitingly close to cliff-walls coming out like giant draperies from Ras al Shaikh. This was the only place we could imagine that Rashad might have come with Said in search of shelter. We maintained an intensive lookout for any outline of house or ship, for any glow from lamps. From the cabin roofs we waved our own lamps and flashed signals with our torches. No response.

No lighthouse on this cape. No spark of light of any sort. Nobody could be there. The little we could make out from shadows discerned on the black rock walls justified the hostile impression left from a distance in daylight. It would be suicide for us to venture in between the dark mountain draperies to follow the curving canyon indicated on the chart, even if it possibly offered a sort of narrow shelter with a precipitous overhang on all sides. If the dhow had ventured in there, one might think they would have managed to place a lantern on the cliffs to show us their hiding place. At a speed of two knots we passed the narrow entrance and left the cape, Ras al Shaikh, behind.

I crawled into the cabin again and made another entry in the diary. Our time was only 5.30 p.m. but it was pitch dark. Now we had no idea where our lost companions could be. Captain Said must have taken a completely different course from what we had assumed. There was no other place to look for them between here and the Hormuz Strait. We were rolling so desperately, sailing barely into the wind in the coastal surge, that I got more shadow than light from the tiny petrol lamp swinging from the ceiling and almost hitting my head, and I made a note that although I sat on the floor with widespread legs in an attempt to keep my balance while writing, I would fall over unless I clung with one hand to the cane wall. Nothing could be left loose on board. Hanging on the wall, my binoculars swung out and hit me in the jaw. Shirts, jackets and trousers hanging from bamboo rods on walls and ceiling performed a synchronised show, like a ghostly army of robots doing morning exercise together with clockwork precision. All swung at the same moment, in the same direction, at the same angle. Towels and underwear, buckets and baskets, lamps and watches, all rose from the wall together and swung together, right and left, forward and aft, until they flopped back against the wall in a common clash.

Finding it impossible either to sit or to kneel inside the cabin I crawled out with my lifeline and fully appreciated Norman's acrobatic skill when I heard him shouting to us from the top of the swaying mast that he saw light-flashes ahead. So far they were only rhythmic reflections in the sky of revolving beams from some distant lighthouse that would soon rise above the horizon on the port side of the bow. Detlef shouted back from the bridge that it must be the lighthouse on the other side of the entrance to the Hormuz Strait. We now had to keep it to starboard in order to clear the last, invisible cape of the Arabian peninsula, but later we had to turn at a sharp angle and keep it on the port side as we swung into the open strait.

By now the wind had started noticeably to turn more westerly, fulfilling our wildest hopes. I was convinced that the flow of the sea beneath us had also been turned by the impassable rock barrier and was forced to follow the coastline in our direction. The strain on the rudder-oars had been so strong that the port-side fork began to gape again and threatened to burst Carlo's ropes. To relieve part of the violent pressure before a catastrophe occurred we summoned all men aft and pulled the heavy oar shaft up until a quarter of its blade was out of the water. The tiller of this oar could then no longer be reached from the floor of the steering platform, and a new form of maritime acrobatics had to be introduced just as the rolling was at its worst. The starboard helmsman had to be in charge of the normal steering and shout up to the man on the port-side tiller each time assistance was needed from this second and now most cumbersome oar. In the dark Carlo was to do the proper steering and I climbed up on the bridge rail to reach the other tiller, while slackened woodwork joints bit and shrieked like angry cats in their lashings, and special care was needed not to get a finger or toe caught. With one foot on the cabin roof and the other balancing on a narrow plank tied on outside the bridge rail to steady the oar, I saw nothing but a dancing glow from a lamp Asbjörn had hoisted to the swinging masthead. I knew that even with a flashlight there would have been nothing to see below me but black water, nor anything to grab but the tall oarshaft which was never firm because I myself turned it with the tiller I clung to. It constantly jammed, because it was raised out of position, so with my lifeline around the slippery shaft, I needed both hands to push or pull the tiller which I clung to for support, whenever Carlo yelled for a fast turn of my oar.

At the time it was hard to see the comic aspects of this crazy nocturnal rope dancing. A Sumerian would have depicted himself as standing blind-folded on the back of a bouncing gazelle. But then the lighthouse danced into sight as a bright spark in front of us. Carlo shouted in triumph that he could see it on his side of the sail. Soon it swung so far out on his side that even I could see it from the port side, which meant that by this time we had turned *Tigris* so far away from the shore that all land was now on the starboard side. Under ever more stars we could discern a lofty skyline of jagged crests and pyramids that no longer seemed to come nearer. We were winning. We made it without leeway.

From then on I forgot that I had any problem in keeping my balance, I just concentrated on keeping the bright spark visible as much as possible from the starboard side of the sail. What was left of the two hours steering watch passed with the grand feeling of galloping through a starlit sky on a winged Pegasus that willingly let us decide the course. Then Gherman and HP came fumbling their way up on the bridge to take over the unconventional steering. We began to see ships' lights everywhere. We ourselves were hardly visible to others. Norman had improved this insecure situation by digging out of his personal case a battery-driven flashlight that sent out blips at intervals when he hung it up in the mast. In this wind the cheap local fuel in our kerosene lamps produced more soot than light. To us the masthead blips looked professional and to others must have seemed more impressive than a faint lamp-glow, but Detlef assured us it meant nothing in nautical terms. For that very reason he admitted it might serve as a confusing sort of scarecrow to any vessel coming too close.

A couple of fixed lights, as from houses ashore, suddenly turned up very close to our starboard side, and the rotating beams from the lighthouse were now clear and strong, circling the sky just ahead of us. Detlef was in charge of navigation. Norman came down from the mast, having tied glittering strips of tinfoil to stays and cabin walls; these, as distinct from reeds, bamboo and wood, should show up on radar. We were heading at full speed and with limited control straight into the double lane of the world's busiest shipping channel, because there was no dhow to show us the unofficial passage somewhere behind and between the closely packed islands of the cape.

Never on any sea had we seen so many brilliantly lit ships in motion at the same time as appeared around us at the moment when Detlef ordered a sharp, 90° turn to starboard and the men on the bridge sent us into the main traffic lane of the Hormuz Strait. We immediately received a violent air stream straight at our back and were pressed into a wind funnel between two opposed capes of the same continent, a sort of Asiatic Straits of Gibraltar. The current must also at this time have run like a river out of the gulf. Our speed past the tip of the Arabian dagger was the fastest we had ever experienced with a reed-ship, and the black mountain silhouettes at our side were changing from one minute to the next. With this speed *Tigris* responded to the slightest touch on the tillers and we raced in between the superships which thundered around us as if we were all of a kind.

Things went almost ridiculously well, and with double steering watch and both navigators alert on the roof Carlo and I could steal a few minutes snooze before we were back at the steering oars for our next turn at 2 a.m. It was usually enough to crawl inside the square door-opening to imagine oneself in a low jungle-hut far from the sea. The atmosphere of cane and bamboo was highly unmaritime, but most relaxing. Winds and waves were immediately left behind as the concern of those still on deck; inside was a neutral zone of peace and rest, even if the crests of the billows peeped at us through the door opening almost within reach of a hand. That night was rather special. As I crawled in to stretch out on my mattress beside the door I was as happy as a boy experiencing for the first time the berth by the window of a night train, lying on my side to watch illuminated ships and black mountains passing by like railway-stations in the Alps. Gone was the threat of shipwreck and collision; we were travelling as if on a double-tracked railway line.

I was awakened by Detlef crawling over my legs heading for his own berth. 'We've made it,' he said. 'We're outside.' It was half-past midnight and the night was at its darkest, still young. We were outside? I crawled to the starboard door opening and lifted the canvas cover that someone had rolled down to shut out the many passing lights. It was an unforgettable change of scene. Beautiful. Impressive. The rolling had ceased and the sky was full of stars over vaguely moonlit rocks and hillocks. These were at the foot of tall, wild peaks and mountain ridges, which together formed a fabulous landscape just beside our ship. Turning to the other side we were almost blinded by the revolving beams from a nearby lighthouse that lit up the rocks of the island on which it stood while sweeping sky and sea. No white-caps there. The sea was silent; we could hear ships right and left. No roaring wind or shrieking wood. Peaceful and idyllic. The lights from the ships reflected romantically in the water, as in a sheltered Norwegian fjord. *Tigris* itself relaxed, from sail to steering-bridge, after its record run.

Our speed dropped down to two knots, then to only one. Norman had measured almost five knots in the Hormuz Strait, and to this should be added the speed of the current. We were indeed outside the gulf. We were already sheltered by the cliffs of Oman, but soon discovered that a powerful current still held us in its grip and dragged us away from the Arabian peninsula. We turned still further to starboard, and the speed dropped to half a knot as we sailed southwards into almost complete shelter of the same Arabian dagger that had seemed poised to cut us to bits a few hours earlier, when we struggled northwards, in the very opposite direction, along its windward edge.

'Boys, we've navigated!' was the jubilant exclamation from Norman as we unfolded our chart under a flashlight on the cabin roof to take a decision on what to do next. We now had endless possibilities. And one single problem, but it was a major one. The way into the Indian Ocean lay open ahead of us, but behind us, deep inside the gulf, we had lost Rashad. We had absolutely no idea where the wretched dhow might have gone. Our best guess now would be that Said had steered for one of the tiny Arab emirates. The wind was feeble but perfect for us to get clear of land and steer out of this gaping Gulf of Oman into the Arabian Sea and Indian Ocean. But we could not abandon Rashad, the youngest and therefore probably least experienced of us all. His parents would despair if we had to send a message that we had 'lost' him and had no idea as to where he could be found.

The dhow belonged to Oman and had to come through the Hormuz Strait to reach its home port in the capital of Muscat. If it kept away from the shipping lanes it would have to steer in between or very close to the Oman rocks. Our best hope was to wait in sheltered water where we were, far enough from the coast to be safe from the rocks in case of changing winds, and yet no further out than to be able to see every vessel coasting close to land.

We lowered the port-side steering oar back into position, adjusted loose ropes, and had very easy steering, with much of the night still ahead of us. While the others returned to sleep, Carlo and I were back on the bridge sharing the 2 a.m.–4 a.m. watch, and we agreed that this was the most beautiful moment any of us had ever had at sea. The soaring peaks and rugged mountain silhouettes that sheltered us made an unbelievably impressive setting. They were even more picturesque from this side, their profiles designed to delight the senses, and they stopped the wind and calmed the waves they had so wildly agitated on the other side. To be back on the bridge in these transformed surroundings was like a happy dream after a turbulent nightmare. We had succeeded in escaping through a hole in a fence; the walls of the gulf were behind us. We were in another world now, a world with different waves, another wind. Somewhere outside us was the monsoon area. The monsoon blows regularly across the Indian Ocean as if set in motion by clockwork, turning like a pendulum to move in opposite directions every half-year. The winter monsoon blows from north-east, from Asia to Africa; the summer monsoon from south-west, from Africa back to Asia. Ideal conditions for primitive craft.

Somehow even the big modern ships we had dreaded inside the gulf seemed friendlier out here. A brightly-lit luxury cruiser with coloured lamps in garlands on all decks passed us and made us feel as if we too were on a pleasure trip; the last revellers were probably still in the bars. We certainly envied nobody. It was great to be afloat on a Sumerian ma-gur.

After breakfast Asbjörn inflated our tiny rubber dinghy and Norris and Toru went out to film. The palm-stems in the bow were found to be as tight and solid as when we had jammed them in before we left Bahrain. Most of the men climbed overboard and had a swim, but kept a grip on one of the ropes of *Tigris*.

It was close on noon when someone shouted from the bridge that a small boat was in sight, emerging from some hills which proved to be detached islands. The vessel changed course and came towards us. It was a dhow! For some long moments we were all kept in suspense. Then Norris exclaimed that he thought he recognised Rashad through the binoculars. We all looked closely. It was Rashad. There was Captain Said, too. Everybody. Our dinghy was still in the water and HP was quick to meet the dhow and bring our lost friend back to the reed-ship where he belonged. He almost kissed the bundles as he climbed back to us on *Tigris*, where happy hand-shakings and embraces awaited him from left and right. It was just great. Rashad was a splendid fellow. We all wanted to hear his adventures, and he just came back at the right time for us to gather around the lunch table. Carlo produced a steaming risotto and salami. Gherman opened tequila, Yuri vodka, and I Norwegian aquavit. There were triple reasons to celebrate, and first of all we toasted Rashad's return.

Rashad was almost overcome with joy at all these festivities in addition to the happy ending of his involuntary captivity. The six men in the dhow had lost contact with us after Sirri island, when Said hastened away after finding himself illegally in Iranian waters. He had later been afraid of coming back against the weather, as the sea was wild and the dhow in a shaky condition. It was leaking, and things got worse when the bolts connecting the only pump snapped one by one until water poured through everywhere. The pitching and rolling became indescribable, and the splashing from side to side inside the square fresh-water tank became so violent that the seams opened and four-fifths of the water supply ran out into the hull. The excessive pitching caused by all the water running back and forth in the hull made the propeller rotate almost as much above water as under, with the result that the pistons started to grumble and create engine trouble. Finally, also before they left us, the steering had become hazardous; the metal casing of the rudder lever broke at the seams and the wooden parts began to waver about. They had not meant to leave us that evening we came to Sirri, but the next day the waves had become still worse and they had no choice. Rashad had argued desperately with the captain and the crew to keep a watch for us, but they had been forced to save their own skins. Everyone on the dhow had been dead tired by the time they had come beside the oil rigs, and they had headed as fast as they could towards Ras al Shaikh, without any charts. Captain Said had explained that he steered 'by his heart'. They had first reached the wild mountain coast more or less like us, and followed it until they found a narrow cleft in the rock wall. Here they had steered in and entered a channel where it blew so hard between the walls that *Tigris* would have been unmanageable. But they had turned right into an inner cove where Rashad felt he had entered something out of a fairy-tale. Where the rock walls ended there was a sort of scoop in the mountains and here was an almost prehistoric village, still inhabited, stepped in terraces between the cliffs. Even in this sheltered place the wind penetrated and blew in wild gusts as if from bellows; but they were able to anchor beside some small fishing boats and were totally invisible when we passed Ras al Shaikh looking for lights. They had repaired the rudder and what else they could with wire and wood, but all that needed welding would have to be done in Muscat. They had come out this morning at six o'clock to look for us, and, failing to see us they had steered north and left the gulf by turning in a tighter curve than that of the main shipping lane. Finally they saw us and everybody had been amazed at the speed we had made. The food on board had been fish, rice and curry, and the company had been good, except that they were all close to exhaustion from pumping and repairing.

For Rashad, to be back on the sturdy *Tigris* was like returning from a floating bathtub to a stabilised luxury liner, and never had we seen such extravagance at sea as what Yuri produced from his personal case when the sun set behind the hills of Oman: Russian champagne and caviar, astronaut bread and turkey-a-la-Space, with moon-cheese and a whole variety of Sputnik tubes from which we squeezed our mouths full of pastes, creams, jams, desserts and juices – all the pocket-size dishes that make up the menu of Yuri's countrymen when travelling away from Planet Earth. Dr Yuri Alexandrovitch Senkevitch was a serious space scientist, occupied with living conditions in capsules when he was not floating about at water-level with us on prehistoric raft-ships. Whether ma-gur or spacecraft, none of us would deny that night that there was still a lot of fun to be had on Planet Earth. We enjoyed the Sumerian view of a thin sliver of moon as we squeezed astronaut mouthfuls between our jaws and celebrated the fact that we really had three very special reasons to make the most of the evening:

Rashad was back with us. We were safe outside the gulf. It was the last day of the year!

CHAPTER 7

We Search for a Pyramid and Find Makan

IF A railway engine were to come in through my door while I sat at breakfast I would be greatly surprised. But not more surprised than when the bow of a ship came in while I lay in my bed.

It was no dream. I was not asleep. I had awakened earlier when I heard the sound of an engine approaching in the night and a hoarse voice shouting something from a distance in a hostile tone. The voice of a stranger. I recognised the familiar voices of Norris and Rashad shouting back from the bridge as if in despair: 'Keep off! Keep off!'

It was 2.30 a.m. and stars could be seen through the starboard door opening until a searchlight started to play arrogantly on the starboard wall. I was at once wide awake.

We were somewhere off the coast of Oman. We had been filled with warnings about this area. Small boats were said to have been sacked and looted by unidentified modern pirates in this part of the Arabian Sea. The papers had recently reported how a Danish couple in a small yacht had been stripped of everything on board except a meagre ration of water by which they survived. Whoever were out there in the dark could not know that we were eleven men on *Tigris*, and that there was an exit on either side of our tiny bamboo cabin. I was just about to shake the others awake and sneak out through the port side opening, where we could lie in ambush, when I stopped to listen to the hostile voices, this time very near:

'What is this?' The angry question was shouted in English with an Arab accent.

'A ship,' Rashad shouted back, almost indignantly.

'Then what are the two big cubes you carry on board?' The searchlight played again on both our bamboo cabins.

'They are cabins! Keep off! Eleven men from different countries are asleep inside!'

But Rashad's warning was of no avail. The starlit sky was blackened as the pointed tip of a bow filled the starboard door opening, and in the same second everyone was awakened by a violent shock in the mattresses.

'Get away! You're breaking our ship!' Rashad yelled in despair from the bridge, while the rest of us responded to the assault with a regular war cry.

Yuri's legs were along the door opening where the bow entered, and half-asleep, hearing Rashad's voice, he raged at him through the cane wall: 'Tell them to scramble away, these are international waters!'

'They are not,' Rashad retorted angrily. 'We are right up under the Oman coast, and besides, you don't tell people to scramble away when they are pointing a machine-gun at you!'

We now began to understand the situation. But not so our uninvited visitors. I have never seen eyes as big and white with fear and bewilderment as when we caught sight of the black face of the uniformed Omani policeman at the wheel of the vessel the moment he made full speed astern and left the door open for us to crawl out like angry dogs from a kennel. We had been called upon by a patrolling police vessel with three armed men from Oman's coast-guard. The crazy blips from our masthead, intended to keep ships away, must have had the opposite effect. They had arrived for an inspection. But coming close enough to see the golden reed bundles of our Sumerian ma-gur the policemen had completely forgotten to steer or stop. No vessel like the one showing up in the searchlight's beam had sailed in local waters since long before the days of the Prophet. Too confused to steer, they rammed us amidships so that reeds and bamboo shook. Their surprise did not diminish when a roar as from the falling Tower of Babel, in nine languages, arose from inside the two 'cubes', and bearded savages, angry as lions, crawled forth everywhere. The three stunned policemen backed their vessel away faster than they had come.

Never will they see again so many angry men swarming out so fast on hands and knees from two tiny bamboo cabins. Half asleep, but boiling over with mingled fear and fury, we all hurried to the starboard bundles expecting to find the side of *Tigris* damaged beyond repair. Confronted with eleven men raising their fists and roaring in a multitude of languages, of which they gathered the meaning at least in Arabic and English, the three bewildered patrol-men just went on backing away until they disappeared in the dark behind our accompanying dhow.

The last we heard was a sudden new outburst in Arabic; they got their voices back as they passed our companions, whose vessel was duly marked as from Oman. We were sure we heard them swearing. But no, said Rashad: 'They are accusing their own countrymen of keeping company with a vessel full of *Shatans*.' He did not have to translate that familiar Semitic word.

If the berdi reeds had been as dry and brittle as they were when we built our ship in Iraq, this intended police inspection would have put an end to our expedition there and then, in the calm waters of Oman, barely south of the Hormuz Strait. But what happens to papyrus happens to berdi too: when sun-dried first and then soaked, the fibres gain an incredible strength. Fortunately for us, the rough sea during the last days in the gulf had wetted all reeds and canes so thoroughly that the *Tigris* bundles had become as tough as compact rubber fenders, and the plaited cabins flexed like leather shoes. Even in daylight we could find no damage either above or below the waterline, apart from slight displacement of a few deck planks and bamboo ribs, all easily pulled back into place.

The sight of the police boat's bow inside our cabin door made us more alert to the possibility of a similarly unexpected encounter with a bow of vastly superior dimensions. Such bows had thundered past to the right and left of us for forty-eight hours as the old year went out and the new year began. In fact, after two nights and two days feeling like snails between feet in a crowded ball-room, this was the first night we had spent in calm waters, away from traffic. When night fell on us after the first day of the year, there was not a ship-light to be seen from horizon to horizon, and the sea along the coast was so calm that stars were reflected in the water. Relaxed and secure, we had crawled to bed only to wake up to a collision.

It had not been without doubt that we had decided to lay our course down along the east coast of Oman. Rarely have I relived the sentiments of early travellers more intensely than when we emerged from the narrow Hormuz Strait and had the choice of where to steer. We had no predetermined itinerary. The Gulf of Oman opened before us like a funnel as land ran away from the strait in two directions, east and south. The greatest temptation in our case was to avoid further threats from cliffs and coastlines by steering straight for the open Indian Ocean while the wind still blew in that direction. But that is hardly what the first ma-gur on explorations outside the gulf would have done. The fascinating appearance of both the shores running in different directions outside the strait made them most tempting to explore for anyone who, like ourselves, had never been there before. They fired our curiosity, even though we modern sailors knew from geography that one coast ran on to India and the Far East, while the other ran to the Red Sea and Africa.

Carlo was the only one who voted for leaving the menace of rocks and shores and sailing straight into the open ocean as fast as we could. He felt restless and wanted us to move on while the wind was good. Norris pleaded for a chance to go close inshore along the Arabian peninsula, since the rugged peaks and fierce cliffs of Oman were among the most spectacular formations any of us had seen from the sea. The strange formations offered him some really impressive sequences for the expedition film. Norman supported this itinerary. He said it would take us down to the port of Muscat, where we could overhaul the rig and the steering system before we set off into the open ocean. Others, and I for one, were fascinated by the alternative possibility of following the coast of mainland Asia.

Iran, the former Persian Empire, was now well within sight on the other side of the Hormuz Strait. Distant blue ridges hovered inland behind friendly coastal hills and hummocks. They seemed to beckon a tempting invitation to follow them as they undulated eastwards in parallel rows until lost between sea and sky in the direction of Pakistan. In the broad daylight blue mountains merged into blue sky, but they stood out clearly in the early dawn when the sky behind them was fiery red and pregnant with a hidden sun. From left to right this coastline gradually petered out, and just before the silhouettes of the continental formations completely disappeared in the eastern ocean they took on bizarre shapes such that we could never quite determine whether they were those of fantastic rock islets or cloud formations. It would have been extremely tempting for any early explorer to follow this enticing shoreline eastwards.

I was particularly tempted myself, since we knew those hills would guide us to the coasts of Pakistan and India, the former realm of the Indus valley civilisation. This had been one of the three major civilisations of the ancient world, second to none in age and importance after those of Mesopotamia and Egypt. Throughout the last five millennia an untold number of ma-gur must have been among the great variety of ships that had followed this continental coast to and from the Hormuz Strait. Archaeology has proved extensive contact between Mesopotamian ports, like Ur and Uruk, and the mighty city states of Mohenjo-Daro and Harappa in the Indus valley, with the gulf island of Bahrain as an intermediate trading centre. This was the sailing route Geoffrey Bibby had challenged me to attempt in a Mesopotamian vessel; a route that tempted me almost irresistibly.

But we turned our reed bow to the south instead, and followed the Arabian peninsula. This might not have been the first choice of prehistoric explorers. The barren, jagged mountain walls of Oman refused access to all but birds, with hostile cliffs rising straight from the sea. Yet these cliffs seemed to get less forbidding towards the south, where they gradually fell off into rolling hills. No early civilisation comparable to that of the Indus valley had ever been known to have flourished in Oman. Yet I had a very special reason to cast my vote with those who favoured following this shoreline in the direction of Africa. Firstly, according to all weathermen, the winter wind in this area ought to blow from Asia to Africa and not turn in the opposite direction until spring. Secondly, and this was something I could not quite get out of my mind, some unconfirmed rumours of archaeological discoveries in Oman had reached me the day before we set sail down the rivers of Iraq. An unknown messenger, a German reporter with moustaches as big as the handlebars of a bike, had called on me with exciting news from Fuad Safar, the Acting Director-General of Baghdad Museum. According to dependable sources, a Sumerian ziggurat, a stepped temple-pyramid of the type so far never discovered outside the river valleys of Mesopotamia had been found in the sand somewhere in Oman, on the Muscat side outside the Hormuz Strait.

I had refused to believe it. It sounded like a joke invented by the journalist just because we had painted a ziggurat on our sail. But the German swore he was only bringing the message he had been urged to pass on. The noted Baghdad archaeologist had been quite excited, he said, and insisted that we should try to visit Oman. It was the first time, he stressed, that a Sumerian structure had been discovered outside Iraq.

I mentioned the strange message to my men. 'Too good to be true,' said Norman. This was my own feeling too. Pyramids are rare and far between. It is not the sort of thing one stumbles upon in the sand. Potsherds yes, but not pyramids. In the Old World none had been found outside Egypt and Mesopotamia. It would be too much of a coincidence if the first Sumerian ziggurat should be discovered in a distant land bordering the Indian Ocean, just as we hoisted sail hoping to go that far in a Sumerian boat. Although reluctantly, we all laughed it off and tried to forget the message.

But every once in a while, when Norman and I were together on the two tillers at the steering platform, he would acquire a day-dreaming expression and then let slip: 'Gee, wouldn't that be something, to find a Sumerian ziggurat in a land facing the Indian Ocean!'

The topic was fully revived at Bahrain, when Geoffrey Bibby took us to the ruins of his Dilmun temple-pyramid. It had, he stated, all the main characteristics of a Mesopotamian ziggurat. It was compact, stepped and sun-oriented, with lateral stairways and a temple on top. Nothing else like it existed outside Mesopotamia. That is, apart from ancient Mexico and Peru. Bibby, in fact, referred to his temple as a 'mini-ziggurat'. He had even found Mesopotamian artifacts at this Bahrain temple. Bahrain was almost halfway to Oman. I ventured a bold question to Bibby: had he heard rumours of a Sumerian ziggurat found recently in Oman?

Never. He had heard nothing of the sort.

If Bibby had not heard of it, with his key position in gulf archaeology, then the whole story had to be invented. We tried to forget it once more.

But Oman maintained its magic grip on both Norman and me. Outside the gulf, when he wanted to coast southwards and over-haul our rigging in Oman, I suspected he had not entirely forgotten the rumoured ziggurat. It was admittedly a major reason why I discarded the unique chance of sailing on a visit to the Indus Valley, a decision I took with much doubt and a heavy heart.

The idea that we could do with a general overhaul before we took off for distant lands was not entirely unfounded. Our superstructure had suffered more from two days of rolling in the wakes of supertankers than from the stormy waters of the unsheltered gulf. The mountains of Oman had given us shelter from the storm raging in the gulf from the very moment we had rounded the cape outside the Hormuz Strait. The sea should therefore have been just as calm as the air but for the frantic traffic. We woke up in the very midst of the shipping lane, with ships on all sides, most of them tankers. We were so relieved at having sailed clear of reefs and cliffs, and at suddenly being in shelter, that we did not fully perceive that this was the most crazy place in which to rock about in a reed-ship with a foul wind that barely permitted us to move.

In this exposed position we celebrated New Year's Eve. The next night our heavy sail hung down completely slack. Most of our progress came from a strong current that pulled us parallel to the coast of Oman in the direction of Muscat. Our kerosene lamps and the faint blips at our masthead seemed like glow-worms compared to the bright electric lights on all the big ships that passed us. For this reason one of the helmsmen had to run up in front of the sail every three or four minutes during the night, to watch for arriving lights and to play with our strongest flashlight on the sail to disclose our own whereabouts to such ships as still had human beings on lookout.

On my watch a huge and brilliantly lit oil platform was slowly towed past us on its way into the gulf. I counted lights from up to twelve ships at a time; they came and went and rushed so fast that we swung in their wakes as in a hammock, and with a fast and crazy rhythm that we had never experienced in any ocean. Ocean swells give a sturdy reed-ship a slow and pleasant motion, soothing to the most restless mind; even the short and choppy seas of the stormy gulf dug up troughs wide enough for an almost decent rolling as compared with the hasty, nerve-racking rocking in the man-made wakes of the superships. We felt maltreated, as if put into a cocktail-shaker, or as if galloping without reins on an unsaddled bull. Everything inside us seemed shaken to bits; we were vexed and harassed by tumbling about on deck or in bed, unable to set proper sea-legs or lie down without rolling over like barrels. The short intervals of calm sea between sudden wakes, cross-wakes, backwash from other wakes, and the next set of wakes, made the fitful rocking the more disturbing.

Bridge-poles and mast-legs started a regular stilt-dancing in the sockets that held them to the reeds, and without wind-lift the heavy sail rolled with the masts and helped strain and stretch all stays and guys and other cordage securing rigging, cabins and bridge to the reeds. Every time a huge tanker chased by at twenty knots the sudden jerks, throwing us from side to side, were so rough that we feared something would break loose from the deck bundles. The two heavy rudder-oars again hammered left and right inside the steering forks, until either of these thick blocks split into two and had to be adjusted with new rope and wooden wedges. The brief encounter with the police boat hardly improved this slightly jerky condition of our woodwork, but in calm sea the straddle-mast and bridge-poles stood as steady as any man setting sea-legs.

The day after the collision we lost sight of the wild mountains of northern Oman, but in the afternoon we came close enough inshore to see low land with a few scattered tall trees. The landscape was so flat that the invisible waterfront almost certainly had to be a long beach, with offshore water shallow enough for us to anchor. But before we made up our minds to steer towards land, Norman shouted from his corner in the main cabin that he had succeeded in getting two-way radio contact with a coastal station. We were told by the authorities in Muscat to keep away from the coast of Oman until we received permission to land. Another message was relayed from the BBC on behalf of the consortium: Norman was told not to use his own amateur radio for any kind of message. He was not to give our position to the radio hams. If he failed, as hitherto, to contact land on the special wave-length of the consortium's transceiver, he should give no position and tell no news to the radio amateurs except 'all well'.

This prohibition made Norman furious. 'What if we get wrecked?' he asked and tore off his earphones.

It was still broad daylight when the dhow stopped alongside the floats of a large fishnet which this time we carefully avoided. The men in the dhow began pulling in the net, and to our surprise they started picking out splashing fish that were caught in the meshes. This done, they tied a plastic bag to one of the floats and threw the net back into the sea. They had placed three *dinars* inside the bag as payment for the fresh seafood and beckoned to Rashad to come over in the dinghy for our share.

Shortly after, steaming fresh fish was scooped out of Carlo's big pot, while Toru prepared some raw cubes soaked in soya sauce, Japanese style. It was a most pleasant surprise to those on board who had never tasted raw fish. Toru himself looked as if he too had come out of a frying pan, with rashes and blisters all over. As if to console his own discomforts, he had been sitting hectically stroking the feathers of a big, bushy-headed kingfisher that had landed on board almost exhausted. Toru had been the first to dive overboard that day, into a sea that proved to be packed full of a species of small jellyfish. These tiny, glassy coelenterates were suddenly around us in endless quantities, waving coquettishly with their violet skirts and trailing long stinging filaments behind while shamelessly mating everywhere on the surface as if intent on filling the world's oceans with their own kind.

Man, too, had left evidence of his own effort to dominate sea and land. The surface of the open water outside the entrance to the gulf was covered by a thin rainbow-coloured oil slick. We had expected more tar balls, but solid oil clots were small and far between. There were more to be seen a decade earlier, when our unexpected observations from *Ra I* and *II* urged us to send a message to the

United Nations that the Atlantic Ocean was becoming polluted. Since then deliberate discharge of waste oil from tankers has clearly diminished. But we did see some giant oil tankers pump out their sludge with the ballast water in unbroken cascades just before they turned to enter the Hormuz Strait. This was probably a convenient area, since the now existing regulations would otherwise have had to be broken at greater risk inside the gulf.

A school of at least a hundred porpoises was seen tumbling and leaping high around us in the thin oil slick. And there was no lack of plankton, though invisible as the stellar heaven until the sun went down. But in the starlit night the plankton sparkled like fire-crackers around the rudder-oars as they cut the water. At intervals we saw large flashes deep down in the sea, as if somebody signalled with a small lamp down there or struck a match that burnt for half a second. The night was now so mild that we steered in shirt-sleeves. The new moon once more rode like a Sumerian ship above the calm waters; a month had passed since the last time. The squeaking and creaking of *Tigris* had completely died down, with the unfortunate result that it was possible for HP to point out that Thor, Gherman and Yuri were snoring.

We saw few of the poisonous sea snakes, although this was supposed to be one of their main breeding grounds. But in the mornings we began to find the first dead flying-fish on deck.

Although the patrolling police boat never came back, it was clear that the Oman authorities had been notified of our whereabouts. And Captain Said was clearly worried, as if he had been appointed our guardian by the visiting police. He was now just as panicky about our coming too close to the shores of his own country as he had been about his own trespassing into Iranian territorial waters when his dhow had been in serious need of repair. Obviously, we had come too close to shore for his comfort, and he insisted on holding us in tow, almost as prisoners, when we approached the off-shore islets of Suwadi, where we anchored side by side right up against the cliffs.

There was a great temptation to go ashore, but we had received new radio warnings from Bahrain not to attempt any landing. All the afternoon we had seen a beautiful white beach, fringed by scattered palms and other trees, with calm water in front and the shadows of blue mountains far beyond. The distant range of mountains must have been the wild, steep ridges we had followed in *Tigris* in the opposite direction on the other side of this great peninsula. All the mountains in this part of Oman seemed to have been swept towards the side facing the gulf and the setting sun, leaving wide plains on our side, which faced the sunrise and the open road into the ocean. Here, indeed, on this friendly open beach, an ancient explorer would have steered his raft-ship in to go ashore. We were aching to do the same, but the last radio contact had given us strict instructions first to clear our passports and obtain landing permits in Muscat, way down the coast.

With our binoculars we could see a large number of small boats pulled up all along the white beach behind the islands. They were too far away for us to distinguish any details, and I had yet to learn what I would have missed if the Sultan of Oman were to refuse us access to his carefully guarded Sultanate. The few vessels that came close enough to be seen clearly showed no details of outstanding interest.

As the evening drew near, the long beach as far as we could see became alive, and numerous small fishing-boats, all with motors, left the beach and swarmed like ants into the sea. A few canoe-like rowing-boats came around our island to tend to fishing-nets. In one shaky craft were two old men brought by a young boy who wielded a natural branch with a make-shift paddle-blade lashed to the end. A charming old rascal with eagle nose and long white beard offered us live fish at a modest price. Lacking local cash we showed him a far too large note in Bahrain dinars and a still larger one in Qatar currency. He grabbed both and told Rashad they were going across to the dhow to learn the values before they made their choice and paid us back the change. Once they were around the dhow they suddenly put three pairs of oars into the water and rowed away behind the island so fast that they would have won any boat race.

When we were left alone again that evening there were fish and sea-birds everywhere around us. A couple of turtles raised their heads like periscopes above the calm water and looked at us. And more than once a huge creature we never saw, probably a whale, barely touched the surface and loudly took in a deep breath before submerging, leaving only ripples to mark the spot.

What a marvellous place. The islands formed a cluster with friendly slopes and bright beaches along the sheltered channels between them, but with cliffs almost 300 feet high on the side facing the open sea. We all itched to swim ashore. But Captain Said begged us for his sake not to touch the shore with our feet; he now admitted he felt responsible, as the police had seen him in our company in Oman waters.

Norman suddenly made excellent contact with a shore station

again. Once more it was Bahrain Radio. We sent an official request to Gulf Agency in Muscat to help us obtain permission to land. We were told that the harbour authorities would give us their answer next day, but that in no circumstances would we be allowed to come ashore anywhere except in the port of Oman's capital, Muscat.

There was no oil slick off the Suwadi islands; only some tiny tar balls and bits of plastic drifted by. But when Toru again insisted on diving to the bottom, this time to film us weighing anchor next morning, he came up and said he could see neither rope nor anchor at seven metres depth, for the sea was full of small white particles. We all put goggles on and had a look. It was like watching a calm snow-drift through a winter window. The whole mass of sea water was on a slow move past our anchor rope in the same direction we had sailed. The current was made visible by billions of tiny white shreds and morsels too minute for us to identify, but looking very like dissolved breadcrumbs or pulverised papier-mâché. Where it originated was anybody's guess, but the current came from the Hormuz Strait, as we had done.

At 8 a.m. we hoisted sail and left the islands. Said did not object to our sailing on our own, but he never went out of sight. The wind was feeble, from south-west, and gave us some leeway towards the open sea, but the favourable local current and a new topsail Norman had devised and hoisted on a bamboo boom helped us to keep a steady course parallel to land. Only some big rollers came in from the sea at irregular intervals and in groups of two or three at a time, clearly distant salutes from the big tankers now well outside the eastern horizon.

We followed a long low coast with the blue mountains barely visible in the background and were passing the town of Barka when Norman again made contact with Bahrain Radio and received a new message from Muscat: we were not permitted to land, but the matter was now being discussed at 'high levels'.

At 3.15 p.m. a large patrol-vessel bearing the word 'Police' and the name *Haras II* in European letters caught up with us from astern. A friendly officer waved and shouted: 'Are you all right?'

'Yes, thank you,' I shouted, and waved back from the bridge. But my waving turned to frantic gesticulations when I saw the heavy vessel turn to come straight for our side like a charging rhinoceros. For a moment I thought it was a joke, perhaps a humorous reference to the behaviour of the other police vessel further up the coast. But I was soon to learn that this was no joke. There was apparently something magnetic to police inspectors about our widest point, where the door-opening of the main cabin gaped towards visitors. Or perhaps our entire vessel looked like an indestructable fender of bundles and ropes, and that a head-on arrival was the usual way of boarding barges and floating platforms. Certain is it that a bow appeared in our cabin door for the second time. While we on *Tigris* ran about on deck and cabin roofs yelling and waving our arms in despair, *Haras II* seemingly took a good aim and rammed us with great force precisely where we had been hit the last time. HP rolled over on his back as the bow of the police boat hit him in the stomach. He and Yuri were sitting side by side in the low door opening and both were pushed inside, where Norman sat alone in his corner with ear-phones on his head. Norman did not believe his own eyes when he saw a large bow closing the doorway. Fortunately the gunwales of the police vessel were higher than the side bundles of *Tigris*, so the ship ran up into the six strong backstays to the mast and was stopped by a network of rope and bamboo devised by Carlo to reduce the danger of tumbling overboard in heavy seas. Bamboo, canes and rigging squeaked and creaked while masts and cabin wavered under the impact.

When we had recovered from the shock and had ascertained that our ship was still in one piece, we began to look for the police-boat. Our wild behaviour must have seemed utterly uncivilised to the truly courteous visitors, who perhaps had only come to welcome us but now escaped at full speed in the direction of Muscat. After all, their confidence about our solidity had proved justified by a test, so why had we chased them away with such an impolite war dance?

Norman received a new message from shore: we were told to wait in international waters outside Muscat until next day; it was not yet decided whether we would be permitted to come ashore.

'Tell them we are on a reed-raft,' I said to Norman. 'We'll drift away. It is too deep for us to anchor in the open sea. Ask why we cannot come into port.'

Norman conveyed the message, and added that we sailed under United Nations flag. A moment after he took off his ear-phones: 'Believe it or not,' he said, 'they say it is because we have a Russian on board.'

The dhow accompanied us unexpectedly close to shore, as if Said was convinced that we would no longer escape. There was scarcely a habitation to be seen, but at one place we saw what most resembled a walled medieval burgh with defence towers and parapets partly hiding big, Arabian-style buildings. One of the Sultan's incredible seaside castles. His main palace was in Muscat. Soon afterwards the lowlands ended and bizarre mountain formations again reached the water's edge. By now we were close enough to Muscat to see ships at anchor and others heading for port. We were glad to take a tow from the dhow for fear of more collisions. Shortly, countless city lights were lit on the coastal plateaux and in the valleys. Way ahead the night twinkled with navigation lights and ships in increasing numbers, so many that we began to feel uncomfortable. Muscat was clearly a busy modern port.

Again we were reminded that we were living in a changing world. The Sultanate of Oman formed a large portion of the Arabian peninsula, some twenty times larger than Kuwait, yet it had until quite recently been one of the least known territories in the world. The country had been closed to all foreign visitors until seven years ago, when the present totalitarian ruler, Sultan Qaboos, put his own father in prison and began to modernise the country. He started by building roads and allowing the first automobiles to be imported. We were yet to learn, however, that in spite of every evidence of economic boom and building activity in the immediate vicinity of Muscat, tourists were still not admitted to any part of the country. The only foreigners of any category allowed ashore were those approved by the Sultan personally.

As night fell on us the contours of a large rock island with a lighthouse rose ahead. In front of it was the cargo ship anchorage with several vessels awaiting their turn to enter Muscat harbour, or perhaps some were too big to get in. In a renewed radio contact we made it clear that unless we were at least permitted to anchor among the cargo ships, the current would carry us far beyond Muscat with no possibility of returning next morning. We were finally permitted to anchor with the other vessels. But neither Said nor we could reach bottom in deep water outside, so when the dhow headed for the harbour entrance we quietly followed and ended up by ourselves inside, at anchor in the midst of a most picturesque group of dhows. Dhows from all surrounding countries, although in the starlight they looked like the silhouettes of a fleet of Viking ships.

It is hard to say who were most surprised at break of day when we and everybody else in the harbour woke up to discover what was next to us. Black, brown and pale-faced dhow-sailors from Africa, from the Arabian peninsula, from Pakistan and India gazed at us, and we at them, as we tumbled out of our respective sleeping quarters to the tiny balconies with a round hole in the floor that hung outboard beside the stern of all our small vessels and permitted us to squat unseen by others, but for our heads, which were offered a perfect view in all directions. Theirs were like wooden barrels, painted in vivid colours like the rest of their wooden hulls right up to the ornamental scrolls of the lofty bows. Ours were only round screens of golden reed matting, to match the rest of the ship. But we had two. One on either side of the stern, to avoid queueing in emergency. It was a splendid retreat with maximum privacy at sea, but in harbour it took some time to get accustomed to spectators in surrounding barrels, gazing at our faces and shamelessly studying our droppings while imperturbably letting their own splash into the calm water from a considerable height.

Some dhows had entered after us and we were completely shut in; but Said's was far away. Still, the fascination of the surroundings was not primarily the exotic Afro-Asiatic watercraft at our sides, but the majestic scenery enclosing all and forming the harbour with its amazing blend of old and new. Big ships were seen behind the dhows in this large new harbour, named Port Qaboos after the Sultan himself. Few ports can be more picturesque. From the modern breakwaters, docks and warehouses dark lava rocks run up to the walls of a medieval Portuguese fort perched on naked cliffs and dominating the whole harbour. Above the multitude of masts, the ragged ridges soared naked and black against an unpolluted blue sky, their shapes eroded into draperies and spires. A sweet fragrance as from mixed incense and tropical spices wafted into the harbour from a row of tall, white-painted Arab buildings at the foot of the cliffs.

The dark rock above us had barely begun to turn mustard-coloured in the first rays of the sun when motor-boats came racing between the dhows to lie to at our side. First came a helpful Swede, representing the Gulf Agency, who told us that all concerned had now approved our landing; it was now up to the Sultan in person to sign. Then another extremely polite police officer arrived in a launch which we carefully guided to our bundles with half a dozen of our thickest bamboo poles. He left after some sort of brief courtesy visit. Some hours later we were boarded by a most cordial Scottish customs officer, who sat down at our table for a long good-humoured chat and then left with our 'crew-list' to make photostat copies for the immigration authorities who were yet to come.

Carlo was preparing lunch when a large police boat entered the harbour and the uniformed men on board spotted us between the dhows. We got our bamboo poles ready, but three police officers transferred themselves to a tiny motor-boat before they came up to us with extreme caution and climbed on board. Two of them were Indians in large turbans, and the police captain, from Delhi, was as pleasant and congenial as his Scottish colleague from the customs. These visitors, too, sat down for a long and interesting chat and then left, confirming that our landing permit would be signed the moment the Sultan entered his office; it was waiting on his desk.

In the afternoon a small boat from the West German consulate came near and the German consul told Detlef that he would contact the UN representative in Oman and also the American Embassy. Then another police officer boarded *Tigris* and wanted our crew list. I told him the list had gone ashore with a colleague of his, but this was unknown to our new visitor, who suggested that the previous one was probably from the 'detectives' and not from his police. I typed up another crew list in five copies, and to my surprise this pleased the police officer so much that he agreed to tow us in alongside the mole, where we could use the toilet facilities of a small green metal shed of unmistakably long-time service.

Our hearts were in our throats and our bamboo poles were ready for push-off when the police-boat towed us zig-zagging between the dhows and delivered us to a concrete dock. We already had our mooring lines ashore when our agent turned up and shouted that this was the wrong place. The right one was a mole set at right-angles to the first. The police-boat made a full turn and we all rushed to the other side of *Tigris* to use our poles for a smooth docking there; we had to protect the cross-beams on deck fore and aft, which jutted out beyond the berdi and could be broken. It was a crazy manoeuvre, but Rashad was already ashore on this second pier with one mooring line when the English harbour-master came and now ordered us to a third place, just behind a three-masted school-ship on the first pier. This was too much for the interplay between *Tigris* and the police-boat, the space being restricted to the angle between the two moles, and we ran with our stern full force against the concrete wall while I was alone in the rear of the vessel with my bamboo pole. I pushed off with all my strength until I realised that the other end of my pole was going to be pressed back through the rear cabin wall. Turning aside, I lost the battle and the thick steering-log astern, the one projecting to either side to support the rudder-oars, hit the concrete wall with a nasty crash, and all the woodwork astern was forced out of position towards the starboard side.

Ignorant of our disaster, some civilian immigration authorities now came on a visit and asked in a most gentle manner if they could 'borrow' our eleven passports just for a few minutes. A Norwegian freighter, the *Tyr,* was unloading on the other side of the same pier, and nobody objected to our accepting an invitation from the captain to a fabulous cold buffet dinner.

The next day was Friday; there was no life in the docks, and we dragged the rudder-oars up on the pier and began adjusting the displaced woodwork astern. As on the previous day a helicopter was circling low over us, time and again. The agent could report that the Sultan had yet to sign our landing permit. It would only take a second when His Majesty came to his desk. In the meantime we had to stay on board.

The next day, four men came on board and politely asked for permission to take a look around. They apparently enjoyed the sightseeing and were all smiles when they came to me and asked if they could also visit the hull. With an equally polite smile I tried to explain that there was no hull, nothing under deck. Nothing but what they had already seen. At first they became serious and afterwards hostile when I began to describe how the *Tigris* was constructed. A bundle-boat. A Sumerian ma-gur. Two of them now openly began to sniff around, peeping under mattresses and deck planks. The other two interrogated me like a suspect. No answers seemed to satisfy them. As it became apparent that they were secret police or detectives sent for a final check-up before we could enter the Sultanate, I fetched a reed-boat model built by our Titicaca Indians, newspaper clippings from other Arab lands, and a letter from the Norwegian Foreign Office. The four men still seemed unconvinced about our lack of a hull as they walked ashore. Nevertheless, an English girl was now sent into the guarded docks to make an interview for the *Oman Times,* and soon afterwards the agent came

47–50. The ruins of Mohenjo-Daro, major city of the long-lost Indus Valley civilisation which suddenly emerged about 2500 BC and as mysteriously disappeared about a millennium later. It had two-storied brick houses, streets with covered sewers, and a perfect swimming pool, waterproofed with asphalt between the bricks. *(opposite & p. 226)*
51. An Indus Valley reed-ship incised on a Mohenjo-Daro seal. *(p. 226)*
52. Carts from the present day Indus Valley are identical with four-thousand-years-old ceramic models from Mohenjo-Daro. *(p. 226)*
53–54. Departure from Pakistan and the continent of Asia. Norman and Yuri hoist the sail; we study our maps at the dining table as we are now able to navigate and pick our own course. *(p. 227)*

and reported that Sultan Qaboos had signed; we were 'under observation', but could visit Muscat and film and photograph anything except the Sultan's new palace.

The Sultan had built his modern Port Qaboos in the bay of the old town Al Matrah, severed from his own main residence and the capital, Muscat, only by a loop the road made around some narrow mountain draperies. The fish-market on the beach and the fruit-market among the old buildings of Al Matrah were neither more nor less colourful than other Arab souks in towns not yet opened to tourism, but the human types were outstanding. If Hollywood had assembled men with such long beards and such remarkable profiles for a Biblical film, I would have criticised the producer for exaggeration. But the Sultan had certainly not staged these characters for our benefit. The crowd seemed dominated by hook-nosed old men in turbans and long gowns, proudly displaying curved silver daggers in their belts, each a master-piece of craft-work. In their foot-length raiment, with turban and full beard, they probably seemed older than they were, though the beards were more often white than black, sometimes split, and often reaching the waist in competition with Santa Claus and Methuselah. Carlo was desperately operating his cameras, for every single face seemed worthy of a picture.

The crowd in the souk, in the narrow streets, and everywhere in this land, reflected that Oman was an old melting-pot of Semitic, Persian, Pakistani, Indian and African types. To me the faces testified to what we knew of Oman history. Oman had been a maritime centre and port of call for sailing vessels from both Asia and Africa since long before the days of the Prophet. The steady monsoon would bring African traders northwards to Oman, and even to Pakistan and India, in the summer; and back to Africa, in company with merchant sailors from Arabia and mainland Asia, in the winter months. The strategic importance of Oman, with its central position in the Near East sailing routes, was quickly exploited by the conquering Portuguese as soon as Vasco da Gama learnt from the Arabs their age-old knowledge of the monsoons.

55–56. Into the Indian Ocean. Once far from land *Tigris* always found space between the ocean swells. *(pp. 228–30)*
57–58. After a storm we had problems in replacing a broken top-mast. *(p. 231)*
59–60. Ocean pollution; a red belt with no visible beginning or end ran from horizon to horizon in mid-ocean. *(opposite)*

Today Muscat is the capital of Oman mainly because of its modern harbour facilities, but the former capital is Sohar, on the open shore further up the coast, which we had passed. At Sohar early sailing ships could anchor in shallow water off the shore, and from there it was easy sailing to the entrance of the gulf, with Bahrain, Persia and Mesopotamia beyond. Exposed on the open beach without sheltered harbour, Sohar was lost to the world when Sultan Qaboos started modernising Oman, beginning with the new capital. He built himself a spectacular dream castle, dominating the former harbour of Muscat, and moved all traffic to the adjacent Port Qaboos. His other initial efforts include so far a major international airport, roads, public buildings and residential quarters above all in and near the flourishing capital.

It was therefore the more interesting to me when the name Sohar was brought up during our first evening ashore, at a dinner party given us by the English-born General Manager of Port Qaboos, Barry Metcalfe, and his wife Kate. Near Sohar, somebody said, there were supposed to be small boats similar to our own. We were just back hot and exhausted after our first day in the souk, and could think of nothing else when the Metcalfes and two of their neighbours made us dance with joy and gratitude under clean fresh-water showers before we sank down, all refreshed, in deep armchairs, balancing plates loaded with chicken curry and with huge mugs of foaming cold beer.

It was all so good that it took me a while to realise that the gentleman who had mentioned the bundle-boats up near Sohar was the noted Italian archaeologist Paolo Costa, Inspector-General of the Directorate of Antiquity in the Sultanate of Oman. In spite of his pompous title, Costa was a most jovial and earthbound person, and we were soon on first name terms and way back in the distant millennia before Moslem beliefs and architecture had reached Oman. Up north there were copper-mines dating from prehistoric periods, said Paolo. There were also underground aqueducts, and the hills of Oman were full of stone towers. They were burial mounds of the same general period as those we had seen by the thousand in Bahrain.

In southern Oman some remarkable prehistoric ruins had now been found on the coast near the borders of South Yemen. But we could not go there, for the Peoples Republic of South Yemen was communistic and the Sultan's worst enemy, so there was an almost constant state of war in the border area.

I could no longer suppress my curiosity about the rumoured

ziggurat, and Norman moved his chair closer as I asked Paolo Costa if it were true that a Sumerian temple-pyramid had been found in Oman.

'Something has been found,' he replied to my surprise. 'Whether it is Sumerian I cannot say, but it shares all its characteristics with a Sumerian ziggurat!'

We wasted no time. By sunrise we were already outside the police gates of the port, and with Paolo Costa as guide and the harbour authority bus for transport we began our view of a part of the world where tourists are not admitted. From the super-modern settlements in the outskirts of Muscat every fifty miles seemed to take us an extra millennium back through the ages until we passed the ancient town of Nizwa and reached the ageless mountain town of Al Hamra, about 150 miles from Muscat. The Sultan's new road had just barely reached there, but electricity and waterpipes not yet. Nor was there any sign yet of transistor radios, pop-drinks or plastic bags. And apart from the fact that all men under forty-five had used the new road to move to Muscat for paid work, life in this attractive town was hardly changed since the days of the ancient Middle East civilisations. The setting was again as out of the Bible, or rather, as out of the Koran.

From the mustard-coloured clusters of tall Arab mud-brick houses perched on naked rock, barefooted women in colourful costumes walked like queens, with pottery-vessels crowning their heads, along the foot-paths to the old aqueducts in the shade of the date-palm plantations. Their tinkling jewellery, like the silver daggers in the belts of the men, looked like collectors' items. Back from the moist soil reserved for growing crops they re-entered the narrow rock streets. Steep, and polished to a shine by hoofs and feet, these much trodden passages climb from the evergreen palm gardens between the sand-coloured houses seemingly into the blue sky. The dark and cool arcades were crowded with robed men sitting or standing, thinking or talking, as unconcerned about the clock as were the calmly ruminating desert goats and little pack-donkeys sharing the shade with them. Never have I seen so many men resembling the popular image of 'Uncle Sam', with mighty beard and prominent nose. Never, except on the pre-Columbian stone reliefs of the unidentified Olmecs who brought civilisation to the aborigines in the Gulf of Mexico long before the arrival of the Europeans. Something in the bearing and the tolerant expression of some of these relaxed men made them look like old and learned sages. Their indulgent comportment as we trampled about in their arcades made us feel like school-boys playing in front of professors assembled to ponder upon the mysteries of life. These men could probably not write. But men just like them had first invented script. Their ancestors had started the busy clock of civilisation ticking and had passed it on to us, who have made it tick a thousand-fold faster and barely have begun to see our own fallacies which we now send out to all customers at the ends of our roads. Al Hamra had been built on rock and the impressive buildings will survive to be seen by generations of tourists who may come to future Oman. But the houses will appear like empty snail shells on the beach, for not even an omnipotent Sultan can drive young men backwards along the road built to the streets of Muscat.

The living past we experienced in some of the still inhabited towns and villages of Oman's mountain valleys gave warmth and meaning to the empty prehistoric ruins we were shown in their immediate vicinity. Most significant to us from *Tigris* were the remarkable vestiges of former activity at Tawi Arja, in the dry river plains of Wadi al Jitti in northern Oman. To get there Paolo Costa drove us in his Land Cruiser northwards along the Batinah plains, the coastal lowlands we had seen from the sea. We recognised our anchorage off Suwadi islands, and continued northwards on a good road until we reached the modest outskirts of the one-time capital, Sohar, about three hours fast driving from Muscat. Here another new road left the coastal plain and led straight inland towards the wild mountains we had first seen from the opposite side, while sailing to escape from the gulf. Rolling hills began to rise around us and we saw to our excitement that all were capped with rows of old stone towers strikingly like those of Bahrain. Costa confirmed that many of these round towers had now been opened; they were burial mounds, and dated from the third millennium BC, Sumerian time, as on Bahrain. Not only were they contemporary with those of Bahrain, they even had the same cross-shaped inner stone chambers. These prehistoric towers almost seemed like cairns set on either side of the wadi to mark the road inland towards the temple-structure to which we were heading.

Before long Costa drove sharply off the road and we bumped along the bottom of a canyon with nothing but camel tracks that wound ahead into a hidden world of dried-up river beds and small alluvial flats where nothing grew except some sparse and twisted thorn-trees with inedible berries. In the whole area we saw not a single house, but in a few places far apart we passed semi-nomad families living in a sort of symbiosis with a desert tree. Their primitive dwellings consisted of simple platforms suspended among the crooked branches of big thorn-trees and walled in like a nest by smaller branches. All the sparse clothing and utensils of the residents hung about on the branches safely above the reach of the goats, and the tree seemed to thrive from the rent paid by the tenants. We were told that even when these roaming tree-dwellers moved into a real house they would leave the floor empty and hang all their possessions from the ceiling and on the walls.

In this barren landscape we reached the biggest of the dry river beds, known as Wadi al Jitti. Like a broad motorway paved with smooth pebbles, it wound through a sandy plain flanked by hills and peaks where worked chips of jasper and a prehistoric stone circle were the only trace of man to be seen between the sparse thorn-trees. Along the inland horizon, still far away, were the tall jagged crests of the west coast mountain chains, rising one behind the other in ever higher rows, as if set to block the passage to both Saudi Arabia and the gulf. In contrast, the wide wadi ran like a flat gravel-road in the opposite direction, straight to Sohar and the open seaside beach from where we had now come. Without a road we drove on until we rounded a black, conical peak that served Costa as a landmark, and on the open plain before us lay what we had come for.

We were still in the Land Cruiser when some large chocolate-coloured boulders, superimposed to form the terraced wall of a partly buried building, struck my eyes before Costa even had time to point in that direction. It was difficult to remain calmly seated until we came to a stop at close range. This was what I had hoped for but had not dared to believe. There was no longer any doubt.

As Norman and I walked up to the structure with Costa, Norris rushed into position with his sound camera to record the first arrival in untold centuries of reed-boat voyagers from Mesopotamia at what may once have been a Sumerian sanctuary. We stood at the foot of a partly ruined, man-made mound, still well enough pre-served to show its main form. While we were gazing at the big, brown boulders in the walls, Costa opened with a solemn speech: 'Unless we can make a thorough excavation of this site,' he said, 'we cannot say whether it is possible to date it to the third millennium BC, but we can say that this huge structure is a unique feature, square, stepped and made of random stones with perfect masonry work; it is in the middle of a plain surrounded by hills and in this position is obviously not a fortification, so we can say that it is certainly a temple.'

Norman and I listened and swallowed the big mound with our eyes as Costa took us to the side where a long, narrow ramp led from the ground up to the top terrace. Gherman was almost beside himself with excitement; this was a stepped pyramid of the type we had so often seen together among pre-Columbian ruins in Mexico. At the same time, the whole concept was that of a Mesopotamian ziggurat. As Costa emphasised, nothing like it was known in any other part of Oman or the entire Arabian peninsula. Huge natural boulders had been used to wall in a rectangular structure that rose above the plain in compact, superimposed terraces, four of which were seen above ground. The four corners pointed in cardinal directions, and the well-preserved, stone-lined ramp led centrally up one side in the fashion characteristic of the temple-pyramids of the sun-worshippers of Mesopotamia and pre-Columbian America. Whoever might have built it did not follow Moslem norms, but it struck me that the concept was also the same as that of the Dilmun temple Geoffrey Bibby had excavated on Bahrain and described as a mini-ziggurat.

It was impossible to say how much of the structure, if any, was lost in the ground, but, barely emerging from the hard-packed terrain, the top of a boulder-wall was visible, enclosing a rectangular temple court extending from one side of the terraced mound. Nobody had yet attempted excavation. There were vestiges of other walls and an elaborate system of aqueducts all around. Even traces of stone masonry on a nearby hill, directly overlooking the temple-pyramid. I was amazed that nobody had started digging.

Costa shrugged his shoulders. There had been no time. With his assistants, two of whom came with us and saw the temple-mound for the first time, Costa was busy surveying archaeological sites all over Oman. This discovery was new; the site was stumbled upon when the Sultan recently permitted a general archaeological and geological survey of Oman. Four years ago the wadi had been visited by a group of mining geologists representing Prospection Oman Limited. They operated with Land Rovers as part of a mineral exploration programme started by Dr C. C. Huston in collaboration with His Majesty Sultan Qaboos. The sites discovered had been reported by Prospection Limited to the Oman Government, but as the report was mainly concerned with mining possibilities, the news of the temple mound had barely begun to leak out. A Harvard expedition, led by J. Humphries, had visited the temple site and briefly reported the existence in the mining area of 'a ziggurat of Mesopotamian type'.[1] The survey was mainly of interest for economic mineralisation. The early Arabs had clearly tried to benefit from the presence of some of these prehistoric mines, but there was still much to be extracted if the long forgotten mines were reopened.

Mines. This rang a bell. A stone's throw from the temple-mound the sun shone on some unusual piles of stone chips or cinders of pretty colours; red, brown, violet, yellow and green. This was slag.

'Remember I told you about the prehistoric copper mines,' said Costa. 'They are here!'

The parts of a long unsolved puzzle seemed to fit logically together, and it made me hold my breath a moment as Costa pointed to a nearby mountain, behind the hill overlooking the temple-mound. It looked like a huge, rotten, rusty-red tooth, almost worn through at the middle. This was no eroded crater. Not the work of nature. What he pointed to was clearly the work of man.

'One of the prehistoric quarries,' said Costa. 'Everywhere in this vicinity you will see evidence of early copper mining activity.'

I began to see a meaning in the strange location of a seemingly Sumerian ziggurat. Fresh in my mind since my researches in Iraq and our days with Geoffrey Bibby in Bahrain were the texts of some of the old Sumerian clay tablets. I began to add two and two together and could hardly keep my suspicions to myself.

Costa told us that an estimated 40,000 tons of slag were scattered about the foot of that one mountain. Prospection Limited crews had discovered a total of some forty-six such ancient mine sites in northern Oman. Costa was itching to take us further inland to another site where an estimated 100,000 tons of slag were heaped in piles. A whole mountain had been removed by prehistoric miners who had turned it into a valley covered by multi-coloured slag to look like a giant painter's palette. Here prehistoric copper miners had been crushing the slag from their many small furnaces.

Bumping through trackless canyons and dried-up river beds we reached what Costa considered the most impressive site in Oman, and we were astounded by the immensity of the prehistoric activities which had transformed the whole landscape into a sort of giant arena or painted outdoor theatre. Of the former mountain nothing was left by the copper miners but a monumental metallic outcrop, formed like a triumphal arch on the low hill at the side of a battlefield filled with multi-coloured debris. Perhaps this majestic gateway was left there on purpose to commemorate the original opening through which the miners had first worked their way into the copper mountain that they had gradually caused to disappear.

I could no longer keep my deductions to myself. 'Makan,' I exclaimed as Costa led us into the great gateway and turned to show us the spectacular sight. 'Yes,' he confirmed, 'this may well be Makan, the legendary Copper Mountain of the old Sumerians. There is no copper nearer to Mesopotamia.'

I recalled how Geoffrey Bibby had taken us down between the walls of the prehistoric city port he had excavated on the island of Bahrain, to show us where he had found all the scraps of unworked copper in the open square just behind the gate in the sea wall. He had emphasised that this was evidence to prove that local merchants had sailed to ports outside the gulf in Sumerian times to fetch this important metal, so dearly required by all Bronze Age cultures. The import of copper had been of paramount importance to the founders of Mesopotamian civilisations, as copper was not locally available in the twin-river country or elsewhere in the gulf.

Perhaps more than anyone else, Bibby had speculated on the origins and transport routes of the Mesopotamian copper trade, and showed how texts on the ancient tablets recorded that the imports had come by sea from a land known to the ancient Sumerian scribes as Makan or Magan. Two tablets slightly over four thousand years old, found at Ur, represent receipts left by a Sumerian merchant for goods received by him from the main temple. One listed sixty talents of wool, seventy garments, one hundred and eighty skins, and six *kur* (nearly two thousand litres) of good sesame oil, as 'merchandise for buying copper'. The second tablet is more specific, garments and wool were received as 'merchandise for buying copper from Makan'.

Bibby had found references to Makan in Mesopotamian inscriptions dating from the days of Sargon of Akkad, about 2300 BC, when he boasted of ships from Makan tying up alongside his quay together with ships from Dilmun and Meluhha. King Sargon's grandson claimed that he 'marched against the country of Makan and personally took captive Mannu-dannu, King of Makan'. And Gudea, a governor of Lagash around 2130 BC, imported diorite from the mountains of Makan to fashion numerous stone statues, and some of these still exist with incised inscriptions recording the fact. But references to 'copper from Makan' or to merchandise 'for the purchase of copper, loaded on a ship for Makan' petered out about 1800 BC, according to Bibby. From then on, he found, there seemed to be no more direct sailings to Makan; now all the copper trade went through the markets of Dilmun. But Makan was still known as the primary producer. There were still listed references to 'diorite: produce of Makan', and 'copper: produce of Makan', as distinct from 'palm-trees: produce of Dilmun, produce of Makan, produce of Meluhha'.[2] Whereas the Mesopotamian supreme gods, and Ziusudra, the survivor from the Flood, were intimately linked with the trading post of Dilmun, there are no mythological texts referring to Makan. The gods of the Sumerians never went there. All references to Makan were commercial and matter-of-fact.

In the days when Bibby's excavations on Bahrain began to give strong support to the theory that this island was Dilmun, many scholars had already started to postulate theories as to the whereabouts of Makan too. One school of thought would place Makan in Africa; thus the noted authority Kramer believed that Makan was probably Egypt, and others had suggested Sudan or perhaps Ethiopia. The reason for these assumptions was a hint in late Assyrian texts. When Assyrian kings around the years 700–650 BC waged wars against Egypt, they left inscriptions placing both Makan and Meluhha somewhere to the south of that country, after having reached Lower Egypt overland from the Mediterranean side. But the Assyrians were presumably ignorant of how far to the south of Egypt these two legendary countries lay, for direct trade between Mesopotamia and Makan had ceased over a thousand years earlier.

The second school of thought placed Makan closer to the Sumerian ports. Woolley, for instance, said that 'diorite was brought by sea from Magan, some point on the Persian Gulf'. As to copper he was more specific: 'copper came from Oman, as is shown by analysis of the ores, . . .'[3] Bibby,[4] too, supported this latter view, partly because he felt that Makan had to be within fairly easy sailing distance from Dilmun. But also because a large number of copper objects from Mesopotamia of the period 3000–2000 BC had been analysed and found to contain a slight trace of nickel. Now nickel is fairly rare as an impurity in copper, but a similar slight intermixture had been ascertained in a single specimen of copper ore coming from the territory of the Sultan of Muscat and Oman, which was still closed in the rigid days of Qaboos' father. The sample was reported to be from 'ancient workings' and found in the valley running inland from the port of Sohar.

I was later to learn, through a letter from Mr G. J. Jeffs of Prospection Limited in Canada, that it had been the brief reference to this one copper sample from Oman in Bibby's book *Looking for Dilmun* that had prompted their company to survey the forgotten mines of north Oman with the consent of Sultan Qaboos, a recent survey that led to Mr Jeffs' discovery of the sites Paolo Costa was now able to show us.

Not until after our return from the *Tigris* expedition did I get further news about the temple mound, when I had a surprise visit from Paolo Costa and his wife Germana to my home in Italy. They had then started to excavate the big mound at Tawi Arja. The centuries of exposure that had eroded the structure itself had not managed to accumulate humus strata on the hard-packed surface of the wind-swept and occasionally flood-washed plains on which it stood, and neither potsherds nor other datable remains older than the Moslem period had been found. But under the dirt on top of the pyramid Costa had found the badly eroded remains of a foundation wall of adobe: a small edifice or erection of some sort had stood on the summit platform, built from big square, sun-dried bricks of the type used in ancient Mesopotamia.

As a cautious scholar he refrained from any hasty deductions for lack of conclusive evidence. There was neither carbon nor written tablets to help date the strange structure. It remained unique in the Arabian peninsula, alone as a huge non-Moslem edifice in an area with forty-six old, abandoned copper mines, the only copper within reasonable reach of merchant mariners from Ur in Mesopotamia and their trading partners on the gulf island of Bahrain.

But even if no Sumerian vase is ever found in the barren landscape of the wadi and the surrounding mines, geography and geology combine to argue with rather conclusive strength that northern Oman was the copper country of Makan to the old Sumerians. There is no competitor for that honour in the gulf area. It is anyone's right to speculate as to the identity of the seemingly misplaced mini-ziggurat at Tawi Arja. An enormous amount of labour had been put into its construction. It was not a fort and not an Arab mosque, but it had all the aspects of a ceremonial structure and one known in the Old World only in Mesopotamia, with a single exception in the recently excavated Dilmun mini-ziggurat on Bahrain. We ourselves had sailed a Sumerian ship from Bahrain to the ocean coast of Oman. In Sumerian terms, we had sailed a ma-gur from Dilmun to Makan. It would seem hard to find a theory more plausible than to suspect that the unidentified structure discovered by the mining prospectors at Tawi Arja had been built by, or for the service of, the sun-worshipping merchant mariners from the great civilisations in the twin-river country, who had come here in large numbers because it was the nearest site for mining and smelting copper.

None of the vestiges of prehistoric ingenuity we saw in Oman impressed us more than the subterranean *falaj*. In an area where a plain of sand and gravel stretched as far as the eye could see we were taken to the edge of an open hole in the ground, which made us recoil because it seemed quite bottomless. A stone-lined shaft descended for at least thirty vertical feet until lost in pitch darkness. The ground around the opening was slightly raised like a small crater with dirt and gravel brought up from below. When we looked over the desert we could see similar craters at intervals in a straight line towards both horizons. We learnt that deep down in the ground these shafts were interconnected by a subterranean aqueduct dug for miles upon miles with such precision that the water flowed in an even descent irrespective of hills or other irregularities on the surface terrain.

Near the source a falaj had to run at surface level, or it might even be elevated, to gain the required declination, and we were speechless when we saw an open aqueduct coming down the hillside above a river, then passing under the river and up on the other side! At the crossing point the water tumbled into the top of a chimney-like stone tower, then passed beneath the river bed to come up again through the top of a slightly lower tower on the other side. From then on it flowed elevated again along another hillside in the direction of the sun-scorched plains, where it was to begin its long and cool journey deep underground. We were told that some of the falaj ran for many miles at an incredible depth below deserts and canyons. Some were maintained by the present Arabs and a few were perhaps even built by them, but the origins of this incredible display of engineering skill and mass labour was lost in antiquity. Whoever had first built them, the falaj of prehistoric Oman gave to me a logical explanation to a puzzle connected with the prehistoric water conductors discovered by Bibby and his collaborators on Bahrain. These Dilmun aqueducts, too, were found deep below the sand, with the same peculiar stone-lined 'chimneys' rising at intervals to the surface. In all likelihood, like those of Makan, those of Dilmun had been built intentionally underground and had not been covered afterwards by wind-blown sand drifts.

But an even more apparent and mobile link between Dilmun and Makan became clear as we drove in Costa's Land Cruiser down the broad and flat wadi that led from the temple site and the surrounding mines to the open beach at Sohar. This old town and former capital lay where a broad river must once have had its outlet in the sea, in the remote millennia before the copper miners had put an end to the inland forests. The enormous smelting activities testified by the remains of prehistoric slag shows that here, as in so many other parts of the Middle East and elsewhere, man has misused his environment by turning woodlands into deserts and rivers into wadis. The smelting had required enormous quantities of wood that once had been available locally. The bottom of the wadi was smooth, water-worn river pebbles. There was no other reason for placing the one-time capital of Oman here but the outlet of a broad water-course from a once most important mining area.

The wide sandy beach stretched for over a hundred miles along the local coast, open towards the ocean. We reached the white sand in front of some modest huts of mud and reed-mats at the outskirts of Sohar. Friendly Arab fishermen were sitting in the sun mending nets. Old women and young girls in colourful gowns, but with black masks on their faces, stood calmly in front of their mat walls instead of running into hiding as in the inland villages. Out at sea was a man in a small vessel struggling with long oars to come back to shore. Soon he entered the moderate surf and came riding straight up on the sand. He pulled his boat ashore, full of glittering fish. His boat was a reed-boat, or more correctly, a boat of slender palm-stems precisely of the type we had seen on Bahrain. There were three more of the same kind pulled ashore where the fisherman came in. The name for this kind of boat was *shasha*. They were now gradually disappearing. We were told that they were used for landing cargo from the dhows which had to anchor outside.

I inspected the shasha with keen attention. They were built just like the farteh of Bahrain, so similar in all details that we found no difference at all but the name. Two Arab nations on opposite sides of the Hormuz Strait had inherited the same type of watercraft, but in times so ancient that they survived with different names.

If modern regulations had not forced us all the way down to Muscat, we would have anchored at a convenient depth off this beach, just as the present dhows and the former ma-gurs from Ur and Dilmun, all of which were too big to come right in to the sand. We could just visualise *Tigris* at anchor in the clean blue water outside this beach, with the local shasha coming out to give us shore-to-ship service in the same way they had served merchant mariners who came to load tons of copper in Sumerian times. *Tigris* had been idle in the polluted water of Port Qaboos for a full week now. The reed-ship, which according to current assumptions would waterlog too fast to leave a river, already had seaweed growing on its bottom as long as Neptune's beard, and crabs and sea-hares were breeding among the bundles. We were in no hurry. We were all fascinated by the rare opportunity of visiting Makan.

It was not until my return to Europe that I reread the ancient *Naturalis Historia*[5] written exactly nineteen centuries earlier by Pliny the Elder, to see if the Romans in their days had any information about Oman or its people. I was not a little surprised when I found Pliny referring to an Arabian people called the Macas (*gentem Arabiae Macas*), and that their territory was found by crossing a strait merely five miles wide separating them from the eastern boundaries of the land ruled by the Persian kings, an area where copper was mined. The narrow strait was quite clearly the one known to us as the Hormuz Strait, and the Macas were thus the people of northern Oman. That the Macas lived in a land known to themselves and to the Sumerians of Antiquity as Makan would make perfect sense indeed.

CHAPTER 8

Tigris *and the Superships: the Voyage to Pakistan*

'PORT SIDE row, starboard rest! . . . Both sides row! . . . Ready . . . rooow! . . . Ready . . . rooow! . . .'

I felt like a galley slave-driver as I stood high and dry in the stern, steering while my friends toiled rhythmically, their sweat pouring and their long oars sweeping in unison like the legs of a centipede.

It was an exciting test, and one I had been looking forward to but never yet attempted. We wanted to row our reed-ship out of port, and, with Toru and Carlo in our rubber dinghy to film and photograph, only eight men were left to man the oars: four on either side and myself at a rudder-oar. To see the men I had to balance high on the bridge railing with a grip on the upper of the two tillers we had fastened to either of the long steering shafts.

'Ready . . . rooow! . . . Ready . . . rooow! . . .' The men put the last ounce of their strength into propelling the heavy ship through the water. The wet berdi, with superstructures and full load, must now have weighed close on fifty tons, so each man had to shift the weight of a floating elephant. If we were to lose control, the slightest gust of wind or tidal current in the harbour would force us into involuntary grips with hulls, mooring lines and anchor chains of the ships large and small docked or anchored everywhere around us. The eyes of spectators watching us from ships and shore probably gave the straining men an extra urge not to give in.

We had been up before sunrise to try to carry out this manoeuvre and fight our way out of the big harbour before the wind rose and other people got out of bed. But somehow the news of our departure had spread faster than we were able to take farewell of all our local friends, and people who could not get in through the police gates had lined up on the main city road along the bay. The crews of all the ships in the harbour were also standing as silent spectators when at last we began to move, pulling our bow away from the pier with the anchor rope and pushing off.

At first nothing seemed to happen. Then slowly we began to move. Very slowly. Scarcely one knot, but enough to give me steerage. The *Tigris*, sail down, crawled out from our dock and wormed its way out through the labyrinth of concrete piers and steel hulls. Filthy flotsam began to circle and drift behind us in the harbour. Three blasts of the siren resounded from each ship in turn as we reached and passed it – salutes from modern ocean craft to a replica of their earliest ancestor as it struggled to reach the sea. Two helicopters circled overhead as if to check that we really left. His Majesty the Sultan in his escorted car chased along the city road to Port Qaboos, where he turned around and dashed back to his Muscat castle.

All this was a stimulus to the toiling oarsmen, and the sudden silence was almost oppressive as we left the last ship behind and the long way lay open to the outer breakwater. It was eight in the morning when we were ready to cast loose, and by nine we began to realise the dimensions of this huge modern harbour, for we still had to clear the outer breakwater with a faint sea-breeze now impeding our advance.

By this time the men, untrained in rowing, began to feel really exhausted. *Tigris* was indeed grossly undermanned. The ancient rock-carvings rarely showed less than twenty oars in the water on a reed-ship of fair size, sometimes twice that many, and we had only eight. With twenty we would have flown through the harbour with three times the speed and little effort.

Before we reached the gap through the final breakwater the breeze had become a problem. We hardly seemed to move at all and the wall ahead seemed to remain at the same distance. Not only did the feeble wind reduce the speed, but as it struck us slightly to one side of the high bow it forced us sideways to port, making it difficult eventually to clear the jetty. The rudders and two boards we lowered fore and aft were not enough to counteract this leeway, and I had to give the four starboard men a rest while the others rowed doubly hard, then we let the rowers change sides.

The eight men were exhausting their reserves and all I could do from the tiller was to keep the rudder turned and yell even louder, like a real slave-driver, but shouting cheers of optimism too. The jetty came nearer. Another hundred yards and we would be level with the long wall and clear its tip, where the sea broke and we would be free, free to hoist our square sail. The minutes were endless. A couple of the men at times splashed their oars like drunkards, ending up with backward strokes.

After an hour and a half we reached the gap where the ocean swells rolled against the star-shaped concrete blocks of the jetty that stretched towards us from the port side. The moment we came abreast of the breakwater the wind became too strong for our effort. We barely managed to get outside, for the four oarsmen on the port side hardly had space for full strokes; and once outside we lost the battle with wind and waves and were slowly forced backwards, stern and rudder-blades first, towards the dentated polygonal blocks. We were seconds from disaster.

'Hoist the sail!' I cried. It was too early, but we had no choice, for the men could do no more. It was a desperate situation, with people suddenly screaming with fear on our behalf from the top of the jetty and from half a dozen small boats that had come to see our departure. We were right up against the fingershaped projections of the breakwater blocks. The stern was so close that from the railing of the steering bridge I could look down at little crabs scrambling away in all directions as the rudder-blades swept past with only inches to spare. An ugly performance. A close shave. A motorboat threw us a rope as the exhausted men led by Norman rushed to hoist the sail. It was up and trimmed at record speed. It filled. In a few moments the concrete claws of the breakwater blocks slipped behind us. We were out in the open sea.

We could only envy the early Sumerian seamen. How easy it would have been if we, as they had done, could have gone in and anchored in front of the Sohar beach. With a few oars we could have turned *Tigris* stern to wind at the anchorage and sailed straight off from the sandy shore.

The faint wind blew from SSE, and there was no sign of the strong NE winter monsoon we could have expected in the middle of January. Our sailing speed was so slow that we found ourselves unable to tack with our small sail, and we had to take the feeble wind in from starboard and sail north again in a direction that barely cleared a lofty black cape jutting into the sea north of Port Qaboos. Jokingly I suggested to the crew that we had been to Makan, and that we ought now to sail back north again to visit the Indus Valley we had missed. In reality we had planned to sail southwards to Africa, hoping to get the winter monsoon at our backs.

We had the tall mountains and rock islands of Oman within sight until dusk. Then we could see only a few bright lights which soon sank into the sea, and as night fell only the faint glow in the sky told us the direction of Muscat and Matrah. But long before dark we began to see a familiar sight: the masts and bridge-houses of distant ships forming an uninterrupted line along the eastern horizon. What had looked like white houses soon rose from the horizon and became parts of large ships, one behind the other. We were once more in the very midst of the shipping lane, trying to cross it. The horrible, brutal and unrhythmic side-rolling in the deep, shortly spaced wakes of the superships began again, and so did the ferocious hammering and unmelodic cat-orchestra special to an irritated reed-ship. On my midnight watch with Detlef we barely escaped collision as a small cargo ship came straight for us and we were both able to turn aside only at the very last moment.

The next day, before sunrise, the wind changed from SSE to NNW and we turned to steer 130° with sail filled and at good speed. In the afternoon we adjusted our course to steer just clear of Ras al Hadd, the easternmost cape of Oman, where the Arabian peninsula forms a right-angle corner and falls off in the direction of the Gulf of Aden.

Only now did we begin to realise that a serious problem was brewing on board. Norris, usually happy and cheerful, had suddenly become sullen and quite beside himself. We all knew why. On our last day ashore, travelling in Paolo Costa's Land Cruiser through the prehistoric coppermine area, we had lost our way in the network of wadis and canyons, and the bumping over rocks and gravel banks had been so rough that something in Norris's specially constructed sound camera was shaken to bits. He had been fiddling with all his spare parts almost without food and sleep until it became clear that nothing could be done except in a laboratory. Before we sailed from Oman he sent an emergency cable to the consortium. We comforted him with the prospect of borrowing a film camera from Toru or Gherman, who each had one, although theirs could admittedly not record sound. But this underestimate of Norris's professional needs made him even more depressed. The whole reason for his presence on board was that the National Geographical Society and WQED had sent him to shoot an expedition documentary with synchronised speech and sound, and a common 16mm camera could not fulfil this requirement.

With his camera unusable, Norris had struggled as one of the eight rowers when we left Muscat, but despite this struggle, when we were safe out at sea and he still had been unable to repair his camera, he wanted to get back to shore and airfreight his equipment for repair. For a moment I was almost glad there was no sound recorder working as most of us had begun to raise our voices. Norris insisted on being put ashore. The rest of us could not care a damn about the film at that moment; with the wind so good all we wanted was to get into the open sea and away from cliffs and moles. So we sailed with latent problems stowed away on board. Now Norman was angry as Norris gave him endless emergency messages to transmit via Bahrain Radio, urging spare parts or new equipment to be shipped by air from England or the USA to Muscat and then by surface vessel to some rendezvous in the Gulf of Oman. I agreed with Norman; it would take hours to get the messages through and we had to get away from the Arabian peninsula. It would be madness to jeopardise the whole expedition for a camera.

The result was that the tall, good-humoured Norris seemed to shrink into himself; he gazed in the direction where land had been as if planning to swim ashore, now that his special services with us had come to an abrupt end. It gradually dawned upon me that this was serious: Norris was growing more frustrated with every passing moment. Norman managed to transmit his desperate messages via Bahrain. The Bahrain operator, Frank de Souza, seemed by now to be part of our group; he was the only person in the outside world who managed to hear our confounded consortium transmitter and come back with a reply. I began to hope that something would happen to help Norris while we were still within reach of Oman, before we sailed past the final cape.

We had successfully wriggled our way through the shipping lane and were beyond it, but we entered it again as we steered closer to land in the direction of Ras al Hadd. By night we felt as if we were travelling across prairies with light from scattered homes here and there in the darkness. But the illusion lasted only so long as the lights were at a distance. While Yuri and Gherman shared the night watch a tanker passed so fast and so near that Yuri jumped from the toilet seat while Gherman waved frantically with the stern lamp.

On the third day after leaving Muscat, the good northerly wind died and came back as feeble gusts from E and SE, slowing down our progress towards the cape and giving Norris a faint new hope. The sea had been moderately polluted since we left port, but now we sailed into a serious oil slick with scattered lumps. We tried to escape the traffic by getting on the shoreward side of the shipping lane, but did not know that Bahrain Radio had warned all ships in the area of our presence in the Gulf of Oman. Before we could avoid it, a large luxury liner seemed as if about to run us down when it unexpectedly changed course as if to give the passengers a closer look at a rare form of watercraft. Next a small freighter suddenly stopped its engine and lay drifting, as if trying to trap us, crosswise to our course, then started again as if in despair when it became apparent that we could not stop like them. Shortly afterwards, yet another ship turned off its course and came chasing straight for us as if intending to ram us, but it turned out to be a small Norwegian freighter, *Brunette*, which circled us twice, then resumed its course and disappeared with three blasts of its siren. A big Russian ship, *Akademik Stechkin*, came for us next and repeated the same manoeuvres, as if dancing around a Christmas tree, while a loud-speaker shouted 'Yuri Alexandrovitch' and asked if we needed anything. We did not, and we sailed closer inshore as fast as possible, to avoid the beaten path.

Fishing with rod and spoon, HP caught a big fish, he said, so big that it broke his line and disappeared with his favourite spoon before any of us could testify to the one metre length indicated by the fisherman. A few hours later Asbjörn dived overboard on a line and came up with a really large fish on his hand harpoon. It was a big dolphin, also known as dorado or gold-mackerel. Moreover, it had HP's precious spoon in its mouth, with a piece of broken line trailing behind. HP had not exaggerated. Soon afterwards, a six-foot shark came circling and headed for the big red buoy we always towed behind for security. The buoy was bumped left and right for a moment, then the shark took off.

On the fourth day we sighted wild mountains above a misty coastline on the inside, while all the ships were now on the outside. For a moment we had escaped the traffic lane. But the wind was turning increasingly east and to Norris' delight forced us ever closer to the mountain coast. That evening the unbelievable news came from Frank in Bahrain that a new camera had been prepared for Norris and would be flown to Muscat, if we could wait. What else could we do when we saw the joy on Norris's unshaven face – and the wind all on his side. We turned *Tigris* about and steered back north, parallel to the coast. The uninhabitable rock walls which emerged from the mist seemed a more compact threat than the widely-spaced steel hulls of the shipping lane, and when the mountain coast grew higher, wilder and nearer as night approached, we forced the reed bow as far to starboard as we could take the east wind on our return voyage up the Oman coast.

The nights and days that followed were among those which none of us on *Tigris* are ever likely to forget. We tried a rendezvous with a fishing-vessel chartered by our agent to meet us twelve miles from shore, outside the isolated fishing town of Sur, hidden between the coastal mountains west of Ras al Hadd. But the fishing smack never found us, and at this position, under the lofty rocks and mountain ranges, Norman's radio failed to get contact with either ships or shore. Not even Frank at Bahrain, on the other side of all the mountain ranges, could hear us now. We decided to beat northwards to Muscat, where modern vessels would have less difficulty in finding us. But we had no means to communicate this decision to anybody, leaving the fishing vessel to search for us in vain.

A name can do injustice to a place, and such was the case with Ras al Daud, which in good old Norse meant 'Landslide of Death' to me, but according to Rashad meant 'Cape David' in Arabic. We sighted this desolate cape at night, vaguely lit by a stellar heaven and draped in a ghostly mist from the sea, and as we sailed close I began to feel that nothing I had seen did more deserve the name of Death than this lifeless, sinister cape, standing in the midnight haze like a giant gravestone, with a deep and empty valley, gaping like a tomb, at its left and vanishing into complete darkness under a veil of pale clouds. No life. Nothing moved but mist. There was probably a small invisible beach at the outlet from this narrow valley, and I had been tempted to try a landing in the starlight between the steep rocks, but at close range the place looked so frightening that the navigators agreed with me that we had better stand off from the walls before it was too late.

The cape was already unpleasantly near and we tried to throw out the canvas sea-anchor to reduce our speed. But our wind-drift was already too slow for the sea-anchor to take a hold, hanging straight down like an empty sack. Norman had by now devised a topsail to be hoisted on a bamboo boom, and it was a great success. It gave us better steerage. We could resist the east wind that threatened to send us against the rocks. Much against what would have been our own wishes, if we had had a better choice, we began to steer out towards the busy ocean highway we had tried to avoid. As we began to see the ships around us again and tried to cross the traffic sector, the wind died down and returned only as feeble gusts from varying directions, preventing us from conducting any sensible navigation. We lay in the midst of dense traffic, fighting with sail and steering-oars to get out of the way of fast ocean liners and tankers as they thundered by and left us rocking so violently that the steering platform under our feet seemed ready to tear loose from the reeds and capsize with cabins and masts into the sea. Carlo and Yuri were everywhere trying to secure ropes while the rest of us struggled to get *Tigris* out of this crazy marine highway where an old-fashioned reed-ship did not belong.

By night we felt worse. We could detect at once a speck of light wherever a ship appeared, but not seeing the vessel itself we could never tell if it came on a collision course or would pass at a safe distance. Lights appeared on the horizon everywhere and grew and grew until either a little green or a little red lamp could be distinguished in the general glare of illumination on board. If both the green (starboard) and the red (port) lamps were visible at the same time, then the ship was coming straight for us. We had red and green lights on *Tigris* too, but in the wind the glass sooted up terribly and the flames often blew out. Anyway, big ships rarely had lookouts to see such modest kerosene lamps, and the biggest tankers could neither stop nor turn by the time something was within vision of their bridge. It was up to us to do the sighting and patiently watch every speck of light around us until big enough and near enough for us to distinguish the various lamps. We could then tell, by the alignment of the two coloured lamps, combined with one tall white lamp aft and a low one in front, whether we lay right in the course of the vessel or whether it would pass to our left or right. But by then we could often see the outlines of the black hull against the night sky, with little time left to scramble out of the way if the wind failed. We therefore organised a double night watch to keep a sharp lookout for lights from the moment they appeared, to know whether we had to escape to port or to starboard. It became a routine, but the night was long and we awaited sunrise impatiently, tired of staring into the dark, tired of the incessant battle with rope and canvas, and arms stiff from gripping the oar shafts and legs weary from tumbling about as *Tigris* jumped the angry wakes like a porpoise.

When morning came the lights disappeared but the ships were still there; now we saw them from the moment they rose humming like mosquitos on the horizon, buzzing like wasps as they came nearer, until they passed us like drumming elephants.

Towards sunset on the fifth day the weather changed. Heavy clouds rolled up over the horizon and we heard distant thunder. To the west we could still see the outlines of coastal mountains briefly each time they stood out against the lightning. The night had barely started when a strange combination of lamps that made us think of a Christmas tree floated very slowly past in front of us, crosswise to our course. We marvelled and discussed what they might be, when we noted some other unusual lights from a low and equally slow craft following at a distance behind the first, also intending to cross our course. Both vessels moved so very slowly that there would be time to pass between them. And that is what Gherman and I tried to do when Detlef came up on the bridge and I showed him the Christmas tree. The next second Detlef realised what it was. We threw both tillers hard over to starboard and barely managed to skim around the low stern of the second vessel, which proved to be a heavily loaded, unmanned barge. As a captain of merchant ships Detlef had read from the many superimposed mast lights on the first boat that here was a tug towing a barge on a two hundred yard cable.

I crawled into the main cabin to get some sleep before my own night watch. I was already accustomed to lying calmly and relaxed with eyes closed, even if a natural reaction would have been to sit up and look out each time a ship came near. My ears registered the rhythmic drumming of distant engines that slowly grew in strength and then as slowly died away as ships came and passed us. Then a terrified yell of warning came from the steering platform over and behind my head, followed by the intense droning of a big ship's pistons as it approached us dangerously fast, suddenly so near that the noise chased all of us out on deck like scared mice. I barely had my head out of the port-side door opening when I saw our ochre-coloured sails brightly lit by something big and noisy just on the opposite side of the cabin. As I rose naked to my feet with a grip on a mast stay, everything seemed in chaotic movement and I felt as if I were travelling at night through a city street, almost blinded by the many lights from a tall black wall. A fully lit cargo ship rose out of the dark so close that for a few seconds all seemed part of one unit: the two bamboo cabins and the black wall. Only for a moment; then it was torn away again before we had a clear idea of what we had seen, just time enough to hang on as our bundle-boat was tossed aside by foaming cascades.

Those of us who had managed to crawl out on deck fast enough had barely had time to exchange exclamations of horror and relief, when a new yell came from Yuri on the bridge aft:

'Crazy! Another one!' Yuri's voice was drowned by the thundering pistons of a second steel wall following the first and racing into vision right alongside *Tigris*, sending us sidelong again with its wake, as if ploughing us casually out of its way.

'That was a container ship,' commented Detlef dryly. 'They often make thirty knots.'

We all remained on deck for a while, commenting on the poor visibility without moon or stars. The weather did not look promising. We decided to take down the new topsail. It began to drizzle. It was soon time for Toru and me to take over the steering. Hesitantly the others crawled back into the two airy bamboo huts and rolled down the canvas sheltering the cane walls. It began to rain heavily. Soon the cold shower became a cloudburst. We could see nothing, not even our own rig. Toru and I had donned waterproof suits, good for rain and spray, but not for skin-diving. The water fell in torrents and the noise smothered our voices. We were drenched from neck to toe in the cold rain that percolated our waterproof gear, and were standing in shoes filled like flowerpots. It was almost comic – but not quite. I could barely distinguish Toru's Japanese features if I aimed the beam of my flashlight at him, although he was standing almost within reach on the other side of the narrow steering-bridge.

Now I could hear something. I shone the light into Toru's face to see his expression. He nodded back as he began to hear it too. Pistons. In the rumbling cloud-burst we could scarcely hear our own voices, but the deep rhythmic thump from some approaching supership could not be mistaken, and it came fast. Carlo, supposed to be asleep inside the main cabin, could already hear what we had heard through the thin wall of cane and canvas, and he shouted out to the two of us on the bridge. We were not less uncomfortable than he, but we alone could take action. What should we do? We were indeed masters of the rudder-oars and could turn left or right. But which side would take us out of trouble? Whichever way we turned might be the one where the ship was coming and bring us under its bow.

Stiff with cold and apprehension, we could only hang on to the oars with the wind straight at our backs. The thumping grew unbearable; it came straight down upon us with the wind and woke everyone in the two cabins. Our world was compact darkness, we could not see the sail, but knew it was filled with wind and rain as we took the weather from the stern. I could not even turn my head to look behind us, for the rain was slashed full force into eyes and face by the following wind.

'Hear the engine!' Carlo shouted in despair through the cane wall. Then we heard no more from our companions as the droning rose to a crescendo and the ship caught up with us. A second later I was overcome with joy. 'Yes,' I shouted back through the cane wall. 'I hear it, and I smell it too!' Then I added: 'I can even feel its heat!'

The giant passed by with thundering pistons, so close that we could really smell the oily air from the engine room, and for a moment I even felt a warm radiation from the colossal steel wall that slid by my side, almost invisible in the rain, although right alongside our starboard bundles. In the dark, behind a compact curtain of falling water, I was left with the unclear impression of lights passing high above my head at mast-top level. Up there someone with radar and automatic steering had seen even less of us than we of them. Detlef could tell us from experience that in heavy rain even a radar is affected and shows nothing at all in distances less than a mile. Even our strips of tinfoil were of no avail in this weather.

The screen of rain thinned abruptly, so that we could see our bow in the night mist; at the same time we also caught sight of two clusters of lights close ahead. Two more ships. They looked like two galaxies of bright stars moving one after the other from right to left across our course; we were just about to pass between them at a distance we could not quite grasp, due to confused visibility. But in the next second we realised that these ships were shockingly near, both moving incredibly fast, with identical speed and constant separation. No tug towing a barge could move that fast. For an instant Toru and I were bewildered at their speed, then simultaneously we flung ourselves against the tillers to force *Tigris* to starboard as far as we could go. An unlit ship's side had come into view straight ahead; it joined the two speeding galaxies in their common rush. All three clung together; then it dawned on us that they were in fact one ship. A huge supertanker, heavily loaded, with lights fore and aft only, slid past our bow, its incredibly long and low midsection, with at least a hundred thousand tons of oil, barely visible above the waves. With a strong following wind *Tigris* responded quickly and turned port side to in time for the black tanker to pass alongside so close that from the steering platform we gazed up through windows and portholes and saw details of walls inside the cabins. We could have recognised a face in the bright electric light inside, but nobody seemed awake.

We had had enough. The giant tankers were left racing in their own track off the coast of Oman as we made use of a fresh south-east wind to sail further out and away towards the north-east. Out in the unfrequented part of the sea we felt as human beings again, in an environment shared with fish and birds; we were no longer Lilliputians in a world of robots without flesh and blood, where men on a clumsy ma-gur had no security nor any right to be.

We heard thunder until morning. Then the wind changed to NE and forced us to trim the sail and alter course. All in vain, however, as violent gusts of wind seemed to wait only until we were correctly trimmed before they entirely changed and struck us from the opposite side. The rain returned in heavy squalls and the rudder-oars jammed as the buffalo hides and hemp-rope lashings that held them in their forks at deck and bridge level swelled and tightened. At 10 a.m. we had a glimpse of a hazy sun between thick clouds, and Detlef estimated our approximate position. We had struggled our way back up north and were 226° off Muscat at fifteen miles distance. To our despair the consortium transmitter did not contact anybody now. Neither Muscat, Bahrain nor any ship. At great risk we had managed to come back outside the harbour of Muscat, but nobody was aware of it. The agent with Norris's replacement was probably still looking for us in the sea outside Sur.

Now we all wanted that confounded camera, and for three hours Norman was climbing up and down in the waving mast to hang up various lengths and forms of antennae. Then suddenly Frank came on the air from Bahrain and got our estimated position. We learnt that someone from the Gulf Agency was looking for us in a motor launch off Ras al Daud, where we actually had been. Frank would now try to get Norris's shipment brought back by Land Rover from Sur to Muscat, if we could try to remain where we were. By noon we turned the bow all around from south to north as the current down the coast was strong. We were seriously afraid of getting closer to Oman, for this would mean re-entering the shipping lane.

At noon we suddenly made radio contact with a ship in Muscat harbour that telephoned the navy coastal station that had a twenty-four-hour watch. Our agent was alerted; he had already given up his search for us off Sur and now had the camera in Muscat. This was almost too exciting. Particularly for Norris, who for nearly a week had felt like a lumberjack without an axe. The manager of the agency, Leif Thoernwall, promised to bring us the camera himself if we came closer to Muscat. In renewed squalls of rain, but with a main wind direction from SSE, we kept moving ever closer to the dreaded shipping lane, but as another night fell we had not seen a sign of a rendezvous ship. Nor could we any longer make radio contact with Muscat or Bahrain, although Norman suddenly heard a voice giving landing instructions to an aeroplane in Hawaii.

He tuned up his receiver, and we all heard an American voice say: '. . . maintain 200 knots and keep in contact!' We could only join Norman in a roar of laughter at the tragic-comic message.

As I went to bed the lights of Muscat could be clearly seen reflected in the night sky. The wind had calmed for a moment. We came close enough to have ships unpleasantly near, then turned around and sailed away from land. The previous night's experiences were still too fresh in our minds. We spent this night in drizzle, safely outside the traffic, seeing a few ship's lights far away and lightning flashes followed by thunder over Iran. As we crawled into sleeping-bags or under blankets we all commented on how good it was to be away from the brutal wakes of the superships. Detlef confessed that a few times during the last couple of days, he had lain down feeling a bit seasick, and I, who never feel the sea, had experienced pain as if my stomach was being tossed about inside me.

Next day, at 10 a.m., Norman managed to make contact with Bahrain, and we learnt that a tugboat had gone eighteen miles from Muscat the previous day, looking for us, but the weather had been too rough and they had been forced to return to port. More bad weather was forecast, so no one would come out to look for us that day.

The day was so dark that HP had to light a kerosene lamp to read inside the cabin. We had painted the vaulted roofs with asphalt Sumerian fashion, but rain trickled down the cane walls in this kind of weather. Four sharks followed us all afternoon. The sea might have seemed terrible for those who saw it from port, but to us it was a real blessing: long, regular rollers, the heavy rain smoothing off all the choppy wave crests. *Tigris* moved over the water as quietly as the shadow of a feline; all ropes were swollen and held the wood-work and bamboo tightly in place. It was again so quiet at night that, when I woke, I could help HP to testify that Carlo and Gherman were snoring. We could even hear the friendly gargling and tinkling of water cut into whirls by the rudder-oars. The sky cleared for a moment during my night watch, and I found myself steering towards the Southern Cross. In the late morning it cleared again and Norman was able to get an approximate position, and even managed to transmit it to Bahrain Radio. Frank came back and reported that the harbour tugboat *Muscat* had left Mina Qaboos at sunrise with a speed of ten knots, and would try to contact us by radio when far enough out.

That afternoon, at 3 p.m., *Muscat* met us almost nose to nose with an English harbour-master at the wheel and our Swedish agent Thoernwall waving among the Omanian crew. Two cases of spare parts and a newly mounted sound camera to replace the broken one were all transferred to Norris, who was so afraid that it might drop into the sea that he hardly allowed anyone to give him a hand when the precious cargo was handed from ship to ship.

Our agent was so excited and relieved at having found us at last that he opened a bottle of champagne and drank it all himself, toasting merrily to sea and sky as *Muscat* turned and raced back to reach port before nightfall.

We turned our own bow away from the mainland, hoisted the topsail above the mainsail, and steered deeper into the sea. We needed no champagne to share the joy of good old Norris as he sat on his mattress in the forward cabin, nursing his new baby, which started the familiar hiccoughs that meant it heard every word we said. As the sun set we registered a fabulous speed of 3·5 to 4 knots with straight course for the open ocean. Makan was gone. The destination of *Tigris* was unknown.

Unknown, because I had begun to feel a secret desire to change the plans recently made. Our endeavours to help Norris had brought us further north than I had considered possible in the winter season, when the north-east monsoon should be dominant. Norman's new topsail had given us greater speed and therefore wider steering margins. We appeared to have a choice of any part of the Indian Ocean. There was still a chance to visit Pakistan and see the Indus Valley. The temptation became too big, and for two days I jokingly told the men that our next port of call would be the coast of the great Indus Valley civilisation.

On the third day it was no longer a joke. I told the helmsmen to turn about and steer for Pakistan instead of Africa. I had more reason than ever to visit that part of Asia. I wanted to see Meluhha. So far Meluhha had been for me nothing but an unidentifiable name common on Sumerian tablets. I had consulted Bibby's book again and read: 'Dilmun and Makan were always being named together, often in the same sentence, and often together with a third land, that of Meluhha.'[1]

We had been to both Dilmun and Makan. Where was Meluhha? Though I had never had any idea, the name had always reminded me of names surviving in the Malay area, like the Malacca peninsula and the Moluccas archipelago. But now, convinced that Dilmun and Makan had been properly identified, it seemed to me that

Meluhha, by a very simple system of elimination, had to be the Indus Valley region. There was no other important coastal civilisation nearer to ancient Mesopotamia, and there was a wealth of evidence to prove contact between Mesopotamia and the Indus Valley in Sumerian times, so much contact that there was bound to be some reference to that trading partner in the Sumerian texts. If we set the important Sumerian place-name Meluhha aside as unidentified, what then was left as the Sumerian name for the Indus Valley? There had to be one, since it was the only nearby contemporary civilisation with which we know they had extensive contact.

As with the Sumerians, so also with their contemporaries in the Indus Valley: their culture collapsed, their cities were abandoned, their very existence was forgotten among surviving nations until their ruins were discovered and their arts and crafts were brought back into the daylight by modern archaeologists. The Sumerians, before they vanished, had exchanged their original hieroglyphic signs for a cuneiform spelling which modern scientists have been able to decipher. Thus the word 'Meluhha' has come down to us as the name for one of their foreign trading partners. But the hieroglyphic signs of the Indus Valley civilisation, although well known today and easily recognisable, have so far withstood all the efforts of the cipher experts, and we know none of their geographical names, not even the one they used for themselves and their own land. Like fingerprints, the Indus Valley script cannot be read, but wherever it is found it identifies the hands that made it.

Indus Valley seals inscribed with Indus Valley script and decorated with characteristic Indus Valley motifs have been excavated by archaeologists, not only in the Indus Valley, but also in distant Mesopotamia. A quite considerable number of them have been found during excavations in Iraq, all the way from the former gulf sites of Ur, Uruk, Lagash and Susa up to Kish, Tell Asmar and Brak, the latter site inside the borders of present-day Syria. Contact must have been quite considerable for so many and such far-flung fingerprints to be left even to our day. I have never seen a more beautiful Indus Valley seal than one ornamented with rows of Indian elephants and rhinoceros, found by archaeologists at Tell Asmar in Iraq and on exhibit in Baghdad Museum.

Like Makan, some archaeologists have suspected that Meluhha was in Africa, and for the same reason: that the late Assyrian kings, returning from overland campaigns on Mediterranean shores, left inscriptions that Makan and Meluhha were both located somewhere south of Egypt. Thus the scholar Kramer, who placed Makan in upper Egypt, suspected that Meluhha was Ethiopia.[2] Bibby on the other hand, after his excavations disclosing extensive Sumerian contacts with Bahrain, wrote:

But it is difficult to fit an African location to the text of the Ur tablets or to the facts of archaeology. Distance alone was a factor, and though I had never been conservative in my estimation of the distances which trading vessels could cover I could not ignore the fact that the sailing distance from Bahrain to Africa was twice that from Bahrain to India. Ivory and gold, which were products of Meluhha, could come equally well from Africa and from India, but the carnelian of Meluhha could only come from Rajputana in India.[3]

Indeed, when he showed us his excavations of the prehistoric city port on Bahrain, half the size of mighty Ur, Bibby emphasised that he had found an Indus Valley seal just inside the harbour gate. He had also found an Indus Valley flint weight that made him suspect closer trade relations between Bahrain and the Indus Valley than between Bahrain and Mesopotamia, which was only a third of the distance away. Even before he saw the strength and buoyancy of our reed-ship in the harbour of Bahrain, Bibby had expressed his confidence in the early sailors of the Dilmun runs:

These merchants who were raising capital and assembling cargo for the voyages to Dilmun were not investing in some wild argosy to a mythical land of immortality beyond the horizon of the known world. This was routine business; it was the way they made their living. Nor should we imagine that the trading was carried on exclusively by Mesopotamians. Two of the people recorded as paying tithes to the Ningal temple are specifically listed as natives of Dilmun, so there were probably Dilmunite merchants resident in Mesopotamia and Mesopotamian merchants resident in Dilmun, while ships of both countries would be engaged in the carrying traffic. Ships of other nationalities, too, would, in these two first centuries of the second millennium BC, be beating up the Gulf to Dilmun and be beached upon its shores, under the walls of its cities. The ships from Makan would be heavily loaded with their cargoes of copper, while the ships from the cities of the Indus Valley civilization would have cargoes, then as now, of timber (and perhaps, though there is no evidence, of cotton), in addition to their lighter and more valuable stores of ivory, lapiz lazuli, and carnelian.[4]

Bibby even ventured a wild but interesting theory. He pointed out that the Sanskrit-speaking Aryans who had entered India from the north and probably overthrown the original Indus Valley civilisation used the non-Sanskrit loan-word *mleccha* to denote non-Aryans, people who did not worship Aryan gods. He asks: 'Could it be that *Mleccha* was the Indus people's own name for themselves and their country?'

The Indus Valley, until recently part of India, was now the heart and soul of Pakistan. Nobody on board *Tigris* objected to the idea of turning back north and taking the alternative route we had rejected as we came through the Hormuz Strait.

For ten days we sailed north-east with no other company than patrolling sharks, a faithful escort of dolphins, and brief visits from playful porpoises and a few curious whales of larger species. Colourful tropical birds landed on our deck to rest, and, attracted by the broad shadow beneath our bundles, multi-coloured fish swam with us like domesticated animals in quantities the like of which I had not seen since drifting over the Pacific on *Kon-Tiki*. Every day HP, Asbjörn and Carlo diverted themselves and extended our provisions by fishing and spearing two of the most savoury species: the dolphin, alias gold-mackerel, which shone in colours of green, rust and bright gold when patrolling in all its splendour down in its own kingdom, but was flat and disproportionate and faded to become a silver relief of a fish-tail with a bulldog head when brought on deck. And the streamlined rainbow-runner with its slim torpedo shape which merited its name in or out of the water. Triggerfish, pilot-fish, remoras and any other kind keeping us company we tried to leave alone, and we threw them quickly back if they rushed on the hook before the fisherman could get his flying-fish bait out of the way.

This was when Carlo insisted he had heard a grasshopper during his night watch. We refused to believe him, but then I found one in my own bed, yellow-green and as long as a finger. On my own watch I heard one singing near me in the stern and another answering from the bow. Soon we all heard grasshoppers everywhere. We saw them crawling and jumping about on reeds and bamboo, and even taking to wing on short circles over the sea, but always coming back to our marine haystack. They seemed to be happiest among the bushy ends of the reed bundles at the top of the high bow and stern. Having obviously embarked as silent stowaways in Oman, together with a number of crabs on our bottom, they had never seen such a wealth of vegetation in any part of the barren Arabian land from which they came.

The wind was so strong at times that our mainsail ripped and we had to take it down in the middle of the night at a moment when it bulged like a parachute and pulled us ahead with such speed over the swells that I was worried lest any man in the dark should make a false step. For thirty-two hours we rolled about in a strong current with a north-west wind and the sea-anchor out, but unable to steer. We had passed 60°E, the longitude of Ras al Hadd, and when Toru, HP and Asbjörn reported that they had mended the sail and that it was ready to go up, we set our course as close into the wind as our steering with the added topsail permitted. We saw that the sea around us was once more polluted. Bathing was only permitted on a rope. For the first time we found that our side bundles were covered not only with slippery green sea grass and soft-shelled white goose-barnacles with long pink gills waving, but also with large colonies of small conical barnacles of the common type, hard and sharp as shark's teeth, which scratched us badly if we did not take care. Clearly this was another form of stowaway that had found time to take hold in port while still only larvae, because *Tigris*, unlike *Ra*, had spent days in harbour in Bahrain and Muscat. It was equally refreshing to body and mind to hang alongside our floating home and watch how beautifully the bundles maintained their shape and how high they still floated. The midline of the bundles was scarcely below water level when we swam around *Tigris* for a check-up on a calm day seventy-three days after the launching.

The weather forecast that day for the Gulf of Oman and the Arabian Sea had been NE wind at ten to twenty knots, but it was dead calm. Norman now told Frank on Bahrain Radio that their forecasts had invariably been completely wrong. But strangest of all, the famous monsoon had not blown on our sails for a single whole day. The most usual wind came from Iran, but we also had winds that varied between SE and SW. No matter from which side the wind blew, we struggled to keep a fairly steady course between 60° and 90°, that is to the north of east.

We succeeded. On the day after the calm Norman and Detlef both shot the sun in its noon position and placed us at 23°50′ N and 62°05′ E, about seventy miles off the Asiatic mountain coast known as Makran. The barren and inhospitable coast of Makran begins in

Iran as the northern shore of the Gulf of Oman, and stretches well into Pakistan, where it borders on the Indus Valley. Looking at the map it struck me that Makran both phonetically and geographically was surprisingly close to Makan. A single letter as well as the narrow Hormuz Strait were all that separated the one from the other. Could it be that the legendary land of Makan had embraced both sides of the Gulf of Oman and controlled the important passage through the Hormuz Strait, the copper mines being in its southern part and the northern part bordering directly on Meluhha? This apparent possibility fascinated me as we steered eastwards, well off the desolate Makran coast in the direction of the Indus Valley.

Next day again we were still further east and a mere fifty-five miles from the coast of Pakistan. Only a stiff and steady NE wind could now prevent us from reaching our goal, but that should indeed have been the prevailing wind at this time of the year. It was not. Southerly winds began to dominate, and the problem now was not whether we could reach the Makran coast, but whether we could avoid being pushed ashore before we got far enough east to find a safe landing place. This part of the Makran coast was nothing but cliffs and impenetrable mangrove swamp.

We were twenty-two miles off Pakistan and expecting to see land at any moment. Norman consulted our copy of *West Coast of India Pilot*, 1975, and read on the very first page that the charts for coastal navigation were adequate everywhere 'except for part of the Makran coast'. In these parts, we learnt, the charts were based on surveys made in the last century with lead and line, and 'such charts cover areas subject to sudden shoalings and shifting sandbanks or where, as on Makran coast, they coincide with small scale soundings which, because of later (1945) volcanic disturbance of the sea bed, are now of doubtful value'. Also, 'A small volcanic island was observed in 1945, 3 miles sw of Jazirāt Chahārdam, but in 1947 it had submerged.'

It sounded like an exciting area, with many possible surprises. The waters around us were literally full of fish. And an infinity of tiny white worms. Carlo, our professional alpinist, had turned into a passionate deep-sea fisherman and caught six large dolphins one after the other before I could stop him. Yuri got the idea of hanging some up to dry on bamboo rods, to produce stockfish. During a delicious lunch on dolphin with polenta we sighted a white dhow-sail on the horizon ahead. We made a good three knots in that direction, but when the dhow came near enough to see us well, it made a wide semi-circle past us and disappeared. We had at last seen another vessel under sail. This looked promising.

On 24 January problems began to develop. Heavy clouds formed over the Karachi area in front of us, and a changing wind struck us in violent gusts, quickly whipping up a choppy sea. We tried to get the topsail down before it was too late, but the topping-lifts that held it were stuck and it would not come down. At dusk we saw incessant flashes of lightning one after the other in the black clouds over Karachi. HP tied a knife to a short bamboo rod, and after dinner, when the wind got tempestuous and there was no other choice, Norman and Detlef climbed to the footplate on top of the bipod mast, and one held the other as they cut the topping-lifts and brought the madly flapping topsail down. A full moon peeped at us in fleeting gaps between rapidly accumulating clouds, then it disappeared as the rattling, rumbling blitz came over us and over the unseen coast of Makran. We had sailed into really bad weather.

I fell asleep before my midnight watch; then, as I grabbed my flashlight and lifeline to climb up on the steering-bridge with Toru, we found the masts naked: no sails. As Norris and Asbjörn huddled away and left further problems to us, they reported that the wind had swung about in all directions each time an extra black thunder cloud had passed overhead. It had been impossible to undo and refasten sheets and braces fast enough every time they had to trim the sail, and the rolling had been so bad that they had feared the dancing yard-arm would break and the mainsail be torn to shreds. Norman had been out to help them save the sail, and we were now adrift with only the sea-anchor to help keep our stern to the wind. The coast was near and there was an island somewhere, so we had to be alert. The *West Coast of India Pilot* left no doubt that it was risky to be adrift completely without steering control in such treacherous waters. A faint wind began to blow in our favour from the direction of the Hormuz Strait, and after half an hour I woke up enough men to hoist the mainsail in a heavy rolling sea. Three whales came up around us, breathing deeply in the dark, and by our sparse light we noted two big white birds riding high at either end of our vessel. The sail came up, but no sooner was it filled before the wind changed to NE, the proper direction of the monsoon. If it really was the monsoon that had now come to stay, then we would run the risk of never seeing land in Pakistan. Never had our desire to get on to the Makran coast been stronger.

We studied the chart again. The dead reckoning would place us off the fishing village of Pasni. Land was just below the horizon. This was still Makran and not yet the Indus Valley. But the Indus Valley civilisation had dominated this coast. Archaeologists had found ruins of prehistoric forts built by the founders of the Indus Valley civilisation; they blocked access to the only two valleys leading inland from this coast. One was at Sutkagen Dor, which we already had behind us, very near the borders of Iran. The other was at Sotka-Koh, the passage that opens at Pasni, which we now had right inside us. In a report on recent surveys of these prehistoric coastal forts, the archaeologist G. F. Dales had shown that they contained numerous Indus Valley potsherds, and that these were attributable to the earliest period of Indus Valley civilisation. According to Dales, this special type of ceramic ware had been brought from the lower part of the Indus Valley, and the archaeological evidence left no doubt that the mighty city states of Mohenjo-Daro and Harappa, further up the Indus river, had possessed maritime traditions and built these forts at the only possible landing places on the Makran coast to defend themselves from other sea-borne intruders. The testimony of these Indus Valley settlements along the coast strengthened Dales in his conviction that 'none of the great civilisations of the world originated or thrived in a cultural and economic vacuum'.[5]

Such words were more meaningful to me than ever as I lay rereading old notes between our cane walls within a few miles of the Sotka-Koh fort.

'I have discovered Pakistan!' I was torn back to reality by Norman's voice high above the ceiling. He was yodelling with joy from the top of the mast. I stuck my books back in the case below my mattress before I crawled out and climbed up the mast ladder. This was the eleventh day after leaving Muscat. It was crowded up on the ladder, for we all wanted to see this new land. It was 9.05 a.m. on 26 January, and to me it was a great moment. The land Norman had discovered at port quarter of the bow was a table-topped island, first light blue and then, as we turned past, showing bright yellow in the morning sun. Yuri, with an Antarctic expedition behind him, likened it to a flat iceberg.

This was Astola island. We still could not see the mainland coast of Makran because of haze. There was nothing to tempt us ashore, there being nothing on the island but allegedly an abandoned Hindu shrine and an infinite quantity of small poisonous snakes. This island has been identified by some as Carnine or Nasola, visited by Alexander the Great's admiral Nearchus. But Pliny,[6] the Roman historian, refers to it in AD 77 as the Isle of the Sun or the Couch of the Nymphs, 'on which all animals without exception die, from causes not ascertained'. Perhaps the Romans did not know about the snakes? We were sailing in historic waters. Alexander's biographer, Arrian, who had access to the actual log-books of the admiral, tells that when the Greeks reached India after their exhausting inland march through the wild mountain deserts of Persia, they built a fleet on the Indus river and sailed back to Mesopotamia along the Persian coast. This happened in the winter of 326–325 BC, two thousand three hundred years ago, yet it seemed like modern European history to us who came in the wake of ships that had forced Indus Valley rulers to build forts along the coast twenty centuries or more before the days of Alexander.

The weather had cleared, but there were other heavy clouds in the making and thunder in the air. We sailed eastwards along Pakistan's most desolate coast and attempted no landing in an area where neither past nor present man could gain a foothold. No forts were needed here; a brief reference in the pilot book inferred that nature had raised vertical limestone walls intersected with impassable mangrove pallisades. This was perhaps the least known part of the Asiatic coast we could have found. We were to land, but had no idea of where we would end up. As the snake island of Astola sank in the sea behind us we saw no further sign of land before the sun set among changing clouds.

A night of profound impressions and a feeling of high adventure awaited us as we said goodnight and crawled to bed in our two cabins. We all felt excited. Something was in the air. Not only the awareness that a world unknown to us was our port-side neighbour. It was just that things did not feel quite right. We had never experienced a night like that in the open sea. No waves, not a sound except peaceful snoring, grasshoppers singing, and then at intervals the sharp splash of a fish leaping. The atmosphere was just as on a mountain lake. It was as if the heart of the planet had stopped beating. Perhaps it was the silence before a storm. Even the two big seabirds were sitting motionless at either tip of our vessel as if they were stuffed.

I was out more than once, nervously scouting for land, and each time I seemed to bump into Carlo on a similar mission. It bothered me that we seemed to be equally tense, and Carlo really got moody when I rejected his idea of sounding the bottom with a piece of string. 'No need, we are far from land,' I said, 'it's midnight, go back to bed.' 'But look at the water,' Carlo retorted, and leaned over the side-bundle with his torch. The calm water was more white than blue. 'Must be mud from some river,' I proposed. 'We have no string long enough to reach bottom out here.' Somehow my answer offended Carlo, who crawled back into the cabin visibly unhappy and unconvinced. I remained seated outside the port-side door opening, on the long and narrow bench we had lashed on along either side of the cabin. Land had to be on this side, and for a while I strained my eyes looking in vain for something, perhaps the contours of a distant coastline. Nothing. According to our dead reckoning we were still too far from shore. Then I, too, crawled to bed.

At 01.30 I was awakened by a loud conversation between Norris and Asbjörn on the bridge above my headrest. They spoke about land, and as I crawled to the port-side doorway I saw it too. In the darkness of the night, lifted above a low white mist, the wavy outline of a continuous mountain chain was showing clearly, black against a starlit sky. The moon was up, still fairly big. The mountain range left us with the impression of being very high and very far away. The bearings of peaks and passes did not change, although we moved eastwards at a reasonable speed, taking a southerly wind in from the starboard side. We had no maps showing anything other than the main outline of this area, but the brief description in the pilot book left the impression of level limestone cliffs and no coastline as impressive as this. The water around us was still milky. I sat for a while gazing at Asia and listening to the leaping fish, then I headed back to bed.

At 2 a.m. I was out to share steering watch with Toru. The distant coast was still there, almost unchanged. It could only be very far away. Toru and I had no difficulty in keeping a course of 85°–90°, which would take us safely clear of Ras Ormara, estimated at 76°, the only cape jutting far enough into the ocean to impede our safe passage to Karachi.

During our watch the distant silhouette of a coast suddenly began to change appearance. The valleys rose to the level of the peaks and the undulating ridge straightened out to a flat plateau. At the same time the dark range brightened before our eyes to a ghostly grey and seemed moreover to move right up to shouting distance. But there was no echo. Perhaps it was not even land, since parts of the ghost wall rolled up and dissolved like clouds. Damn it. Had we been fooled?

When I was expecting Carlo for the change of watch, Detlef happened to come out instead to answer a call of nature. The milky cloud bank left him unimpressed, but he noted the milky water. 'It is clay from some river outlet,' he shouted up to me on the bridge. That was almost a verbal repetition of my own words that had somehow irritated Carlo, who just now happened to come crawling out with a roll of string to measure the depth before his watch began. Carlo straightway threw the string back in through the door, climbed to the steering-bridge and grabbed the tiller from my hands without a word.

'Don't steer below 85° or we will hit the cape,' I said, and as I saw how unreasonably touchy he was that night, a devil prompted me: 'Don't worry, just clouds,' I said stupidly and pointed casually to the mysterious fog wall on the port side.

I regretted teasing him the moment I had spoken, for I was far from convinced myself. I was so uncertain that I sat down with Detlef on the port side bench, watching the wall of mist that had risen so suddenly to obscure the view of the distant mountains I had seen so clearly with Norris and Asbjörn. I was almost falling asleep with my back against the springy cane wall when something happened to the low cloud bank we were sailing along. Its milky colour began to turn more yellowish, and many vertical stripes or cracks began to show up very distinctly, as if the broken ice-edge of a Greenland glacier was slowly throwing off a thin veil of night mist. The mist lifted and dissolved. The clouds were gone. But left naked in the moonlight were pale cliffs of lime or petrified clay, smooth as crystal, rising to an even height of 600 to 700 feet all along, as far as we could see under the sparkling sky. Obscured by mist and camouflaged by a sea discoloured by erosion from the rock itself, the coastline of Makran had been close beside us for hours. Excessive leeway or a northbound current had brought us out of our intended course after passing Astola island. But these cliffs had stopped our leeway and set us on a straight course. The same elements that had pulled us towards the coast had been halted by the cliffs themselves and forced to turn aside. Both ocean current and airstream had been diverted and were forced to follow the limestone wall, and *Tigris* with them. Out at sea the wind and water had their freedom. In here the rocks were in command. And for once they had arranged everything in our favour; they had even calmed the waves.

What Norris had discovered on his watch were the distant inland mountain ranges of Baluchistan. Their 3000-foot peaks and crests were seen against the stars while we were still so far out at sea that the extensive lowland deserts and the 700-foot coastal cliffs in front of them were yet too low to show up against the sky. As we came closer, however, the seaboard wall rose above our heads, all wrapped in sea mist, and obscured the view beyond.

We were now so busy trimming the sail and forcing a course further away from the spooky cliffs that we forgot to try Carlo's string, and although his instinctive concern was certainly warranted we shall never know if the milky water was due to shallows or to erosion from the chalky cliffs. Certain is it that Makran was only a gunshot away while we were admiring the high ridges in distant Baluchistan.

We managed with no great effort to hold our own and even increase our distance from the long rock wall, and I crawled back into bed. I was awakened a couple of hours later by Detlef shaking my legs violently; 'We can see Ras Ormara now, and it is doubtful if we can clear the cape!'

I felt an icy chill through my bones. It was before sunrise on the 27th and we coasted with good speed along moonlit white cliffs, but just ahead they ended and opened up into a large bay. Beyond the bay a towering promontory, twice as high as the other cliffs, jutted into the sea with wild peaks and precipices that scared the helmsmen from coasting on, the more so since we began to get a very strong onshore wind the moment the cliffs inside disappeared. The wind followed the cliffs. It was again as if nature took the lead and wanted us to turn with the rock walls and flotsam into an unknown bay. Ras Ormara towered in front like a policeman guiding the traffic to follow the curve in a left turn.

Against an intensely red sky that preceded the first light, Ras Ormara reminded me of the lofty North Cape of Norway, dropping into a black sea at noon in midwinter with the sun just below the horizon. But this was more of an Asiatic dragon, drinking from the ocean, its dorsal crest towering against a sky the colour of blood, its jagged tail sloping off towards the inner bay.

We had to take an immediate decision. I wakened Norman, who was on the bridge in two leaps and taking the bearings.

'We will never make it,' he yelled. 'We are steering for sure collision with the cliffs!'

'Let us steer into this bay and drop anchor,' I shouted back. There was no time to lose. All hands were summoned. The topsail went down. The mainsail was trimmed. With the wind at our back we turned into a huge bay. Ras Ormara at our starboard side provided a magnificent sight, the more impressive as the sky behind the dragon began to play in brighter shades, until the sun shone on the ivory cliffs on the other side of the bay while we were still in shadow and could only see rays of silver radiating above the head of the dragon. Three white sails began to show up deep inside the bay. As they approached we hoisted the Pakistani flag, green and white with moon and star, but the sails belonged to fishing canoes which passed us cautiously at a great distance. We could not see where the bay ended.

Toru and Asbjörn went out in the rubber dinghy to film the sunrise and to sound the depth ahead of us with lead and string. 'Five fathoms!' 'Four fathoms!' they yelled back to us. Really shallow for this far out. Here, too, the water was milky. Norman tried to call Ormara, Karachi, Bahrain; he called 'all stations'; he tried anybody who could help us ask for landing permit. Nobody heard us anywhere. We could only prepare the anchors and go in without permission. Without charts we entered an unknown bay with the two men ahead probing the bottom, which now seemed to maintain the same depth. Well inside the bay we lowered the mainsail too, and threw out both anchors at four fathoms. They dragged over a sandy bottom and we noted that the onshore wind was increasing. Heavy clouds rolled up over the ocean behind us and we began to see swirls of desert sand rising into the air on both sides of the bay. A storm came in from the ocean and although the promontory gave us partial shelter, tall waves tumbled in and *Tigris* danced and continued to drag on the anchor ropes. The birds were so tame in this bay that they landed amongst us without fear. One big gannet rode about with the two men in the dinghy and left its visiting card on Detlef when he refused to stop rowing and allow it to perch on his oar. There were obviously no bird-hunters in this area, but we could smell fish half a mile away when the sailing canoes passed.

At noon we were still slowly dragging our anchors and by now the wind was striking us hard. This was not the right weather to risk lifting anchors and start crossing about among unknown cliffs, looking for a bottom that would give a better hold. I sent Norman and Rashad ahead into the bay on reconnaissance, to look for people and try to learn with English or Arabic where we could best ride off the storm. Asbjörn took them in, and with three men in life-vests the rubber dinghy was loaded to capacity and vanished behind every wave.

Hours passed. Asbjörn had been back once to report that the surf was terrific, but they had found a small sheltered corner of the beach under the rocks to the far right. The other two had walked inland with a man who could speak some Arabic. Having told us this much, Asbjörn returned to wait for them, but this time none of the three came back.

The eight of us left on board realised that we could never transfer ourselves to the sheltered corner found by the men in the dinghy. Sand was whirling high on either side of the bay, the anchors dragged, and we began to see a wide stretch of low land ahead, apparently a sandy isthmus linking the lofty cliffs of the Ras Ormara peninsula to the mainland. To the far right, where Asbjörn had indicated, we saw some black canoes pulled up on the sand, but no dinghy. We took the bearing of some lonely palms that came into sight near the tip of the dragon's tail, and as it showed that we were still moving slowly we threw out the canvas sea-anchor, hoping it might again dig into the bottom and take a hold as had happened off Failaka. But here the water was still too deep. As the wind forced us shorewards the four fathoms under us were reduced to three, then two-and-a-half, and then two. From now on fathoms were no use for measuring and we began to reckon in metres. We pulled both rudder-oars up until the blades were level with the bottom of the vessel, and secured them there with ropes.

We who were left on the bundles had a most exciting but nerve-wrecking afternoon, with *Tigris* hanging on in the two tight ropes. Our lives were probably not in danger unless breakers should drag us overboard and throw us against rocks, but we had fears for all our deck cargo and of jeopardising our chances of ever getting to sea again if *Tigris* were washed ashore on this desolate Makran coast. All along the stormy bay we heard the rhythmic drone of ocean surf from the big swells that came rolling in from the open sea. The height of the breaking walls could only be understood by their roar, as they turned their round backs towards us.

In spite of fear and uncertainty, we had time to look around and realise that we had come to a truly exotic place, unlike anything we had ever known. If the setting was both scenic and spectacular, with rocky side-curtains flanking a flat, sandy stage-set against the open sky, the performers were no less reminiscent of the theatre. The first we saw was a long caravan of camels coming out from the rock draperies at the dragon's tail and passing right along the four-mile-wide stage at the water's edge. The turbanned drivers, with bag-shaped trousers and coloured shirts, drove their striding beasts right along the shore where the sand was wet and hard. No sooner had one caravan disappeared among the rocks to the left before another emerged at the right, with twelve or fourteen camels in each procession, always striding across the stage from right to left. From the opposite direction came camel drivers with single beasts, or pairs, all with big burdens of twigs that could only be firewood. With them, or independently, came women carrying large bundles on their heads, robed from top to toe in green, red or other very gay colours. Soon we had them passing us only some four hundred yards away, and with the surf drumming between them and us we certainly felt as if watching from orchestra stalls in a theatre, the more so since none of the people ashore paid us the slightest attention. Not one stopped to look, or even as much as turned a head our way to glimpse a reed-ship arriving tail first, with dragging anchors, from another world into their own.

No one stared, not even the animals, when we came close enough to see goats and dogs pattering along in the long-legged company. We from our deck absorbed all these strange surroundings as we struggled to save our ship and all our possessions. None of us failed to realise that this was the most appropriate setting we could have found for the arrival of a reed-ship. Although everything seemed staged for us, we were simply reliving history.

The village must be somewhere behind the sands to the right, where the smaller groups hurried with firewood and bundles. On this side of the dunes, facing our bay, were only some tiny huts of mud and mats tucked away under the few palms at the dragon's tail. The canoes, too, were on the beach on that side. But we could make out the tower of a little mosque behind the dunes. It was Friday, and it seemed that Muslims from the desert were returning to the wilderness from a visit to the town. The others, with firewood, were coming home before the sun set, for it was already low in the western sky.

Much to our surprise, a large flock of pink flamingoes circled above us in plough formation, their necks and legs extended as if shot from a bow. Then came the thunder and the rain. No doubt a real storm was raging out in the open sea; Ras Ormara clearly gave us some protection. The men in the dinghy were not back. The silence among us on board *Tigris* was emphasised by the rhythmic droning of the surf and the occasional clatter in the thunderclouds. Toru was quietly frying dolphin and onions in the open galley. HP was measuring the depth with his own lifeline. He reported two metres. I double-checked and got 2.20 metres. This would leave us half a metre clearance in calm water. But the water was anything but calm. For a while the surf had been rising and breakers now began to fumble into vertical walls around us and throw themselves with force against the sickle-shaped ship. With the anchor-ropes lashed to the bow, the stern was turned to the land and the purpose of the elegant sickle shape became clear again. The ma-gur was not designed as a river boat. The curve was there to help it ride a stormy sea and let the ship lift its tall neck or tail over the surf like a swan in choppy water. The broad, raised chest of *Tigris* leapt up, split the surf, and as it came down again sent spray to either side, in fact wetting the deck less than the rain.

Carlo grabbed one of the oak oars and an axe, and with Detlef's help began to ram the oar into the bottom beside the bundles. The sand gave no hold; the oar yielded and came out. The sun was getting low, and there were no more people on the beach. Now the ebb tide started and began to suck the water away from the bay. To our despair we saw the beach creeping towards us, getting wider and wider, whilst we had ever less water beneath our reeds. We had the impression of being still completely afloat, hugging the surf with our seabound bow in wild leaps. But almost certainly the landward stern was now touching the bottom in the intervals between the high waves, and this, combined with the seaward suction of the ebbing tide, helped the two small anchors and the buried sea-anchor to hold us in one spot, ridiculously close to the obliterated camel tracks ashore. It had been my last hope that, when the stern ran aground, the drag on the anchors would diminish and we could dance the weather off, half-afloat in one spot, until the storm abated and we could pull ourselves back into deep water.

In this desperate position we sat down at the long table to eat a warm supper, feeling like passengers on a racing speed-boat jumping the waves towards nowhere.

At 5.40 p.m. *Tigris* time the sun set and HP lit the kerosene lamps. Far away someone lit another small lamp among the mat huts at the end of the bay. The rain had ceased. At 6 p.m. we heard shouting from the sea and waved our lights from the cabin roof. Riding with the surf in the dark came Asbjörn, alone in the inflated dinghy; he was grasped by many arms and pulled on board. He had no news of the other two. He had walked inland alone until he found a small house with a tower; he had peeped inside but had learned from a few praying Orientals that this was a mosque, and that the village was far away on the other side of the sandy isthmus. He then came back to the dinghy which he had hidden behind a big canoe, wrote TIGRIS in the sand and a message to the others that if they signalled he would come to fetch them.

Nobody signalled. For a moment we saw another faint light at the opposite end of the beach, then even that disappeared and all remained dark.

At 7 p.m. we all leapt from our seats around the long table. We heard more shouting from out at sea. There were lights. I could have sworn I recognised Norman's voice. We all shouted, but our voices were drowned by the surf and carried inland by the wind. Then we heard many voices and the sound of an engine. We climbed roofs and mast ladders with all the lamps we had and waved and yelled warnings, for we did not want to lure the boat into a trap. It was a small and shallow dhow that came into the range of our lights, full of Pakistanis and with Norman and Rashad in the bow. They came straight for us and near enough for our two friends to jump on to the reed-bundles before any of our visitors realised we were half up on land. Caught by surprise and drenched by surf, the experienced crew of the dhow performed a brilliant manoeuvre, throwing their vessel about and steering at full speed out again until their lights and the sound of them were lost in the darkness.

Once again we were eleven men to share the fate of *Tigris*. Norman and Rashad told us they had walked a couple of miles across the low sandy isthmus until they reached a village by the sea on the other side of Ras Ormara. Rashad explained enthusiastically that the scattered houses along the trail were built from mats with arched roofs, following the very same design as that of the Marsh Arabs. In the village some of the houses were built of stone, and these had been the houses of the school master and the police. There was a real storm out in the ocean and the friendly villagers feared for our lives on the west side of Ras Ormara. They had finally decided to send a rescue-vessel around the cape to come to our assistance. It had been a terrible trip around the cape. What happened to the rescue-vessel after it had brought us our two companions we shall never learn; once it was swallowed up by the night it never came back. Norman and Rashad assured us that the rescue-party, in passing Ras Ormara, had been in greater danger than us, and the friendly fishermen would certainly have their work cut out to round the cape once more and get back to the more sheltered village bay before the storm got worse. After all, they had found us alive, riding our large bundles close to land, unhurt and cheerful. And the dhow captain had shouted to Rashad that this anchorage was safe for us but not for them.

It began to seem as if he was right, although never had we heard of anyone using an anchorage like ours, in the midst of high surf, with the stern almost ashore. There was no longer any visible change in our position, however, nor any sign of an immediate worsening of the storm, so we could only relax and console ourselves with the knowledge that three of us had been ashore in Pakistan. We divided the night-watch and crept to bed in the two cane shacks where we were rocked brusquely asleep on our reed cradle to the diabolic lullaby of the Makran surf.

CHAPTER 9

In the Indus Valley in Search of Meluhha

On Monday, 30 January, the sky over Pakistan was blue and the sea was without breaking crests as the ocean-going salvage tug *Jason* of Norway slid around Ras Ormara and into the large bay on its western side.

The storm that had struck the coast three days earlier had abated, and Captain Hansen of Bergen was standing on the bridge with binoculars, looking for a wreck. There was none to be seen. He searched for derelicts among the fallen rocks and treacherous sandbars at the foot of the lofty cliffs, but in vain.

But peacefully anchored in the middle of the bay was a moon-shaped boat, the strangest watercraft the captain and his Norwegian crew of twelve had ever seen afloat. Not an Arab dhow. Not a Chinese junk. Certainly not a Pakistani canoe. There was not even a name on the vessel's stern, for it was nothing but a sheaf.

At slow speed the 1200-ton salvage ship moved up alongside the little boat and anchored. The vessel they had discovered proved to be abandoned, but secured to the bottom by ropes. There was not a living soul on deck nor inside the open doors of the two woven huts. But there were plates and cups and an empty frying pan on the outdoor table, and shark tails and underwear hung side by side to dry on the cane walls. Stockfish was hanging in the mast stays, and rolls of toilet paper were hanging inside two tiny outboard balconies astern.

The captain and his men began to search the long sandy isthmus at the end of the bay and observed a crowd of people all gathered at one spot on the beach. They were huddled together in a compact ring, watching something, and others were running from inland to catch a glimpse of whatever they had found in the sand. Captain Hansen lowered the landing barge and hurried ashore with some of his men. He had come to this desolate bay on instructions from the Norwegian salvage company Salvator in Bergen, as a result of a message from Bahrain Radio reporting that the vessel *Tigris* of Larvik, Norway, had been wrecked west of Ras Ormara in Pakistan.

The ship's shallow landing-barge hit the bottom before reaching the beach, and the captain and his first officer left the men to hold the bouncing boat as they waded ashore and hurried over to the crowd. They forced their way gently into the inner circle and found themselves front row spectators of a party which included drums, oriental dancing and wrestling competitions in the sand. Guests of honour within the ring were bearded foreigners of different ethnic types, strangely mixed, and they and their hosts were all so amused that the newcomers were scarcely noticed.

I myself was not there and was the next to be surprised. I had been over in the village with Gherman; and returning on foot across the isthmus we hurried so as not to miss the dancing which the village schoolmaster had arranged for our entertainment on the beach. Coming back over the dunes we had already sighted the most peculiar kind of ship anchored beside our own, where we had moved when the weather had calmed down two days earlier. I had never seen such a vessel, but realised I was to meet the captain when Yuri came to meet me with two blond Europeans.

'Do you speak English?' I asked, introducing myself. I could see that Yuri was having fun.

'Captain Hansen,' replied the stranger in Norwegian and added that, yes, he could speak English, but if I didn't mind we could use our mother tongue.

Shark-skin drums and Pakistani songs filled the air and dancers and wrestlers tumbled about in the sand during the improvised beach party as I recovered from my surprise and asked Captain Hansen what he was doing with a Norwegian salvage vessel in such a desolate place as Ormara bay. I told him that in spite of this friendly welcome dance the village police had confided to me that foreigners were not allowed to come to Ormara. The village people recalled only the visit of one foreigner before, and he had come overland to collect stones.

Captain Hansen replied that he came with permission from the coastguard in Karachi. His Norwegian salvage-tug operated all over the world. At the moment it was working in the Indus delta, pulling grounded vessels off the mud-flats in the labyrinth of channels between the mangrove swamps. They had just come back to base in Karachi and rescued an old Greek ship on fire outside the harbour when they received radio instructions to salvage *Tigris* in west Ormara bay. 'But coming here we find your boat at anchor in the bay with everyone dancing on the beach!'

We tried to think out what might have happened, since we ourselves had sent no radio message when we sailed into the surf of Ormara bay. Norman was in fact ashore with Rashad, and when he came back at night to ride the surf at anchor with us, we needed nobody, and nobody could have come to our help fast enough anyhow. Perhaps it had been this very silence, and the fact that a major sandstorm was known to have swept from the Arabian peninsula to Pakistan that day, that had led to speculation that *Tigris* had been wrecked. That night, however, my entry in the expedition journal had read:

Terrible onshore wind now howling in the masts as we dance in the surf with wild splashes. So we go to bed fully dressed with an inferno of waves lifting us up and down, lightning in the sky, wind and water so noisy that the grasshopper in the galley can scarcely be heard. Dhow said before it left that this anchorage was safe for us but not for them. They would come back tomorrow or try to send somebody else. Insh Allah. It is not a pleasant moment for me and my men as we now go to bed. Right now the stormy wind is howling. Will our anchors hold? Insh Allah.

Next morning we woke up to find ourselves in a calm bay. We were in shelter; there was now an offshore wind. At sea the waves ran white. I had been up and out by 4 a.m., aroused by the change that I could feel even in my sleep. It was again this strange silence, as if we were back on the rivers of Iraq; suddenly there was no surf, no more jumping, only a slow rolling all along the broad bay on either side. But in the distance I heard a sound as of a great waterfall, coming from inland. Clearly a strong surf was running against the narrow isthmus from the other side and the weather had changed direction completely. The storm now struck the village bay east of Ras Ormara. The dhow would have been better off had it remained on our side.

The cliffs now protecting us stood like the walls of a glacier rising from the sea. The shallow water was more milky than ever, although the tide had risen and left only a narrow beach a couple of hundred yards further away. I found Toru in the galley and he offered me a cup of coffee. This was too good to be true. I crawled back into bed but could not sleep. As the clouds were chased away I could see the moon and the familiar Plough and Pole Star.

The sky was clear and the wind still offshore when the sun rose. For once all eleven men could crowd around the deck table at the same time. Gherman served porridge and fresh eggs, which he had preserved by painting each one with oil before departure. A friendly group of Urdu-speaking fishermen brought along by Carlo volunteered to help us to a good anchorage out in the middle of the bay. Since Norman and Rashad had seen the village the day before, they were left on watch while the rest of us took shore leave. The dinghy ferried us two at a time into shallow water and we waded on to a broad and beautiful beach. Miles of white sand and not a human soul but ourselves stepping on shells and sponges and scaring little crabs into their wet holes as we trampled barefoot to the dry sand, where we put on our shoes.

It was good to see *Tigris* riding high and to remind ourselves that we were now ashore in Pakistan. We were still in part of what we call the Makran coast today, but to the Sumerians this might have been a transition area between Makan and Meluhha, since the westernmost of the Indus Valley military installations were behind us.

The sand beyond the reach of the waves was full of potsherds. There were hardly any very ancient remains to be found on this low isthmus, however, because the sea level, as in Mesopotamia, had been higher in Sumerian times. The sandbar had probably been completely submerged, leaving the high rocks of Ras Ormara as an island. Local geology was fascinating: Asbjörn discovered that the limestone cliffs were full of perfectly round balls of soft chalky stone the size of oranges which, when broken, revealed the most beautiful fossils of fancy shells, clams and worm-like creatures. The men collected large quantities as souvenirs. But the hard-packed crusts of sand forming the low isthmus were quite different from the limestone cliffs and rose in a series of arched terraces parallel to the beach as a result of the successive formation of new beach-lines each time the ocean had subsided. This was important, for, as Dales had pointed out, it meant that the ancient forts of Sutkagen Dor and Sotka-Koh had been built right beside the former edge of the sea.

As we reached the upper level and started walking across the sandy isthmus the surface was so full of potsherds that it seemed as if the population had broken all the jars it possessed as soon as the merchants from Karachi had brought them modern metalware. In fact, the first thing we saw were small groups of women escaping from us across the sands in their foot-length robes with tin jars on their heads. A well had been dug not far from our beach and the women came from far and wide for its water.

As Rashad had pointed out, the houses we passed on our way were so similar to the characteristic reed dwellings of the Marsh Arabs that it was difficult to see how people living in a desert and people living in swamps could build homes so similar without a common heritage. When we finally reached Ormara village, on the other side of the dunes, even the mat-houses were rectangular, with gabled or flat roofs like the more recent stone huts. Otherwise the village gave us a fascinating taste of a community that reflected deep roots with sparse grafting from abroad; its environment was sea, cliffs and desert. Inland was all barren wilderness, with clean white sand-dunes heaped like thick snow-drifts to the very edge of the settlement. There was no overland road to Karachi, only camel tracks and, in the opposite direction, a jeep passage westwards to the prehistoric fort of Sotka-Koh and Pasni. While most of Pakistan became increasingly affected by the hubbub of modern Karachi, Ormara and its surrounding desert, like the swamps of the Indus delta, offered great natural obstacles to cultural revolution. Ormara patiently awaited its turn to take the leap into the twentieth century. In the meantime, life continued closer to my image of a self-sustained, civilised community of the third millennium BC than anything I had ever seen except among the fine Marsh Arabs of Iraq and the timeless mountain town-dwellers of Oman. Here I felt, as more than once before in isolated communities, that the now utterly interdependent outside world, with its problems of petrol and military prestige, may be shaky, but should it ever collapse, as did the empires of Sumer and the Indus Valley, the camel trains of Ormara will still go on loaded with palm leaves and canes for house construction, and fish, dates and vegetables for human survival. The proud women will still find a way of draping themselves in coloured robes to sail over the dunes for milk and water; the men will still haul in the fish and till the plots between the palms.

But today intrepid traders were doing their best to convert the life-style of the Ormara fishermen into one of cash economy. Ormara already had an outlet for its products in the merchant mariners who came in dhows from Karachi, not far along the same coast, and from Colombo on the distant island of Ceylon. The local industry was shark fishing. The result so far for the poor villagers was more flies than money. Never had any of us seen such incredible quantities of flies: at first we brushed them away and concentrated our attention on the charming little village, with its three tiny mosques and a market place straight out of *The Thousand and One Nights*. We passed through a mob of camels, donkeys, dogs, cats and chicken and, with a horde of children at our heels, were met by bearded men of mixed Pakistani and Arab physiognomy, most with turbans, but one with a marvellous silver fez on his head. Some had enormous hooked noses, while others had flatter, more negroid faces, the same remarkable blend we had first seen in Oman. When we asked if they understood Arabic or English they shook their heads and spoke in Urdu.

They all received us with smiles and signs of welcome, except for one bushy-bearded villain who drew his curved sabre, pointed the tip at his own stomach and then swung it over our heads. He then disappeared into the crowd with little reaction from the rest, except that they tried to make us understand that he was a religious fanatic by pointing to themselves and to the mosque saying, 'Allah! Allah! Allah! Allah!' Then they pointed after the savage-looking man, waving their bodies from side to side with eyes closed as in trance, and mumbling spitefully, 'Alli-allo, alli-allo.'

Not much wiser from this instruction in local religious practices, we passed the shopping street, hardly larger than our own reed-ship, with about half a dozen small alcoves raised a yard above the sand on poles. Each was a shop the size of a doll's house with an open front wall where the staff of one sat cross-legged, with the merchandise before his legs. I took the inventory of the smallest shop: seven carrots and five potatoes. Most had saucers or small bowls with green, grey or white grains or seeds in front of them, but one had biscuits, Arab cigarettes and bits of crude soap and candy. The most elegant of all was that of the long-bearded tailor, whose needle seemed to strike the woven mats of ceiling and walls whenever he started sewing. In front of him we almost stumbled over a man who squatted in the lane selling tea made on an open charcoal fire. Not a single woman could be seen in this exotic crowd, except beautiful girls below the age of six or seven and old crones with green stones dangling from nose-rings.

The flies were everywhere in abnormal quantities, but their numbers seemed modest compared with what we were to see as we reached that part of the little village beside the sea. Here were a few open yards or sandy fields fenced in with woven, man-high, palm-leaf mats, but if we stretched on tip-toe we could peep over the fence and witness a barbaric sight. Right under my nose was a man standing knee-deep in a mud-hole in the ground, trampling upon fish bodies as large as himself, while everywhere around him lay colossal sharks' heads gaping and staring in all directions. An old man was standing among the heads with a bucket, pouring water into the hole where the youngster was eagerly stamping on the beheaded sharks in the muddy water. And when he had trampled enough he dragged the bodies out of the soup and threw them on the ground, which was everywhere covered with soiled fillets, fins and heads. Most of the heads were of sharks and giant rays whose evil little eyes and great grinning jaws dominated the scene. Another barefoot boy moved around with a long rake, sorting out or turning over smaller fillets, raking them about into orderly rows and piles as if they were hay. In the midst of it all stood a long-legged camel with two huge baskets which were slowly filled to the brim by two men sorting out dry, sand-crusted fillets, one apparently selling and the other buying.

On all my sea voyages we had occasionally eaten shark, and it is a good meal if slices are soaked overnight in water to extract the ammonia from the cells. The shark is a primitive fish without organs to eliminate the urine, which thus enters its own blood. We now understood what these men were doing: they were pressing and washing the ammonia out of the giant shark fillets before they were dried and exported to distant markets. The camel driver was buying it as a delicacy for his own tribe somewhere in the Makran desert or beyond, and two dhows were anchored off the village beach, waiting for their loads. 'Colombo' the villagers had said as they pointed to the dhows, and this was confirmed by the school-master. Many hands were to share the profits from these sand-encrusted shark fillets before they were masticated in Sri Lanka to the south of India. This direct oversea trade with distant Ceylon by small boat seemed to rest on old traditions, and made much impression on me.

We had seen many sharks in our wake on the way to Ormara, but few as large as most of these. There must have been an almost endless quantity of the man-eating monsters in the open ocean outside Ras Ormara, to judge by the grotesque field of yawning heads, some piled up and others strewn about within the fences. But it was not the sharp teeth of the beheaded predators that most impressed us, it was the mask worn by those left alone: a close mosaic of shiny bluish-black beads. We could scarcely believe that normal house-flies could pack together so densely in such incredible numbers until Carlo bent down with his camera over a big head. It was stripped to its skin in a second, while Carlo almost fell over backwards, shrouded in a buzzing cloud. Wherever we looked, on the fence, on the fish-piles, were compact blankets of flies, sometimes seeming to crawl on top of each other in search of a vacant place to feed. And the odour of generations of old fish was wafted for miles on the warm air and must have tickled the scent organs of flies and set in motion a mass-migration to Ormara from every part of Makran.

Perhaps it was this strong stench that benumbed my senses and made me impervious to lesser smells by the time we reached the open beach and found two of the fishermen building themselves a new plank canoe. The wood was white, but the other canoes drawn up after use were dark brown and seemed to be covered with a thick coat of hard, waterproof varnish. The boat-builders saw me examining this covering and made it sufficiently clear that they had made the varnish from shark liver oil. I put my nose to the canoe and sniffed, but could detect nothing. I recalled that Arab fishermen in Iraq had told me they had used shark oil in former times to waterproof their dhows, and they suggested that we should do the same with *Tigris*. I had refused. I remembered the horrible odour of stale fish oil in warm weather, and was afraid we would be driven from our own ship when we started to stink like rotten seafood. But these Ormara canoes were waterproofed with shark-liver oil and had no smell. The Gilgamesh epic of the Mesopotamian king who sailed to Dilmun was again brought to my mind. He mixed oil with his asphalt when he built his reed-ship. We had found that pure asphalt cracked on reeds. Perhaps it should have been mixed with pitch and oil, probably shark-liver oil. With our reeds cut in August we would float for months still, but with such a cover there might not be any absorption at all. We were learning from people with centuries of experience, and were at any rate doing far better than during the first fumbling experiment with *Ra*.

I counted more than fifty dogs full of fish entrails lying out-stretched as if dead on the wet sand of the broad tidal flats. The sea had withdrawn again. It was good to know that *Tigris* was now afloat, independent of the tides in the bay on the other side. The drag on the ropes during the storm had been so great that the anchor shafts had bent, but the grip had held, probably because the stern had touched the bottom. After this first visit to the village we hurried back to make sure that the two men left alone on board had no problems. On the way across the isthmus we met to our surprise a young lieutenant of the Pakistani coastguard, who was camped with his men in four tents at the tip of the dragon's tail. He was polite from the moment he started his interrogation, but overwhelmingly friendly when he learnt that the name of our little 'dhow' was *Tigris*. He had a radio in his tent and had received a message from the Coastguard Command in Karachi that 'a large ship named *Tigris*' was cruising in the waters of Makran, and that he should not make any difficulties but provide all assistance needed! He was personally responsible to his superiors in Karachi for the whole Pakistani part of the Makran coast.

We had not known that the Makran coast was out of bounds to foreigners, and we never learnt why, as we could not have imagined a friendlier reception anywhere. The coastguard lieutenant, the schoolmaster, the police, and an expert from Karachi who was organising local fishery, all spoke English and became true friends. The Urdu-speaking population also did their best to show us hospitality, inviting the men into their huts and treating them to gifts of fresh lobster. Some came with eggs and goat's milk.

The confusion concerning the landing of our 'big ship' could not have been cleared during the discussion between the Ormara tent and Karachi headquarters that day. The lieutenant went back to his field transmitter to report that *Tigris* was here. We went back to our own bay, where Norman sat in his cabin corner desperately trying to make contact with any part of the outside world. He at last heard a voice with a garbled message to the effect that a ship was on the way from Karachi to Dubai and would stop at Ormara to deliver 800 gallons of water to *Tigris*! Norman did his best to yell into the microphone that we had more water than we needed for another three months, and eight hundred gallons more would sink our ship. But nobody seemed to hear him.

When we told our story to Captain Hansen of *Jason*, he assured us that he had not brought us any water nor was he bound for Dubai; he had come with orders to salvage *Tigris* and bring us to Karachi. And he was ready to give us a free tow any time we wanted.

We got together to discuss what to do. After our experience with *Slavsk* we feared a tow more than any storm. What we had planned was to stay at anchor in west Ormara bay until the weather had stabilised and then sail to Karachi. We wanted to travel overland from there to see the inland ruins of Mohenjo-Daro. The barometric pressure was still very low and we intended to wait for a wind that would neither wreck us on Ras Ormara nor send us away across the Indian Ocean before we had seen the prehistoric city in the Indus Valley. We could also take the camel road to Pasni and travel inland from the old fort of Sotka-Koh, but this would mean leaving *Tigris* unguarded for a long time. We would have more time to see the ruins if we accepted Captain Hansen's offer to tow us for one day along the coast to the harbour of Karachi, and we agreed. So as the musicians picked up their drums and followed the large crowd back to the village the two vessels in west Ormara bay weighed anchor. The broad and powerful *Jason* took the lead and *Tigris* followed like a toy on a string, and before we rounded Ras Ormara we were already swallowed up by the darkness.

The colourful life of timeless Ormara village was still vivid in our memories a week later when we entered the prehistoric ghost city of Mohenjo-Daro, deep within the Indus Valley. We had left *Tigris* tied to a buoy in the busy modern port of Karachi, where the Pakistani harbour authorities had given us a hearty welcome and the Navy Command had offered to guard our vessel in front of the Naval Academy while we went off on our inland excursion. Sani, a young guide from the National Museum, took us by minibus on good roads through various Pakistani towns to the ruins of Mohenjo-Daro, 350 miles from Karachi.

Mohenjo-Daro means simply 'Mound of the Dead', and no one knows what the real name of the city might have been during the thousand years it was full of life, from about 2500 to about 1500 BC. In the beginning of the third century AD Buddhists arrived and built a tiny temple on top of the ruins. The site was still nothing but a huge mound of sand and debris in 1922 when archaeologists were attracted to the place and began digging. Until then the mere existence of the underlying civilisation was unknown. Scholars had barely begun to suspect that the Indus Valley had housed a very early civilisation. The work which was begun after the First World War by R. D. Banerji and other members of Sir John Marshall's expedition was followed up by many others and has so far un-covered 240 acres of an urban settlement some three miles in circumference. Nevertheless, large sections of what was once a metropolis of the first order are still deeply buried and unexamined. The surrounding plains, once irrigated fields with dates, figs and crops of wheat, barley, cotton, peas, sesame and other garden products, are now barren stretches of sand, scattered with dusty tamarisk-bushes. The broad Indus river, once flowing past the city wharves, has since withdrawn. Even the climate has become dry and less favourable. Like the deserted ruins of Ur and Babylon, Mohenjo-Daro now lies far from any water. The ruins of former homes and shops stretch for a mile in the direction of the calm Indus, which was originally the life-preserving artery of the city and an unpredictable threat at the same time. Excavations have revealed that the city was rebuilt seven times following floods, before it was eventually abandoned for some unknown reason, for we do not know why the whole population suddenly disappeared in about the fifteenth century BC. Many of the skeletons found among the ruins show signs of a violent death in combat. Sir Mortimer Wheeler and other authorities have suspected that although climatic, economic and political deterioration might have weakened this firmly established civilisation, its ultimate extinction was more likely to have been completed by deliberate large-scale destruction by the Aryan invasion from the north. The fertile plains that supported the Indus Valley civilisation for a thousand years were then conquered by Aryans who arrived, not by sea, but by the high inland mountain passes. These invaders were the Sanskrit-speaking people who subsequently used the non-Sanskrit loan-word *Mleccha* to denote non-Aryans, a term which Bibby tentatively identified with Meluhha and considered a possible name for the original Indus civilisation.

Whether Mleccha to the Aryans or Meluhha to the Sumerians, the founders of the metropolis known to us as the 'Mound of the Dead' have left vestiges that make a profound impression on any visitor. It is a monument of the age-old unity of mankind, a lesson in not to underestimate the intellect and capacity of other people in places or in epochs remote from our own. Take away our hundreds of generations of accumulated inheritance, and then compare what is left of our abilities with those of the founders of the Indus civilisation. Counting the age of humanity in millions of years, we begin to understand that the human brain was fully developed by 3000 BC. The citizens of Mohenjo-Daro and their uncivilised contemporaries would have learnt to drive a car, turn on a television set and knot a neck-tie as easily as any African or European today. In reasoning and inventiveness little has been gained or lost in the build-up of the human species during the last five millennia. With this in mind a visitor to Mohenjo-Daro will be left with the impression that the creators of this city with all it contained had either in record time surpassed all other human generations in inventiveness, or that, like later Aryans, they were immigrants bringing with them centuries of cultural inheritance.

There is only one opinion among scholars: the city of Mohenjo-Daro was built according to pre-conceived plans by expert city architects within a rigidly organised society. The planning of the city was one adhered to, but never surpassed, by subsequent town planners in Central and West Asia for four thousand years. Even at the beginning of our own century few of the lesser towns maintained an equally high cultural standard.

The first impression on approaching the ghost city is little different from that of a Sumerian town: a central elevation with a temple surrounded below by a labyrinth of streets and ruins. Originally, however, this temple did not belong there; the Buddhists fifteen hundred years ago had unfortunately modified the summit of the original structure and built their own shrine in the form of a circular mud-wall with a number of monk's cells on the nearest terraces below. But the original foundation of the summit sanctuary was square and stepped, and the general impression was that of a terraced pyramid hill converted to suit the monks.

As we walked between the brick walls, some of which have survived from tall, two-story buildings, we automatically lowered our voices as if in respect for a sanctuary. There were no large temples, no spectacular palaces, but evidence of a level living standard, although the largest houses seemed to have been clustered in the central part, up along the terraces towards the elevated shrine now dominated by the Buddhist ruins. The original planners of the city had started with a clear concept of what they wanted, laying out parallel streets from either side of the oblong central elevation. These streets, running north-south and east-west, were straight and broad, some regular avenues up to thirty feet wide, dividing and subdividing the city into rectangular blocks, each 400 yards long and 200 or 300 yards wide. The walls were built of burnt brick by skilled masons. Most of the houses have their doors and sparse window openings towards the narrow side-lanes, accessible only to pedestrians, as if to avoid the noise and dust from the main streets where, judging by local art, wheeled carts pulled by pairs of oxen, and occasionally by elephants, would pass with building materials and merchandise. What impressed most was the evidence apparent throughout of a high sanitary concept. Although there were no impressive palaces or other structures that indicated the megalomania of a totalitarian emperor, the common citizens were provided with a complex system of sewers: the streets were lined with brick-built drains, and at intervals the large brick slabs covering these channels were provided with proper manholes by which the sewage workers were able to remove debris, some of which the excavating archaeologists found piled beside a manhole.

There were also fresh-water conduits among the houses, and a big public bath or swimming-pool, possibly built for cleansing rituals. It measures 38 feet by 22 feet, is 8 feet deep and it was waterproofed by a double wall of bricks set in asphalt, with another layer of asphalt one inch thick between the walls. Other than the well-designed pool, too deep to stand in when full, there was a pillared hall obviously intended for ceremonies, and another building with extra solid walls and a cloistered court of unknown purpose. But there has been a wealth of discoveries from within the city to give life to all these empty walls. The former residents had left utensils, images and ornaments of enduring materials ranging from basalt and pottery to bronze, gold and precious stones. All that was not perishable has been recovered by careful excavation. Among these discoveries are the inscribed and beautifully decorated seals of the kind that had also found their way to ancient trading partners in Mesopotamia and the gulf islands of Failaka and Bahrain. Whilst we cannot read this script, the miniature illustrations of deities, anthropomorphic beasts and mythical scenes incised on the seals leave us with the impression of an art style as well as a theocracy and cosmology strikingly similar to those of the Sumerians, and to a lesser extent to those of ancient Egypt too. Numerous small pottery figurines and a few excellent bronze statuettes depict normal human beings and everyday life. Small ceramic models show women kneeling to grind flour for bread, while bronze figurines represent others adorned with jewellery and elegantly posed as for a dance. There are pottery models of men with pairs of oxen pulling two-wheeled carts, and bronze miniatures of beautiful chariots. Some of these figurines, like the ceremonial ceramic bird running on two wheels, are so like those of the two other great civilisations of the same epoch as to confound the experts.

One can easily imagine the life of the people behind the abandoned utensils and waste found among the ruins: the farmer's digging tools and a sample of his figs, grain or cultivated cotton; the fisherman's beautiful bronze fish-hooks; the merchant's elegant bronze scales, with precise flint weights, and his variety of dainty seals; the potter's masterpieces in ceramic, ranging from vases ornamented with coloured motifs to effigy jars shaped like birds, beasts and men; the painter's mortars; the carpenter's cutting tools; the jeweller's necklaces, arm-rings and other ornaments wrought in gold or made from precious stones; and not least the metal-worker's astonishing products in bronze, from statuettes to hand-mirrors, created with expertise by the advanced process of *cire-perdue*. Even the gambler is represented with dice indistinguishable from our own and game-boards with proper pieces to be moved on the squares. In marked contrast, the soldier has left little evidence of his former presence within the city boundaries, indicating an un-warlike, mercantile and agricultural society, counting on the strength of its outer defence positions with forts at strategic places along the coast.

Skeletal remains from the Harappan cemetery have given no answer to the obscure origin of the Indus civilisation. F. A. Khan[1] claims that four different physical types were buried together. The majority of the Harappan population, he says, consisted of a 'Mediterranean' type of moderate height, with long head, narrow and prominent nose and long face. But there was also a second type, long-headed, too, but more powerfully built and of tall stature. The third type was short-headed and the fourth was typically Mongolian.

No matter how many ethnic groups had joined hands to form the Indus civilisation, how could the citizens of Mohenjo–Daro and Harappa manage to out-distance all other peoples of their time so abruptly and so completely? Perhaps this question will never be answered by local excavations in Mohenjo–Daro and Harappa. Perhaps these cities were founded by already civilised settlers who came from Amri or Kot Diji, two other mounds packed with antique debris towering above the plains nearer the coast in the same Indus Valley. We visited both, but little was left except a compact mass of broken adobe, potsherds, chips, bones and bits of datable charcoal. From the meagre information to be gained from these sparse fragments science begins to hope that these mounds may provide important clues to the real origin of the Indus civilisation. Extensive digging has brought to light shattered remains of an older culture of equally remarkable character, roughly datable from 3000 BC, or slightly before, to about 2500 BC, when Mohenjo–Daro and Harappa were founded. If this is a mere coincidence it is most remarkable. It will give the Indus civilisation a formative date coinciding with the foundation of the Sumerian and Egyptian dynasties. It is not surprising that simpler minds are ready to accept the wildest of all theories, that civilisation at that time dropped down on our planet from outer space.

When our guide, Sani, led us between the ancient walls of Mohenjo–Daro, the broad streets were empty and the roofless ruins around us looked naked and cheerless, gaping towards the sky. But in our minds we filled the avenues and side-lanes with visions fresh in our memory. If anything, life in Mohenjo-Daro had been still more colourful, still more attractive and advanced than in Ormara. It was easy to imagine the tumultuous movement of crowds in the main streets, while others sat lazily in the shaded side lanes dozing, chatting, or playing games. One could almost hear the noises and scent the smells of hump-necked cattle, hay and spices that filled the air between sunbaked walls at the height of the day; tanned men and women hurrying by with beasts and burdens; rumbling carts with palm leaves for thatching or with cotton for the weavers and food for the market; fishermen with baskets on their heads; bakers with their warm bread; women with eggs, pastries and tropical fruits; crying babies, singing birds; the clinking from the smithy; bleating beasts and shouting children running through the side-lanes while others sat indoors with priests teaching them the script we have never managed to decipher.

We had to multiply our impressions of Ormara a hundredfold to recreate the life of Mohenjo-Daro with its vaster dimensions, broader streets and higher standards. Ormara had never been designed. Each little hut had been put up at the convenience of the fisherman who had built it for his own family. That village, like most others, had been created by natural growth over centuries and had led to no major invention. No sewage system. As we walked through Mohenjo-Daro, from the lower town up the stairways between the buildings clustered around the central tower, we knew, from what fifty years of excavation had revealed, that this was the empty shell of a community once ruled by a priest-king, regarded as a demi-god, just as were the rulers of Mesopotamia and Egypt. These streets, carefully oriented to the sun, had seen his wise men walking about among the common crowd: high priests, scribes, astronomers, architects, inventors. The elite of this society, together with the rulers of the equally large and no less important city of Harappa, five hundred miles further up the river, had created a common empire that had left its ruins and monuments through-out the plains of the Indus Valley. Their territory rapidly grew to stretch a thousand miles from north to south along the river, and twice that distance along the ocean coast, from somewhere in the cliffs of Makran to the jungles of southern India. Perhaps all this land was Meluhha to the Sumerians. To us it has become known as the Indus or Harappan civilisation, since the first and so far most significant sites revealing its former existence were found in areas so named today. Excavations have shown a remarkably uniform and unchanging culture in Harappa and Mohenjo-Daro, from the time these simultaneous cities were founded until they died out a millennium later. Indeed, instead of progress there was a clear deterioration in construction details which some scholars have suspected as due to rapid reconstruction following devastating floods.

The mystery of the Indus Valley civilisation is not so much why it disappeared as how it began. As in Egypt and Sumer, this, the third of the large and contemporary early civilisations, lacks clear local roots even at Amri and Kot Diji, and simply seems to have chosen the fertile banks of a major navigable river as new home for a branch of an old powerful dynasty. Civilisation in Mesopotamia, too, began with settlers who knew how to cast bronze from imported copper and tin, and who buried four-wheeled carts with a troika of harnessed bullocks in royal tombs full of other evidence of a level of civilisation which neither Babylonians nor Assyrians later surpassed.

In searching for the home of the inventors who first gave us a script and the wheel, we find that the Indus civilisation was not only strikingly similar to those of Mesopotamia and Egypt, but also suspiciously contemporary. The difference of a century or so in such important human mile-posts is negligible when today we have to revise former thinking and measure human accomplishments on a scale of over two million years. With the total collapse of accepted dogmas concerning man's antiquity as a species, we have to move on tiptoe and not be blindly wedded to existing assumptions on the age and spread of civilisation. We know too little. And new discoveries are constantly changing a still unclear picture. What we

61–62. Inside the main cabin the author lists the night watches and pins them on a wooden board watched by Carlo. Birthday party with Yuri's caviare, celebrated around the deck table lit by the cameraman's battery lamps. *(opposite)*

63. The sun rose and the sun set as months passed by, and *Tigris* was still floating high. *(pp. 294–5)*

64. The international crew of *Tigris*. Left-hand column: navigator Norman Baker, USA, doctor Yuri Senkevitch, the Soviet Union, and alpinist Carlo Mauri, Italy, all veterans from the two *Ra* voyages. Above the author at the tiller is student Hans Peter Böhn (alias HP), Norway, and below, student Asbjörn Damhus, Denmark. Above *Tigris* is scuba-diver Toru Suzuki, Japan, and below, student Rashad Nazir Salim, Iraq. Right-hand column: sea-captain Detlef Soitzek, West Germany, industrialist Gherman Carrasco, Mexico, and cameraman Norris Brock, USA. *(pp. 296–7)*

know for sure is that the founders of the Egyptian and Sumerian dynasties began to leave their traces on their river banks – and to illustrate the large reed-ships of their ancestors – just when the Indus Valley was settled by a third civilisation. All three established themselves in the three major river valleys centring upon the Arabian peninsula.

The Indus Valley excavations, like those of Mesopotamia and Egypt, have far from exhausted the surprises in store for those who still seem to think that history began with Columbus or maybe the early Greek and Romans. Although attention was naturally first turned inland, where Harappa and Mohenjo-Daro were first found, it has gradually become clear that the people of the Indus civilisation had a predilection for waterfronts. The American geologist L. R. Raikes[2] reconstructed the sea level and major river courses at the time of the ancient Indus culture and found that practically all the sites of this civilisation would have been on the sea or on a river. Scholars have recently taken an increased interest in the seaboard sites, since the importance of shipping and maritime trade to the earliest Indus people has become evident. A pioneer in this respect has been S. R. Rao,[3] a noted Indian archaeologist who has uncovered a large number of the Indus Valley settlements, and gradually moved to the ocean coast. He says:

> Until recently it was generally believed that the Indus civilisation was land-locked . . . Recent explorations have, however, brought to light several Harappan ports giving a coastal aspect to the Indus civilization and suggesting a brisk sea-borne trade between the Indus people and the Sumerians in the late third and early second millenniums BC . . . Thus, the entire coastline of

65. No lack of sea-food, as marine life was attracted to the silent and broad-bottomed reed-ship. The ink of a squid tested by Detlef; Yuri dries fish and savours a piece of dolphin; Asbjörn with basket-full of rainbow runners; flying-fish were picked up on deck; turtles were caught at sea with remora-fishes attached to the breast plate; the red grouper was a surprise gift from a passing trawler. *(p. 298)*
66. The sail gave us speed and steering possibilities. The crew manoeuvres the sail above the forward cabin and the deck table. *(p. 299)*
67. Rowing in the ocean adds to the sailing speed and keeps the sailors fit, in this case Detlef and the author. *(p. 299)*
68. A dangerous calm trapped *Tigris* in a current bound for a forbidden coast. *(opposite)*

Kutch, Kathiawar, and South Gujarat, covering a distance of 1,400 kilometers, was studded with Harappan ports in the second millennium BC. Some were already established as early as the third millennium. . . . no inland station of the Harappa culture is as early in date as Lothal. . . . The largest structure of baked bricks ever constructed by the Harappans is the one laid bare at Lothal on the eastern margin of the township to serve as a dock for berthing ships and handling cargo.

This prehistoric port with its large brick-built dock is perhaps the most thought-compelling discovery with bearing on the Indus civilisation. Built about 2300 BC, it consists of an enormous excavated basin enclosed by thick embankment walls of baked bricks. The well-preserved remains of these baked brick walls still stand 10 feet high and are reinforced behind by a mud-brick wharf 43 to 66 feet wide. The basin is about 709 feet long and about 122 feet wide, and was designed to take ships about 59 to 65 feet long and 13 to nearly 20 feet in width, which is in excess of the size of *Tigris*. Rao states that two ships could pass simultaneously through the forty-foot wide inlet gap in the embankment, and he concludes:

The high degree of engineering skill achieved by the Lothal folk can be understood from the ingenious way in which they could regulate the flow of water into the dock at high and low tides. They could ensure flotation of ships in the basin by sliding a door in the vertical grooves of the flanking walls of the spillway at low water. Excess water was allowed to escape by keeping the spillway open at high water. In no other port of the Bronze Age, early or late, has an artificial dock with water-locking arrangements been found. In fact, in India itself, hydraulic engineering made no further progress in post-Harappan times.

What did the Indus empire have at Lothal? Rao shows that this was the warehouse of a rich rice, cotton and wheat-growing hinterland. Besides, the port had its own bead factories and was also an important centre for ivory-working. Carnelian was imported from inland, but finds of tusks and elephant bones indicate that elephants were reared on the spot. The principal exports of Lothal, says Rao, were ivory, beads of carnelian and steatite, shell inlays and, perhaps, cotton and cotton goods. With Lothal as a major port, the Indus civilisation qualifies well as Meluhha from where the carnelian beads, ivory and steatite reached ancient Mesopotamia, directly or indirectly by way of Bahrain. Rao also uncovered a terracotta head of 'a bearded man with Sumerian features' which to him indicated contacts to the west, and in a merchant's house in Lothal's bazaar street he excavated 'eight gold pendants similar to those found in the Royal Cemetery at Ur'. All this, he says, adds weight to the eighteen seals of Indian origin already found at Ur, besides those found at Susa, Kish, Asmar, Hama, Lagash and Tepe Gawra. At Lothal he even found a circular steatite seal neither wholly Indian nor Sumerian in workmanship, but almost identical with those excavated by the Danes from the Dilmun level on Bahrain. Dealing with this 'Persian Gulf' seal in a special report,[4] Rao concludes:

Indus seals and other knick-knacks could not have travelled to Ur, Kish, Lagash, Tell Asmar, Brak, Diyala and further beyond but for a flourishing trade in which merchants from India and the Persian Gulf took an active part. The use of special types of seals in different regions suggests the existence of merchant middlemen who maintained accounts, documented contracts and despatched sealed packages of goods. Their main centres were southern Mesopotamia, the Persian Gulf and the west coast of India extending into the Indus valley on the one hand and as far as the Gulf of Cambay on the other.

A few model boats were also excavated at Lothal, crudely made from terracotta, some flat-bottomed and others with keel, all possibly votive or good-luck objects such as the small terracotta, asphalt or silver boats common in Mesopotamian excavations. Commonly crude and symbolic, there is nothing in these small pottery boats to indicate dimensions or details of a true sea-going merchant ship. Since the real ships were not of clay but of perishable, buoyant material, we can only judge from the magnificent port facilities and the finds of a seal of distant island type that merchant mariners of Lothal had proper, long-range watercraft.

Since we ourselves had come to the former territory of the Indus people in a berdi ship of Sumerian type and had found bundle-boats surviving *en route* both in Dilmun and Makan, we were probably more curious than anyone who had come here before as to what kind of sea-going ship the Indus people had possessed. As if intending to help posterity with a hint, a citizen of Mohenjo-Daro had left behind another seal, another tiny chip of durable stone, which happened to be just what we most needed. Tiny as it was, it filled a huge gap in the puzzle we were trying to disentangle with the *Tigris* experiment. I had seen objects excavated from the Indus civilisation in major museums in Asia, Europe and America, but it was in a small field exhibit at the foot of the Mohenjo-Daro ruins that I was to get the greatest surprise.

'Here is a ship!' said Toru as he pulled me by the sleeve while I was admiring a stone image with shell-inlaid eyes, just as in Mesopotamia. He dragged me over to the next showcase, where we had to stoop down and press our noses against the glass to see what was there: a tiny yellowish steatite seal, a rectangular piece of soapstone no bigger than my own thumb. The label merely said: 'Seal with house boat'. And there was the boat indeed, engraved on the flat surface. The little stone had been broken in two and carefully mended by its finder. The symbol chosen by the original owner was that of a ship, a sea-going ma-gur of typical sickle shape, designed to ride the ocean swells. Cross-lashings characteristic of a reed-craft were clearly shown, with double rudder-oars astern and a deck cabin between two masts. Both seemed to be straddling bipod masts, and at least one clearly showed the rungs of the mast ladder of the kind copied by us on *Ra* and *Tigris* from Egyptian designs.

Rarely can the artless scratches on a tiny piece of stone have impressed seafarers more than this relic from an ancient mariner or merchant, left for thousands of years in the refuse of Mohenjo-Daro. The former owner could not have fancied that every little furrow he had cut out into his little piece of steatite would be studied with such care and cause so much excitement. But we were not the first to see the implications of this find. In his report on the actual discovery, Ernest J. H. Mackay wrote:

The vessel portrayed on this seal is boldly but roughly cut, apparently with a triangular burin, and is apparently not the work of an experienced seal-cutter; hence its interest, because, probably in consequence of inexperience, the motif is not a stereotyped one. The boat has a sharply upturned prow and stern, a feature which is present in nearly all archaic representations of boats; for example, the same type of boat appears on Early Minoan seals, on the Predynastic pottery of Egypt, and on the cylinder seals of Sumer. In the last mentioned country this type of boat was used down to Assyrian times.[5]

Mackay realised that the parallel hatchings indicate lashings of a boat 'made of reeds like the primitive boats of Egypt and the craft that were used in the swamps of southern Babylonia'. He stated that this seal indicates a type of vessel that was in use in ancient Mohenjo-Daro, and proposed that its owner was perhaps connected with shipping of some kind, for so much attention had been paid to detail.

Not much attention was paid to Mackay's discovery, and he found only one other and much cruder ship scratched on a pottery sherd, which at least showed a strongly arched hull, double-mast with a pair of yards, and a rudder-oar with tiller in the stern. But scholars who have subsequently referred to the discovery agree in their verdict: 'The boat from the seal is without doubt a papyrus boat',[6] and 'This is a splendid seal . . . there is an echo of prehistoric Egypt. . . .'[7]

Did the ancient Indus people actually have access to papyrus? Papyrus was an African plant. The early ships of Egypt were indeed made from papyrus. At that time papyrus grew in abundance all along the Nile and in wide areas of adjacent Africa. It spread to the Mediterranean side of Asia Minor, where rare patches of papyrus plants have survived into modern times, where reed-ships are incised on walls of ancient caves in Israel and on Hittite seals at Gaziantep, and where the Prophet Isaiah (18.ii) speaks of reed-ships arriving with messengers from Egypt. Papyrus must have been known in Corfu at one time, where fishermen made bulrush-boats until recently and called them *papyrella*. It has survived since antiquity on the island of Sicily, while reed-boats (but no papyrus) have survived on nearby Sardinia. It was brought by early seafarers beyond Gibraltar and planted on the Canary Islands, where Romans to their surprise found papyrus growing in the rivers.[8] Who brought root-stocks of this difficult fresh-water plant to those far Atlantic islands is not known, but the Phoenicians were there. They left reed-ship designs on one of their most beautiful vases recently found on the ocean floor off their former Atlantic port of Cadiz in Spain, but' no papyrus. At the former Phoenician port of Lixus, on the Atlantic coast of Morocco, the Berbers built reed-boats until recently, but of an inferior river-grass through local lack of papyrus. Papyrus on the Canary Islands has long since disappeared, and when the medieval Portuguese rediscovered the islands a couple of millennia after the Phoenicians, the blond guanche islanders did not know how to build watercraft of any material whatever, though they made plank coffins for their mummies and practised cranial trepanation and other arts that clearly revealed their former contact with sailors from the far corner of the Mediterranean. And together with fragments of tripod vases of Phoenician origin, archaeologists have found terra-cotta seals indistinguishable in type and decor from specific Mesopotamian seals; they are on exhibition in the Gran Canaria Museum, together with a selection of typical Mexican terra-cotta seals to show their striking similarity.

Clearly papyrus was a most useful river plant to ancient deep-sea navigators, and attempts to plant the tubers might have been made with varying success in areas where there is no papyrus today. The marshes surrounding the oldest Mexican pyramid, at La Venta on the Mexican Gulf, are covered by vast stretches of a plant with a flower and stem that are to an amateur's eye indistinguishable in every respect from papyrus. Botanists are now inbreeding the plant, suspecting that it might be a descendant of papyrus, *Cyperus papyrus*. This important question was recently brought up at an ethnobotany seminar by Professor Donald B. Lawrence of the Department of Botany at the University of Minnesota, who wrote:

'Rediscovery' of Giant sedge *(Cyperus giganteus)* on the Gulf coast of Tabasco State, Mexico, at the famous Olmec culture site, home of 'America's first civilization', and recognition of its similarity to Papyrus, have suggested to this speaker its possible derivation from that Old World plant, its possible introduction to the New World by man in ancient times (3000 or more years ago), and its possible importance for the establishment and development of the Olmec Culture.[9]

There seems to be no record or trace of papyrus growths in Mesopotamia. The first reed-built ma-gur must therefore have been lashed together from berdi, as we had done. And as we were still afloat in Pakistan, berdi had proved suitable for building sea-going craft.

It was not very far from the desert landscape around Mohenjo-Daro to the green belt where the Indus winds its long journey to the mangrove swamps and the sea. From the little museum by the ruins we wasted no time in getting to the river banks, which in former times had run past the walls of the city.

'Reeds! Hey, it's berdi!' Norman was first into the marshes to pull up a plant. Indeed, it was berdi! The typical oval cross-section composed of the crescent-shaped stems of adjoining branches with spongy fill and large-celled, waxy walls was not to be mistaken.

This was the plant from which we had built *Tigris*. Whether it was brought by man or by nature was merely a botanical problem. There were vast stretches of it, and wherever we were in the Indus plains, as soon as there was a vacant area with moisture, we found berdi growing wild. This then, and not papyrus, had been the reed used by the Indus people for building bundle-ships of the same type as those in Mesopotamia and Egypt. Merchants travelling between Ur and the Indus in prehistoric times had ample opportunity to repair or even renew their vessels if their visits should last longer than the reeds would float.

On our way back to *Tigris* in Karachi harbour we took our time and found ourselves once again in a world where folk ways 5000 years apart live side by side and for us blended into a wealth of profound impressions. While the jet-age is well established in cities like Karachi, with everything money can buy in the form of press-button aids and computers, side-roads show the traveller all stages in the few long leaps from the Indus Valley bronze age to the present. Clearly the Aryan invaders in their early days did not strangle or eliminate the Indus civilisation before it had left lasting benefits throughout the territory it dominated. We saw tribes from the highlands of Baluchistan, the land we had first seen against the sky when approaching Makran from the sea, who had come to settle near the river. Their carts, drawn by pairs of oxen sharing a common pole, conformed in the smallest details with the pottery models from Mohenjo-Daro. Since these hill-people had never seen the museum, their carts must have followed a tradition which the Aryans had never disturbed. Some of the perishable wooden wheels were so beautifully carved with relief ornamentation that they could have been used for chandeliers in any modern home. Their houses, as in Ormara, were reminiscent of those of the Iraqi Marsh Arabs who, too, had simply maintained an ancient house-form well known from illustrations on the Sumerian seals. Within these homes we found women sitting on the ground grinding flour with stone hand-mills, as shown in the pottery models of Mohenjo-Daro and Ur. Ceramic bowls had given way to purchased metal ware; otherwise Mohenjo-Daro was reflected in all they possessed. Their animals were those first domesticated by the founders of the Indus civilisation, and their cotton clothing was made from a plant first cultivated, and on a loom first invented, by them. Even the grain used for their bread was the species first grown by those pioneering benefactors of whom they had hardly heard. They had been satisfied with this inheritance and had added nothing for four millennia until now when they were about to absorb the twentieth-century way of life.

A man was sitting by his grass-house hand-pressing dung from his hoofed animals into small tidy cakes which he piled up to sun-dry for fuel. I asked why he and so many other men in his tribe painted their hair and their big bushy beards red. We were merely told that this was their custom.

There was berdi and khassab along all creeks and dikes, and we saw both plants harvested for the building of complete reed huts as well as for the thatching of adobe homes. Most houses, and all towns, were built of sun-baked adobe bricks. The prototype of all old inhabited towns was clearly Mohenjo-Daro, right down to the characteristic brick-covered drains.

The art of adobe is as old as civilisation in the Indus valley, and in Mesopotamia and Egypt too. Where not already introduced by these culture founders, it was spread by the Arabs later, for it is a simple and ingenious building process perfectly suitable to a rainless land where no other building material is found.

We stopped to see village people in the Indus Valley making their adobe blocks. They made them in the same large size and in the same wooden frames I had already seen used in Iraq and Mexico, and when the mixture of selected soil, straw and water began to dry they removed them from the frames and baked the blocks in the sun. Independent invention of such a simple procedure would not be at all surprising in desert countries like those along the North African coast from Egypt to Morocco, but it is surprising to find adobe blocks used by the founders of Mexican civilisation on the jungle coast across the Atlantic. Adobe was used by the Olmecs, the unidentified founders of American civilisation, when they built their sun-oriented temple-pyramid in the coastal swamps at La Venta, where timber and reeds abounded. And adobe blocks were also used by the pre-Incas of Peru when they built their pyramids along the coast in the form of sun-oriented ziggurats with temples on top. The Cerro Colorado pyramid on the north coast of Peru covers 4,000 square yards of ground and some six million adobe blocks had to be manufactured before the early architects could erect this colossal structure. Despite a complete contrast in climate and environment, the founders of pre-Columbian civilisation in the rain forests of Mexico and deserts of Peru were used to building houses in Old World style from adobe bricks.

When we saw donkeys and camels bringing Indus Valley cotton to market, I could not help thinking again of early Mexico and Peru.

As far as science has been able to ascertain, cotton cultivation that produced a useful species with spinnable lint was begun in the Indus Valley plain and spread to Egypt. Yet, extensive fields of cultivated cotton were found by the Spaniards when they reached Mexico and Peru. Until today it has remained a puzzle to scientists that the vertical-frame loom with two-warp beams, found by the Spaniards in use among the Incas, was identical with that which had been used in ancient Egypt and Mesopotamia. Even the most peculiar, ornamented spinning whirls of pottery, used for preparing the yarn, are sometimes indistinguishable and scholars have pointed out that the final fabrics were in certain cases made into clothing also described as identical on both sides of the Atlantic.

This is not all. Of recent years botanists have had something to add. A modern chromosome study has shown that there was something very special about the cotton cultivated by ancient peoples of Mexico and Peru; it was not at all the same species that grew wild in America and which in fact did not produce spinnable lint. The chromosomes of all Old World cottons were different from those of the wild American species, and the pre-Columbian cotton domesticators in Mexico and Peru had somehow obtained the Old World cotton, crossed it with the wild local cotton and obtained a perfectly spinnable product that had the chromosomes of both species combined in a hybrid with double chromosome number, the only species with both types of chromosomes combined. The botanists now leave the question open for the students of human culture to answer: how did the spinnable Old World cotton, the one with the chromosomes of the species first cultivated by the Indus people, get into the hands of the culture-founders of Mexico and Peru? If seeds had been carried by the wind or by birds, an American Indian must have recognised them for what they were and planted them in his fields before any plants had time to grow wild, and with the idea that if the lint grew longer than that of wild cotton he could invent the spinning wheel to make yarn, and the loom to work yarn into cloth.

There is something amusing about the desperate desire of so many historians and anthropologists to reserve the first possible crossings of the Atlantic to the Spaniards and the Vikings. That is, to the Europeans who only reached the Canary Islands two thousand years after the Phoenicians who got there from Asia Minor on successive voyages of exploration and colonisation. There is almost a touch of religious fanaticism in the attempts by the western world to see America as a European creation, completely protected by sea until the local barbarians were found by civilised Christian pioneers. We should try to be more open-minded. The art of navigation, literacy, even the symbol of the cross and the religion we carried to America, we had first obtained from Asia.

With our heads bursting with historic data and new impressions, with a photographic harvest of the most spectacular Arab forts any of us had ever seen, and with Sani tempting us untiringly to visit ever more mosques and Moslem marvels, we suddenly found ourselves squeezed barefoot and exhausted in amongst the packed crowds of another historic sanctuary. To the sound of drums and the smell of incense he elbowed us in for a glimpse of a sacred coffin; this was the thirteenth-century tomb at Shawan of Shabaz Qulanda, who had brought Islam to the area. A man with remarkably long arms was standing high up, collecting paper flowers from a crowd who could not get close enough to the coffin. My curiosity was aroused when he began to walk through the crowd as on stilts, and I forced myself into the vacuum behind him to see how he did it; but to my surprise his trousers reached almost to the floor. A strange joke in a tomb, I thought, a man riding on someone's shoulder, both covered up by a cloak. But then I saw the largest human feet I had seen in my life and the largest hands; indeed even the head was abnormally big, with huge eyes and lips. In the sombre atmosphere of the old tomb it was as if a legendary giant was slowly staggering about with us, and I followed him around three times before I took courage and invited him outside into the sunlight. When he offered me his hand it was like grabbing a ham. He stooped to get out through the temple gate, and when we gathered around him to get his measurement he stood a good 7 ft. 10 in. barefoot. We saw many remarkably tall people in the Indus Valley, but never anyone like him.

For such a flying visit I thought we had done quite well in Pakistan, until we got back to the National Museum in Karachi, when the Museum Director, Taswir Hamidi, asked me if we had been to Hassan Wahan. We had not even heard of the place.

Hassan Wahan was a village on a lake connected with the Indus river not far from Mohenjo-Daro, where people still made pottery of Mohenjo-Daro type and led a life very much as in that city. But there was also a large number of wooden ships on the lake, where the descendants of an ancient people lived. The fishermen who owned these boats adhered to a very peculiar custom: they lived on board with their families and all their possessions and never left their floating homes. The Museum Director had himself spoken to a man more than a hundred years old who had never been ashore.

I was itching to go inland again and visit Hassan Wahan, but *Tigris* was waiting for us in an incredibly polluted harbour; the ocean was our challenge, not a lake, and we had to get back on board, hoist sail and leave Asia. But the friendly Pakistani archaeologist had more to tell me that made me forget the surviving river dwellers. He fully agreed that the early Indus people had contact with Mesopotamia; and showed me objects in his museum that left us in no doubt. Among them he pointed out what he called a 'Gilgamesh' motif on an Indus Valley seal, where the familiar Mesopotamian hero stands bushy-haired between two erect lions. Gilgamesh! The legendary King of Uruk, who sailed to Dilmun to visit the land of his earliest progenitors. The fame of the Gilgamesh myth had spread beyond Bahrain and had reached the Indus.

To me there was no longer any doubt. I agreed with those scholars who identified the Indus region with Meluhha. Meluhha could be nothing else. Dilmun, Makan and Meluhha belonged together. With our ma-gur we had visited all three. And we were ready to embark on further adventures.

CHAPTER 10

From Asia to Africa; from Meluhha to Punt

DEPARTURE from Asia. A cold wind from the north blew through the fissures in the cane wall as I woke up at early dawn and checked that everybody had come on board during the night. This wind was good. The men were all there, back from the last shore leave probably for a long time. How long, nobody knew.

There was never a dead moment in Karachi harbour. Ships everywhere, and small boats moving between them all night. At dawn Pakistani fishermen passed us in small dhows and saluted with broad smiles. They were amazed at our ship. Soon the harbour authorities came on board with the necessary port clearance papers.

'Next port of call?'

'Unknown.'

'We have to know.'

'Then put down Bombay.'

A family of emigrants from the Shah's Iran, owners of the shipping agency Cowasjee & Sons, arrived with friends in a small yacht and threw us a line. They had volunteered to tow us out of port. As we started to move, the smell of rotten eggs came up through the two open slots in our bottom. The same thing had happened as we rowed out of Muscat. Modern harbours are incredibly polluted. We were afraid of the effect on ropes and reeds. Perhaps some of the outer berdi was fermenting. Fortunately we knew from our departure from Muscat that this horrible stench would disappear as soon as the reeds were washed clean by salt ocean waves.

We counted thirty-eight ships at anchor at the mouth of the harbour. *Jason* was one of them, and Captain Hansen was on the bridge. With his loudhailer he wished us *bon voyage*, and told us he had come back the day before from another mission to the swampy mangrove labyrinth of the Indus delta. They had gone to salvage a Greek ship grounded in the swamps, but pirates had already stripped the vessel completely before *Jason* arrived.

Clear of the anchorage we hoisted our sails. Our helpful companions recovered their tow-line and returned to port. The last words the bearded skipper of the yacht shouted back was a generous offer to purchase *Tigris* after use. If we cabled him whenever and wherever our voyage ended, he would send a ship to fetch the reed-ship for exhibition.

Thick smog covered the view of the big city beyond the anchorage. In the last few years Karachi had grown from a mere 700,000 inhabitants to well over five million. There was a sheen of surface oil far out to sea. Two porpoises were rolling up through the slick. A little mouse came peeping up between the reed bundles and then ran into hiding. Perhaps it was the one we had carried with us since we left the Garden of Eden.

We could still see the masts of the anchored ships, like a sunken forest along the horizon behind us, when the north wind died completely. Steering problems ensued. The wind came back in faint gusts, but from the south. The bow turned back towards Karachi. We had to row to get about. By now we had at least learnt enough to hold our own against the wind. The day passed and the distant forest of masts off Karachi was still there, though displaced far towards the west. A tidal current dragged us eastwards in the direction of the Indus delta. As night came we saw all the lights from the anchored ships and the city-glow behind. We threw out the sea-anchor for whatever it was worth.

We had noticed that both bow and stern had sagged slightly during our stay in Pakistan. Carlo took charge of the job of tightening all mast stays, and the ropes holding the cabins to deck, and the flexible reed-ship straightway resumed perfect lines. All the men were in the best of spirits, although most of us had indeed been in better shape physically. Our two out-board seats were constantly occupied. Asbjörn and Norris were the worst affected. I got sudden kidney pains, and Norman had a new attack of his mysterious fever, causing Yuri to suspect malaria, although Norman had been the one who had been taking pills regularly. Norman had such self-control during his repeated spells of fever that nobody knew for sure, until it was confirmed by a laboratory analysis after the voyage, that Norman had actually had malaria. The atmosphere was still markedly one of joy and laughter, the visits to Ormara and Mohenjo-Daro had been great stimuli to us all, while unknown adventures lured ahead. We were all equally eager to get into the ocean and away from land, to which we gave the blame for our temporary troubles. I would never have recalled that a few mosquitos followed us into the ocean, but for the one that remained like a pressed flower in the fourth book of my diary, on the page recording our departure from Asia.

During the night a good sailing wind from the north came back. All lights from ships and land sank behind us. Huge swells indicated interference from a strong current, and there was soon a smell of greenery over the ocean, as from a jungle. The insect-ridden wilderness of the Indus delta was beyond the horizon, but this problematic area, which kept the salvage vessel *Jason* busy, was eventually left behind on our port side. We tried to steer away from it, but the delta was wide enough to be with us for a couple of days at least. So far we simply steered to get clear of all coasts and into the open ocean.

We did not trust the coming of the north-east monsoon. We knew it had failed to blow for the second year in succession. The harbour authorities in Karachi had also told us that in these parts the worst period for storms and treacherous weather was the last days of January and the first half of February. We were to find that they were right. We had left anchorage on 7 February.

The second day at sea the northerly wind increased in strength and, lifted on our way by hissing waves, we sailed away from eastern Pakistan and the north-western coast of India. Next day we lowered the topsail as the wind strength rose to half gale. The mainsail split, and we lashed on two rowing-oars beside the rudder-oars to help steady our course. Sharks began to join us.

On 10 February we had a lull before the storm. The wind blew faintly from the east. We mended the mainsail and began to discuss where to steer next. With this wind the risk of being cast on to the coast of India was greatly reduced. We all wanted to send comforting messages to our families, but Norman was unable to contact anyone with our consortium transceiver, though with his own amateur set he picked up one radio ham in Germany and another in the USA. To our surprise, both reported having heard that we had decided to sail from Karachi eastwards and finally across the Pacific to South America. Crazy! Ridiculous! This was just about the only thing we could not do. We could perhaps reach the Far East, but never cross the Pacific in the opposite direction of the *Kon-Tiki* voyage. All those who had tried had failed, but all those who had followed in *Kon-Tiki*'s wake had succeeded, and even reached Australia.[1]

I had always tried to make the point that, though certain primitive craft were seaworthy, not even a reed-ship could succeed in doing what the early Spanish caravels found impossible: force an old-fashioned sailing craft eastwards along the equator to tropical America. The Pacific Ocean fills half the surface of our planet, and in this unsheltered hemisphere ocean currents and trade-winds are rigorously propelled by the rotation of the earth. In the entire tropical belt, sea and air, set in non-stop movement from Peru and Mexico to Indonesia and the China Sea, are too strong to permit aboriginal mariners to reach America across the Pacific, except in sub-Arctic latitudes, or to enter the mid-Pacific, except from the American side. For eastbound voyagers from the Indus Valley, China would be the end of the line. Any effort by us to sail our primitive reed craft from Asia through the mid-Pacific island area would have failed, just as it would have failed Chinese junks and as it did fail all Spanish and Portuguese caravels and later replicas of prehistoric junks that have tried in modern times to sail due east from China.

For a while the false reports about our intended itinerary infuriated everyone on board as we struggled to sail south with an unsteady easterly wind. It was not the first time news media ashore had fabricated reports about our voyage. The reason was perhaps that the London-based television consortium with exclusive rights to our stories had scant success in distributing the reports we tried to send them with their transceiver, and yet had stopped Norman from telling anything to the hams who heard him. When we had happily waded ashore in Ormara bay, the news media reported our shipwreck. German papers had carried the horrible news that the Japanese member of *Tigris* crew had been eaten by a shark and that the expedition leader had therefore been forced to end the experiment.

To clear up all the confusion ashore and to soothe the growing temper of eleven angry men of the expedition, the consortium sent us a peace messenger. The amiable Peter Clark, whom nobody could blame for the confusion, flew from London to Karachi to straighten out the misunderstandings. When last seen he had barely had time to climb on board and say hello as we passed through the Iraqi river-port of Basra; now he landed just as we were ready to depart from Karachi. He had been commissioned to reassure us that from now on correct news about our voyage would be distributed.

He marvelled at the fact that *Tigris* was just as perfect and undamaged as the day he had seen us pass on the river, and he waved us off to fly straight back to London with the comforting news of what he had seen. Three days later, on the day we got the crazy news about our trying to reverse *Kon-Tiki*'s voyage, we received a panicky question over Bahrain radio from another office of our same London consortium: 'Why had *Tigris* broken its back?'

We all asked Norman to shut down his radio. We had to take it as a joke. If our shore-based contacts could not communicate across a corridor, how could we help across the oceans?

That day *Tigris* had been afloat for three months. The cigar-shaped twin bundles curved as much below as above water and still floated higher than *Ra II* did after three weeks. None of us now had any fear of *Tigris* losing buoyancy quickly. None of the crew had wanted to give up in Pakistan, although they could have had air tickets home, and although we were to leave Asia with an unknown destination. Everyone knew that the risk of running into a storm was overwhelming, and there was a chance of a cyclone. Nobody feared that the bundle-body might be ripped apart by wind or waves. It would take something like a steel hull or coastal cliff to destroy the springy, yielding bundles.

The air had turned misty. There were rain showers during the day and new thick clouds were building up around us. It was warm and oppressive. That day I wrote in my diary:

Watch set back again one hour at noon. As I lie on my mattress I can see the dancing ocean on both sides of the ship simultaneously through the two door-openings; when it rolls like these days I see the water rise up above the door and above the cabin roof on one side while it disappears under the deck on the other, and vice versa. I cannot remember that we had so much rolling on the two *Ra*'s, but then we always sailed along with the elements and not across them; now we steer for given destinations or to avoid dangerous coasts or shipping lanes. *Where do we sail now?* Nobody knows. Madagascar, the Red Sea, or any African coast in between seem most likely and most tempting.

Around the deck table we began to discuss our future course. Norman had just finished reading a book about a yachtsman who spared no phrases when it came to describe the horrors of the Red Sea, and he strongly advocated that we set course for the Seychelles, described as a modern island paradise. Yuri, on the other hand, had no greater desire than to sail straight for Kenya; his dream was to visit the great animal reserves. Carlo favoured steering for the Red Sea; the waters of the ancient Egyptians ought to be linked with those of the Indus Valley and Mesopotamia if these people had really known about each other in early times. We had all seen that Baghdad Museum had a whole room full of beautiful Egyptian ivory work, excavated from archaeological sites in Iraq. Gherman had also impressed everybody with his descriptions of the fabulous marine life he had seen in the Red Sea. Otherwise nobody voiced any specific opinion; the first-time reed-ship sailors just seemed to enjoy life on board and were happy merely steering into the infinite blue. It was for me to make the decision.

The expedition had taken a somewhat different pattern from what I had anticipated. At one time I had seen the main purpose of the experiment was to test the buoyancy of a Sumerian ma-gur built from berdi cut at the correct time. I had visualised sailing with no predetermined goal as long as the berdi kept afloat. If it did not sink we could cross the Indian Ocean with the monsoon at our back, and if it still kept afloat we might even sail down the coast of Africa and perhaps again cross the Atlantic to tropic America. That last leg from South Africa would have been the easiest of all, for we would have had the winds and currents with us all the way, just as on the drifts with the two *Ra*. But instead we had followed the trade routes of Sumerian merchants. We had spent so much time visiting prehistoric remains in Bahrain, Oman and Pakistan that my finances had started to run low. Besides, the buoyancy test of *Tigris* had more than stood up to the time requirement for any straight, long-distance voyage. The ability to navigate to given destinations had become more of a challenge than conducting long-distance runs that had been demonstrated possible by previous expeditions. We had so far voyaged between the legendary Dilmun, Makan and Meluhha of the Sumerian merchant mariners. Across the Indian Ocean lay the Horn of Africa, Somalia, considered by all scientists to be the legendary Punt of the Egyptian voyagers. If we could reach that coast also, then we would have closed the ring. Then we would have tied all the three great civilisations of the Old World together with the very kind of ship all three had in common. We had linked Africa with the New World before, with the same kind of ship, and the New World had in turn been shown to have access to the mid-Pacific, where Easter Island was the nearest speck of land, with its stone statues and vestiges of an undeciphered script which some scientists claim has a strong resemblance to the Indus Valley writing.

317

Our discussions around the deck table were interrupted as distant lightning flashes suddenly leapt across the sky, splitting the black clouds above the mast with a violent clatter. The wind had turned from east to south and increased in strength. We were outside all traffic lanes and had seen only a single sailing dhow from India that day. Those of us who were not on steering turn preferred to crawl into our cabins and roll down the cotton canvas. The heavy clouds soon began to spill like torn goat-skin bags ripped open by the flashes, and good drinking water again cascaded down upon the *Tigris* as we were swallowed up in complete darkness. It was indeed good to know we were alone in these waves. Inside the cabin we could hear every word shouted between the two men on steering watch. I could in fact thrust my finger between the canes at any place and feel the thin cotton canvas that had been pulled down, yet the roar of ocean and wind so deafened the men on the bridge that they could not hear a word of what we shouted from inside.

Thus began the night described in the first pages of this book, when I asked myself if I would have embarked on this adventure had I known that such a moment awaited us. This was the beginning of a roaring gale that swept the Indian Ocean and reached the eastern part of the gulf with fierce sandstorms in Dubai and Abu Dabi. Our little boat was tossed about like a toy vessel between and over mountainous waves, when those of us who still slept were awakened by the familiar call: 'All hands on deck!' It was Norman's voice with an undertone of despair cutting through all the turmoil outside.

In the faint light of swinging kerosene lamps, colliding wave crests rose around us like small volcanoes smoking with foam and spray. When the thick topmast broke and half the mainsail dragged like a water-filled parachute in the sea, heavy as a small whale, we all feared the worst that might have happened: that we could have lost the rigging and been left adrift on a heavy raft-ship with only the rowing-oars to propel us. The straddle-masts could have torn loose from their lashings to the bundles, but we knew by now that the twin-bundle ship was too broad and sturdy to turn bottom up. Indeed, either half was too heavy to be lifted into the air and too buoyant to be forced down under water; capsizing was no threat. Unable to hear an order from stern to midship, each man filled his place where most needed, and all hands together managed to empty the sail and pull it on board. Neptune got nothing from our deck.

All night the storm raged. The wind howled and whistled in the empty rigging and the canvas-covered cabin. Like Noah we waited inside cover for the weather to abate. The rudder-oars were abandoned, lashed on. There was nothing to do but wait. It was an incredible comfort to all of us to know we were on a compact bundle-boat and not inside a fragile plank hull. No worry about the vessel springing a leak; no need for bailing. But without sail our elegantly raised tail was of no avail, and the ungovernable vessel just turned side on to the weather, with breaking seas tumbling on board by the tens of tons, up on the benches, everywhere. But next moment all the frothing water whirling around the cabins was gone, dropping straight down through the sieve-like bottom. And the bundle-boat rose from the sea like a surfacing submarine, glitteringly wet in the lamp-light, and sparkling intensely with the phosphorescent plankton trapped on board. No wonder that this simple kind of self-bailing craft was the first to give primitive boat-builders the security to venture upon the waves, and the one that paved the way for further progress in more sophisticated and demanding maritime architecture.

By morning the storm abated. The wind turned to ESE. Sporadic rain-squalls continued, and the warm easterly wind again brought a distinct smell of jungle. A warm, spicy, botanical aroma of a tropical rain forest was wafted into our nostrils, from India far beyond the horizon. What was more amazing, the wind had brought a rain of insects across the sea. Several beetles, ants, a few big moths, a dragon-fly and some fair-sized spiders came down from the sky to join our old company of grasshoppers who for some time had been silenced by the storm. Seven sharks had taken to our sides in the rough weather, and showed no sign of leaving. The side-on rolling to the sea without sail was terrific after the storm. I timed it; two and a half seconds between each violent jerk, when it felt as if we dropped brutally down into a deep trench. The step-ladder in the empty mast looked like a naked skeleton against the grey morning sky. It was hard to keep balance around the breakfast table. We praised the presence of the mouse as the food that spilled during meals was not easily swept up from between the thousand berdi stalks beneath us. The stowaway crabs, too, were growing in size and number and helped keep the ship clean, operating like tooth-picks between the reeds.

Clearly we had not yet had time to rid ourselves completely of all continental bacteria. Soaked by seas and chilled by rain and storm, four men caught colds and were confined to quarters by Yuri, who said that they all ran temperatures.

We hoisted the mainsail to the foot of the broken topmast and succeeded in maintaining a fairly even southerly course, although the north-east monsoon never showed up. The wind varied between ESE and SW, and at times we had to fall off and turn all about to fill the sail from the opposite quarter and then continue our interrupted course. We were masters of our progress. We could have done better with more oar-blades or lee-boards in the water, but we were indeed no drift voyagers. We held our own in contrary wind, and even managed to force *Tigris* some few degrees against the wind direction. The best we managed, according to Norman's observations, was to advance 80° into the wind. We tested broad tables and rowing-oars lashed to the side bundles as lee-boards, and we tested similar boards lowered as centre-boards or *guara* in the open slots between the twin bundles fore and aft. Any such addition to the under-water keel-effect already provided by the large rudder-oars and by the longitudinal groove between the twin bundles helped reduce leeway. An interplay of guara fore and aft had an immediate effect on the course, but although of paramount importance for manoeuvring South American balsa rafts, and probably even South American reed-boats, the big rudder-oars sufficed for guiding our vessel, and the boards gave the same keel effect whether lowered between or outside the bundles.

Norman and HP spent much of the four days after the storm lashed to the top of the waving bipod mast. It was no easy job to hang on up there and chisel out the broken base of the topmast which had been mortised deep into the wooden block that joined the two legs of the straddle-mast. The hardwood topmast, thick as a man's leg, had broken crosswise and splintered lengthwise, so no piece could be re-used. On Carlo's suggestion Asbjörn skilfully adzed one of the big ash rowing-oars into shape, and it was wedged in as a new topmast, allowing us to gain speed again on our way southwards in constant struggle with changing winds.

The men recovered quickly from their colds, and the spirits of us all seemed higher than ever. Nobody would leave this vessel until it sank under our feet.

The second day after the storm the sky was blue and the sea calm. As we rose from the lunch table Norris, a head higher than the rest of us, exclaimed in surprise: 'Look at that! What can it be?' Ahead of us on starboard side was something like blood topping the small waves along the blue horizon. We climbed the mast. A narrow belt ran like a painted river through the ocean as far as the eye could see.

Asbjörn and HP inflated the rubber dinghy and went ahead of us to check what we were about to run into with our reeds.

The two scouts came back and reported that the coloured liquid was as thick as paint but that it did not seem to be a chemical product. We sailed alongside for a while and then ventured to steer across it. There was a strange fishy odour that could be smelled even within the cabin. The belt was rarely more than a fathom or two wide, but it stretched from horizon to horizon in a straight line from NNW to SSE. We criss-crossed through the orange-red belt, unable to determine what it really was. I had seen a similar 'red tide' of algae caused by pollution off the coast near Rome. But there was no coast here. How could this distinctive belt keep its narrow path like a red carpet laid out for untold miles across the open ocean? We sailed along it all afternoon and saw neither beginning nor end. There were occasional small patches of red on either side of the band, but the general picture was a clearly defined gold-red river across a blue sea. A sample glass could fool anybody into accepting it as thick orange juice but for the fishy smell which gave it away. A closer inspection revealed that the coloured belt, and the sea on either side and as deep as we could see, was polluted by billions of tiny, almost invisible, fibres and tufts as from dissolving cotton. There were asphalt-like oil clots, too, but they did not appear to be more concentrated than what we had commonly seen, and they were tiny, not the large tar-balls we had fished up all across the Atlantic. In the red soup were large quantities of short slimy bands of greenish-grey fish eggs, a few bird feathers, and many small dead coelenterates, just as we had seen in thickly oil-polluted areas traversed with *Ra*.

Somehow this red band did not seem the work of nature alone. The sea off northern Peru turns red at intervals of years, a frightening phenomenon called 'the Painter', caused by the sudden death of immense quantities of fish and birds which discolours the water. The marine disaster of the Painter occurs when the equatorial water of the Niño Current from Panama runs abnormally far south and enters an area with a biotype adapted to the cold Antarctic water of the Humboldt Current. But there was no difference in temperature on the two sides of the red belt we crossed in the Indian Ocean.

We turned away from the painted path in the evening and saw it no more. The last thing we observed in that area was a huge hammerhead shark, twice the size of any of us on board, that came chasing towards *Tigris* at a ferocious speed, then rushed like a rocket to the buoy we towed behind, zigzagged with undiminished speed around us once more, and then shot off, fin above water, along the coloured belt.

Thunderclouds and lightning continued to circle around the horizon for a couple of days, particularly in the direction of India, but as we entered the second half of February all storm clouds disappeared, just as predicted by the harbour authorities in Pakistan. Yet the monsoon did not show up. Its continued absence might have disastrous effects on the rainy seasons in Asia and Africa. We sailed into a blue world, all sky and water, a planet where all land was hidden deep below the sea. No ships, no planes. For days and weeks we were the only human beings, but not the only life.

We did not see a single fishing-boat in this ocean. There was supposed to be nothing to catch in the local waters, which were sometimes described by oceanographers as a marine desert. Perhaps this supposition was correct. But if this was a desert, then our reed-ship was indeed a floating oasis. Whatever roamed about in the surrounding emptiness must have seen our shadow as we drifted silently like a cloud over the sunlit blue sky. Day by day the swimming company around us increased. One marine species after another turned up, multiplied, and the lifeless sea became reanimated until in the end we felt as if floating in a packed fish hatchery.

To the swimming creatures joining us on our southbound course, and to ourselves whenever we dived overboard and looked up, *Tigris* was a small floating island, a green meadow turned upside down. Broad and jovial, with no propeller, with a cosy valley where other ships have a sharp keel, we waddled along, all vegetable matter, waving with spring-green sea-weed and with a speed appealing to the drowsiest creatures of the sea. Tiny crabs had no difficulty clinging to the reeds, and felt at home between the long sea-grass, like fleas in the beard of Neptune. No grass was greener, softer and more pleasing to the eye than the uncut marine lawn growing upside down on *Tigris,* with goose barnacles growing like mushrooms, and mini-crabs and minute fish crawling about like beetles and grasshoppers in a meadow. Why should not all kinds of homeless vagrants be allured to the only garden floating in the blue?

We never tired of hanging with our heads over the side bundles or trailing on a rope outside, watching the changing world on the yellow reeds. By now the white goose-barnacles, small when we left Oman, were as big as flat pigeon eggs, each standing on a long black leg and rhythmically unfolding feathery crowns like yellow wings waving to bring food and oxygen into the broken egg. Our speed made them a permanent spectacle, a large procession of waving banners, surrounded by tiny, legless common barnacles, hidden like mini-tents in the greenery. Between them crabs the size of fingernails patrolled mechanically like armoured tanks, but quick to take cover between the reeds if they saw our giant heads coming down from heaven. In a few places the sun glittered as if upon gold wire wound up into balls, which proved to be marine worms that never unwound and performed no other visible movement than to grow slowly in size. But nothing was stranger than the sea-hares. We had never had them travelling with us on previous raft-ships; now two types were crawling slowly about, feeding on our sea-grass. The smallest was as yellow as a banana and as big as a thumb; the other was somewhat larger, fatter, and of a greenish-brown colour. Except for two long protuberances on the head, reminding one of a hare's long ears, these boneless gastropods are far too clumsy, slow and ugly to merit their name. To us they were rather mini-hippopotami. The dull motion, the clumsy snout, the warty skin of the broad unshapely body were those of a pocket-size hippopotamus as they crawled up to and over the waterline, and slowly wavered their heads, trying to determine where to move next. When scared they would lose all shape and identity, crawling into their own skin, first resembling a cat moving in a sack and then becoming a lifeless fig.

Except for the crabs, none of our submarine passengers ventured above the wet part of the reeds. Perhaps it was the flying-fish that lured the little crabs up on deck as time passed. We did not always find every flying-fish that sailed on board during the night, but the crabs located them between bundles and cargo. The little two-armed, eight-legged rascals posed themselves merrily on top of the titanic helping of sea-food, scaled clear a convenient portion and began serving themselves greedily with both hands.

Our little kitchen-garden did not serve the eye alone. It was Detlef who discovered that by harvesting a pot-full of the biggest goose-barnacles and boiling them with garlic, Iraqi spices and some dried vegetables, he had invented the best variety of fish-soup we had ever tasted, particularly when some bits of flying-fish were thrown in.

Flying-fish are a blessing to any raft-ship voyager in warm waters. As a delicacy they are second only to breakfast herring, and as bait they are superior to the most expensive artificial tackle. The variety we encountered in the Indian Ocean was not the largest I have seen, being only about seven inches long, and when only two or three landed on deck they did not suffice as breakfast for eleven men. But a single one put on a hook would straightway catch a dolphin three feet long, which sufficed as dinner for everyone. And when the old secret of how to lure flying-fish on board was shown to the keenest among the fishermen, Asbjörn, master of our ship lanterns, lit all our kerosene lamps and placed them outside the cabin walls at night. Flying-fish began raining on board like projectiles. Day by day, as we travelled southwards, our morning harvest increased. When we had a late dinner by lamp-light flying-fish sometimes shot across the table, hitting us left and right, tumbling into pots and pans, while fish-scales that marked collision points had to be brushed off canvas and cabin walls. Several times we were awakened by a cold fish landing in the bed, struggling with long breast-fins that could not give them take-off speed for flying away, which in the water is done by the tail. More than once I was awakened by a wet creature wriggling down my neck as I lay naked in my sleeping-bag right inside the door, and one night Carlo, my neighbour, sat up greatly amused when I could not find the visitor that had danced all over my bed. Next morning he found it dead in his own sleeping-bag. There were days when the morning cook was able to serve each of us with three fried flying-fish for breakfast. The morning harvest had mounted to forty by the time we stopped counting and accepted whatever flew on board as part of a normal routine in what our German companion Detlef, versed in the stories of Münchhausen, termed the Schlaraffen-See (the Sea of Luxury).

The first fish to rush for a hook baited with fresh flying-fish was the dolphin, the surface hunter and raft companion of all warm seas. Again, the multi-coloured fish dolphin, *Coryphaena hippuras*, also known as dorado, gold mackerel and mahimahi, must not be confused with the little whale of the mammalian family *Delphinidae*. Fish dolphins had followed us before, notably in the Pacific. We had only two or three with us as we left Pakistan; they never went for any hook, and only Asbjörn and Detlef had success with spears. But from the day we had the first flying-fish to put on the hook we had dolphin dinner whenever we wanted, and, no matter what we pulled up from the sea, next day the number of dolphins swimming around us had increased.

On 16 February I made a note that I had never seen so many fish at sea. On the 18th Gherman and Toru swam under *Tigris* to film the variety of fish species that had joined us by then, and among them were twenty dolphins, in spite of all our fishing. Next night their numbers had passed thirty. With our flashlights they were easily distinguishable from other fish in our company. We kept on fishing, and at intervals I kept adding a note to my diary that never had so many fish swum with us in any sea as now.

At midnight on the 26th I crawled drowsily out of the cabin to take over my steering watch, and got a veritable start as I faced a completely spooky sea. Toru was out before me, playing with his torch over the surface, and his beam did not reveal the usual black night ocean. Just below the surface ghost-like, lifeless bodies stood side by side everywhere, keeping exactly our speed and course, yet motionless as if they were mere reflections of something on board. Dolphins. But never had we seen them in such numbers, and they did not circle about in lively fashion as we were used to seeing, but just stood there like an army of white ski-troopers, escorting us motionless and effortlessly as if gliding down-hill on black snow. Normally we swam with the friendly dolphins, but this silent, uninvited entourage was packed so close and in such impressive numbers that I should have hated falling into the water.

As morning came the dolphins spread out over the surrounding sea. But as soon as the sun set they came back to us. Two nights later their number had passed three hundred. Colourful in daylight, they became pallid ghosts again by night. Broad in profile but slender across the back, they stood evenly spaced, with freedom for their outstretched breast fins, and always on port side, and always with our bow a short distance ahead of their own front line, as if to be sure that we led the way. No matter what speed we made, theirs was the same. By day they were lively, swift, and even seemed playful. In rough seas they would amuse themselves wagging their tails at full speed to rush right up to the highest peak of a wave top, then calmly surf-riding down the steep side.

As the month ended a note in the diary says:

> We are so much part of the marine environment now that I regard the dolphin school as domesticated; they always change speed and direction to go with us whenever we make a change. And at port side gunwale they swim so close that the tailfin often cuts the surface beside us, and with a flashlight in their faces we can see every detail. In daytime they often swim with their white lips open as if to collect plankton; we can then see their pink tongues. In daytime also, the breast-fins, spread out like on a butterfly, are shining so intensely in light blue that they seem to be illuminated and can normally be seen before the fish itself becomes visible. While pale silver at night and when dead, this

parrot-coloured chameleon of the sea is otherwise almost brown on its back, with grass-green head and yellow-green tail-fin, the rest of the body in various tones of green that alters and moves with the direction of sun and shade. A beauty to look at that none can help admiring or tire of, even though we see it all the time. They even leave us with an impression of being a friendly and faithful marine herd, by following us as dogs and sheep follow a shepherd ashore.

We sometimes sailed into an area with a side current rich in plankton. At night the sea was phosphorescent with microscopic organisms that gave a faint glow to anything touching them, fish or reeds, while some less numerous, bigger plankton twinkled as individual sparks. Sea and sky could sometimes be confused. The dolphins then showed up without flashlight as pale, elongated clouds drifting over a pitch-black firmament, and the phosphorescent micro-plankton made each fish send out a faint light even in its wake, giving them a long extra tail as if they were giant luminescent tadpoles. When we used a fine-meshed net to catch the star-like sparks that danced about independently, they proved to be copepods: tiny jumping shrimps with big black eyes.

The wind strengthened and we sailed faster. The only sign that the dolphins also increased their speed was at night, when their luminescent wakes were drawn out in length, making them look less like tadpoles and more like electric eels. It was an unforgettable spectacle that sometimes kept us seated on the port side bench gazing at the nocturnal show long after our steering watches ended. It was a fire-dance each time some fishes decided to change places in the procession, their long undulating comet tails crossing and interlacing in beautiful patterns.

At the first sign of daylight, when the procession broke up, the few that remained would circle around and under us as if awake after a night's sleep. We saw the others at a distance, usually in pairs or three at a time, jumping sideways out of the water in pursuit of glittering flying-fish. For days on end we had the company also of a few long-flying boobies, and it was a common sight to see these sea-birds fly low in front of a hunting dolphin, ready to dive and snatch the escaping flying-fish as soon as it took to the air, leaving the bewildered dolphin to scuttle about below looking in vain for the fish that never came back from the sky.

We wondered much why the dispersed hunters came back to us at night to take up their sleep-walker procession. Carlo theorised that they had discovered the rain of flying-fish close to *Tigris* at night. They swam so closely packed on port side that with mouths open one of them would always get whatever failed to land on deck. But why always assemble on port side? The moon rose on port side and lit up the golden ship on that side in the early evening before the whole procession fell asleep. There were many guesses, but certainly the dolphins herded together like sheep as if for safety during the night, probably finding *Tigris* a large and friendly protector, each hoping that another of the flock might become the prey and sound alarm in case of surprise attack during their sleep-walking. They had enemies in the Schlaraffen-See that could swallow them whole, the way they themselves could swallow half a dozen flying-fish in one gulp.

Sailing along one night with our host of luminescent ghosts, I was half-dormant myself, leaning against the bridge railing with the tiller in my hand. Brought to my senses by a sudden uproar in the sea beside me, I saw the phosphorescent dolphins rushing about in chaos as if a torpedo had landed among them. And there came the luminous torpedo. The terrific speed of the escaping fish prolonged their phosphorescent wakes until they all looked like undulating sea serpents, but the torpedo grew to outstrip them all in width and length as it shot into their midst and provoked havoc in the pattern of light. When the sea was black again, and the dim lights turned off with the dolphins gone, a ten-foot hammerhead shark swam away, fins above water, restless as if not yet satisfied with whatever he had swallowed.

A minute after the giant predator was gone, the dolphins were back with us to resume the procession. If possible, they seemed for a while packed even closer together, with barely room for their outstretched fins.

The hammerhead sharks were terrific hunters. They commonly came at night when the dolphins could be caught by surprise, but were active by day too. We once saw half a dozen escaping dolphins leaping through the air as if in imitation of flying-fish. The last in the row was thrown vertically into the air by the broad transversal snout of a hammerhead shark, and fell down, out of control, right into the terrific jaws of the waiting monster.

The hammerheads seemed the least trustworthy of all the sharks we encountered. In contrast to most other sharks the hammerheads never really kept us company. They appeared suddenly, patrolled our surroundings, and rushed away. One fine morning I was sitting peacefully on the airy outboard seat on the port side stern, with the reed screen discreetly drawn around me, when someone on the bridge shouted, 'big shark astern!' I happened to look straight down between my feet and saw the ugliest sight probably ever seen in any toilet. The broad, squalid head of a monstrous hammerhead slid slowly into view right below my bottom. Never have I seen the grim expression of that graceless species better. No other creatures can have a more grotesque position of the eyes. The flaring, hammershaped head was incredibly wide in front, drawn out to either side almost like cheeks, but resembling broad moustaches except for the small eyes at either extremity, with big nostrils at their sides. The huge mouth, far behind and underneath, leisurely swallowed up my paper as the displaced eyes seemed to gaze up at me ready for the main course. The prospective man-eater was apparently either nearsighted or as disgusted at what it saw as I was, for it moved on. I did not even dare to lift my feet for fear of calling attention to my near presence. The ten feet or more of grey-skinned muscle slid slowly, very slowly, under my seat, and I could have touched both dorsal and tail fin as they in turn passed under my outboard enclosure, but until the tall tail fin sailed out of the toilet I did not even rise to gather up my trousers.

When next I saw the beast it was behind us playing with the big rubber lifebuoy, and the men were preparing a huge hook with a short chain tied to the strong tow-rope. Carlo had put a small shark on as bait, and he caught the big hammerhead to everybody's excitement. A long battle followed with our end of the rope finally tied to the stern log, whereupon the fish pulled us 20° out of course. The captured shark churned the water, then went deep in all directions, and as it came up under us the rope was caught up in the guara aft, which started to waggle above deck. We pulled the guara up. For a while the giant was hammering at our bottom and our wake began to fill with tiny reed fragments. Finally everything grew calm below, but we could not pull in the rope with all our combined strength. HP had made us a small diver's basket of bamboo and rope, following the design we had invented on *Kon-Tiki*. Feeling braver now with my swimming trunks on, I crawled in when the basket was hoisted overboard. To everybody's disappointment but my own relief I found the big hook empty, stuck deep into the reed bottom of our ship. Swimming around me were only our familiar escort, including smaller sharks of a white-finned type we no longer feared. The hammerheaded monster was gone; it looked as if it had made a joke of its own name by hammering the hook so deep into the bottom of *Tigris* that we hardly got it out. It had swallowed the big bait as payment for the effort.

Since the time when we led a daily life with sharks around and upon the *Kon-Tiki* raft, it has been abundantly clear to me that sharks can be just as friendly and just as ferocious as men in and out of uniform. Even one individual shark can change in temperament from one moment to the next. The entourage of *Tigris* consisted of almost charming sharks. That is, though belonging to a detested and carnivorous species, they contrasted with the hammerhead shark both in conduct and appearance and left us with a feeling of sympathy.

Sharks came and left throughout the voyage, but only after we sailed from Karachi did we really make close contact. Their number varied from day to day, but increased as we sailed, and seventeen were counted in our company the day we fought the hammerhead. Their sizes varied from three feet to the length of a man, sometimes more, and at the beginning it was we who were the predators and they the victims. It was too easy to catch them and no sport fighting them; they hardly put up the battle of a dolphin. And they were far less popular as food, even when soaked to reduce the ammonia. Soon even the keenest fisherman was apologetic to the rest of us if by mistake he took a little shark on the hook. We fed them instead like dogs, with leftovers, fish-heads and bones.

One day we had foul wind and our speed permitted bathing. We found Gherman in the water alone before us, floating calmly on his back with no less than ten friendly sharks keeping him company with no sign of evil intent. From then on most of the men lost their awe for the daily shark company; in fact the complete disrespect was too flagrant at times, especially among the youngest on board, who were left with the impression that only blue sharks, hammerhead and tiger sharks would force them to climb out of the sea.

One day I let myself trail behind *Tigris* in the dinghy, lying with my head under the water to look at two long pike-like barracudas, scarcely trustworthy prospective man-eaters that swam with us for a few days, grinning with jutting lower jaws full of fierce teeth. They always swam side by side and deep below us. I removed the goggles that did not quite fit my expedition beard, had just filled my lungs and was head down to relocate the patrolling barracudas, when something was right up against my hair. I looked up and right into a face that had an expression as surprised as mine. I had seen the broad, flat head of a shark before, but not nose to nose, and had never before discovered that a fish could have an expression. Yet this shark definitely had one, as much as I did, radiating friendly curiosity and mild surprise. Being used to dogs, I was perhaps a bit misled by the tail slowly wagging at the other end of the body, and by the corners of the closed mouth that were drawn down as if in bewilderment at what it saw. I saw a shark, but the shark saw the bottom of the dinghy as a big round turtle with a bearded human head at one end. From that moment I have promoted fish to the reasoning species, not so far removed from the warm-blooded beasts as I had always assumed.

Our friendly coexistence with domesticated sharks led to a couple of near disasters. Carlo was sitting well outside the stern, balancing on the narrow oar-blade, washing his old wound with one hand and hanging on to a bit of rope with the other. He had convinced himself, but not Yuri, that this daily salt-water cure was good for the wound. Before this Gherman had been hanging on the other oar after a good soap wash and was standing aft drying himself when he suddenly shouted a warning to Carlo: 'Shark!' Carlo had long been sitting with one of the friendly little sharks swinging from side to side beneath his feet, and answered, 'I know'.

'Shark! Shark! Get out!' Gherman shouted again, for he had discovered a ten-foot man-eater coming, fin above water at high speed, straight for Carlo, who calmly continued to wash his leg. Then, fortunately, the brute began to behave strangely, wagging its tail and whipping the surface as if in challenge before attack. Carlo looked up and saw the big predator approaching fast and directly towards his leg. With the remarkable strength of a life-long mountain climber Carlo hoisted himself up with one arm and grabbed the side bundle with the other as he swung himself up from the rudder blade. The men aft estimated that another shark-length, equalling another second, saved Carlo from losing his foot.

We had almost reached Africa when Gherman without my knowledge took an unforgivable risk. While all but the helmsmen were dozing in the shade of the cabins after a good dolphin lunch, he donned his rubber suit and let himself out from the stern on a long rope to film the mixed company that swam in our wake. Hanging there with his goggles on, white-fin sharks and other fish approached him and paraded on either side of his camera. Some species were by now so tame that Toru used to swim out with a bag of chopped sea food and feed them with his hand. Hanging alone, far behind, Gherman noted one of the really big man-eaters deep down below. It had already seen him and came in slow circles up from the depths. Gherman had filmed sharks in his own Caribbean

Sea and most other shark-infested waters on our planet, so he knew when one of them was approaching with evil intentions. As calmly as possible he started to pull himself in.

While the rest of us were out of sight and most of us dozing, HP and Rashad were joking and laughing between themselves on the steering bridge, and had no notion of what was going on in the ocean behind them. Circle by circle the shark came higher as Gherman, grip by grip, pulled himself closer to *Tigris*. Nobody was there to pull him aboard. An uncontrolled grip or a provocative sound or movement, and the shark would have rushed to attack. It was there, right below him, as he grabbed the starboard rudder-oar and pulled himself up on the bundles. He was pale, speechless, confused, angry at himself and at war with everybody for a couple of days after his unjustifiable adventure.

Among the inhabitants of the open aquarium under *Tigris* were rainbow-runners and trigger-fish, two species not known to us from previous raft voyages in other seas. The rainbow-runner, *Elagatis bipinnulatis*, was a beauty in lines and colour, slim and speedy as a projectile. It derives its name from the two blue and two yellow stripes that run lengthwise on either side, separating the silvery belly from the dark blue back, and it has a tail of pure gold. Usually about a foot and a half long, this fish has a delicious flavour, like a mixture of bonito and mackerel, and at times the rainbow-runners followed us in such numbers that Asbjörn once was told to stop when he had caught a dozen in a few minutes. More than once we saw the rainbow-runners in the closest company with the sharks; in fact it happened that a man-size shark would swim along the side of *Tigris* with one rainbow-runner escorting him on either side, and they stuck so close to his side that it looked as if all three fishes were fastened together.

The trigger-fish, in contrast, was the comedian of the sea. It was clumsy in shape and looked silly swimming, as if more at home in a glass bowl. This species is in fact supposed by marine biologists to be a shallow-water fish living in reefs. But they surely did not live up to their reputation, for here they came to join us in the open sea and swam along with us ten thousand feet above the ocean floor. The first one we saw was like a strange bubble moving aimlessly on the surface, which proved to be the round white mouth of a stocky, speckled little fish. Short and tall it swam in a comical way, with curtain-like fins on back and belly waving left and right but always at the same time and towards the same sides, like wings out of place. This funny creature had difficulty in keeping up with our speed in good wind and sometimes fell over sideways like a drunkard in its eagerness not to lose us.

There are some thirty kinds of trigger-fish and some of them are poisonous, so we never tasted the two types that kept us company. Thus their representatives had risen to incredible numbers by the time we reached Africa. They flippered along deep and high and competed with the sea-hares in grazing on our lawn. When *Tigris* rolled in high seas they sometimes had the grass in their mouth and, reluctant to let it go, were pulled out of the water, fell on their backs, but waved themselves right side up again to come back for another bite. The largest we measured was twelve and a half inches long, but most were the size of an open hand. When caught they could control their own colours, showing round blue spots all over that otherwise were hardly discernible. But their speciality was to manipulate the trigger that gave them their name. Leather-skinned all over, they have a soft fin on the back and in front of it are two, sometimes three, long spines. The foremost of these is the longest and can be raised and locked in a vertical position, so that the fish may hold itself firmly in protective crevices. With all the force of our fingers we could not bend or unlock this long spine. But if we touched the trigger, the smaller spine just behind the long one, the mechanism would immediately unlock and the long spine fold back. This ingenious mechanism the fish could manipulate from the inside. The trigger-fish mingled freely with the sharks, who would hate to get one of them in its throat.

A few zebra-striped pilot-fish swam before the bow or beneath us, but few as compared with those we had seen in the Pacific, probably because we rarely brought on board the bigger sharks they usually accompanied. One five-foot shark had six pilots swimming in front of its nose, but once we pulled it up on deck we saw to our surprise about twenty of them struggling around its tail trying to come along with the master. Carlo left the dead shark hanging with its tail in the water, and we were amazed to see two pilot-fish remaining beside the tail-fin, performing the most beautiful and remarkable dance. The two performers, of equal size but one greenish-yellow with light brown stripes and the other light blue with dark brown stripes, never stayed more than an inch apart, and swam and swayed with identical rhythm as if one was the shadow of the other.

It was something different, however, to be awakened by the sound of someone blowing his nose so loudly that it aroused even those of us able to sleep through the most exclusive snoring. Sitting up to look at the water beside the open doorway, we then saw the moon shine on something colossal, glistening like a polished shoe at our side, but with a big panting blow-hole that left no doubt that we had a living whale at our bedside. No matter how often we might have seen whales in some Marineland, it is quite different to wake up in intimate contact within the whale's own free environment. Unlike the police boats which ran into our door and shook the ship and rigging, the whale with all its tremendous body-strength never touched the reeds, never bumped into us even in the dark. Yet we often had them rolling up suddenly at arms-length or swimming right under our bundles from one side to the other. Most of the visiting whales were porpoises, rolling up with rounded back and dorsal fin, but many were much larger, with straighter backs.

By day, too, whales would venture up to examine us quickly, while we would stand on deck and look straight down into their blow-holes as half a dozen of these huge mammals slid beneath us. We saw some of them blowing like marine fire-brigades, and among those that leapt vertically from the water, body and tail, we recognised the killer-whale with its beautiful black and white decor and tall dorsal fin. We even saw killer-whales and porpoises chasing together in an area glittering with small silvery fish, and among them the water was cut by fins so tall and sharp that they could only belong to hunting sword-fish.

One birthday after the other was celebrated on board. Never have I seen a raft-ship with so many eminent and inventive cooks. Never would I have suspected that raw fish could be served in so many and tasty ways as Toru managed, Japanese style, admitting that he himself was the owner of a small fish restaurant. Detlef's fish soup, Norman's pancakes, Carlo's rice specialities, Yuri's sun-dried rainbow-runners and Gherman's chili-peppered dried meat were unforgettable, and no one could have guessed that two young Scandinavian students could have produced so many fabulous cakes and puddings out of Arab beans, peas, flour and eggs kept fresh with a coat of oil. Yet the prize went to Rashad, the inventor of pickled flying-fish à-la-Tigris. I stole into the galley one morning and wrote down the recipe: Take two flying-fish, clean and cut them into cubes, ¾ large cup vinegar, ¼ cup sea-water, a dash of olive oil, 1 clove of garlic, plenty of chopped onion, 1½ spoons of sugar, 2 teaspoons salt, ½ teaspoon pepper, and a dash of whatever Arab spices he had brought with him.

Yuri's birthday dinner started with this superb appetiser, continued with dolphin, fried, boiled or served raw in half a dozen ways, then came HP's and Asbjörn's special spaghetti pudding. The two rascals had secretly drilled a hole in our wooden table and placed the rubber hose used for inflating the dinghy inside the pudding. As we were all ready to dig in for the first helping, HP solemnly announced that Yuri's very special birthday pudding needed some drops of vodka to swell, as we had no yeast. He poured while Asbjörn secretly pumped with his foot, and to everybody's amazement the pudding started to swell. None of the miracles of the sea perplexed us more than when the growing pudding began to swell in all directions, until something rose like someone's finger out of the middle of the pudding, and grew and grew to emerge as a long balloon with pudding pouring from its sides. We had barely recovered from our surprise, roared with laughter and begun singing 'Happy birthday', when there were unexpected visitors: black whales surfaced and came rolling straight towards us. We saw only three side by side, but there were probably more, as for a while the huge mammals came up at intervals everywhere. For the first time for weeks the little mouse came running up into view, as if it had a peep-hole down below and for a moment preferred the company of singing men and grasshoppers.

Our intimate association with these large marine mammals would probably account for our inheritance of the largest remorafish we had ever known. Black, floppy and ugly, with an oval suction disc at the top of the head, the remoras are too smart to bother to swim with their own tails more than necessary to change transport. They hitch-hike on hard-skinned, scaleless travellers, like sharks, whales and turtles. Hanging on with the fine device at the top of the head, their mouths are free to enjoy spillings from their host. Most of the remoras hitch-hiking on our bundles were finger size, but a couple must have come to us from a whale, for those that clung to our bottom were as long as an outstretched arm.

Once, to our surprise, Asbjörn lifted a sea-turtle out of the ocean with his bare hands. It seemed intent to come and visit us anyhow. We kept it on deck for a while and dreamed of turtle soup and simulated roast veal, but by unanimous vote we preferred to watch him swim happily away. For ancient voyagers in this ocean sea turtles must have provided a welcome change from the fish and dried food diet. Today they have been almost exterminated, although we saw the periscope-like heads of a few. Three days after Asbjörn's success, Rashad on a calm day swam into the sea and somehow managed to grab the big carapace of another turtle. Turtle and boy proved to be equally good swimmers, and it was unclear to the spectators whether the turtle pulled or Rashad pushed as they approached *Tigris* more below than above the surface. But when the young Arab triumphantly managed to lift the struggling reptile with front flippers above water, victory was his. Both our turtles had a couple of remoras firmly attached, and tiny green crabs crawled about on their carapaces as on *Tigris*. This second captive was so angry at its defeat that it tore reeds and bamboo to shreds with its parrot-beak, and we quickly let it back where it belonged. No sooner did the turtle surface some hundred yards from us when Yuri yelled from the cabin roof: 'A big shark took him!' True enough, we saw the water churning with fins and flippers rotating in the waves. Then we saw no more on the surface, and someone mumbled that we might as well have had it ourselves.

Next time Asbjörn attempted to catch something with his bare hands it was something we had never seen before, and he was less fortunate. The sea was calmly undulating and something strange was rippling the surface in one spot, resembling the emerging fingers of a human hand. Asbjörn rowed in pursuit with the dinghy and was soon there, not knowing that what we saw was Neptune, or rather *Neptunus*. 'I've got him!' he shouted triumphantly as he reached over and nearly fell out of the boat when he saw what he had caught. 'Aaiii!' he yelled in pain as he lifted his own hand in the air and tried to shake off something red and sprawling. A big crab! We looked around, and reddish-brown crabs as big as a fist scurried about everywhere, running across the surface as if it were a mirror, then diving down and disappearing. We had never seen it before, this swimming crab of the Indian Ocean known among marine biologists as *Neptunus*. The crab pinched Asbjörn's finger so hard that it bled and he came back to borrow Carlo's spaghetti sieve. With this ingenious implement he and Detlef caught a dozen crabs that were so furious that they clipped the claws and legs off each other when left in the same pot. Soon the men in the dinghy found competitors. The dolphins that swam with us also went for the big crabs, snatching them on the surface right in front of the dinghy and leaping high out of the water in doing so. We also saw a five-foot shark rush after a crab that paddled at full speed to escape; the shark caught it and then jumped in a terrific twist clear of the water. Never before had we seen a shark jump.

The little rascals with the name of the ocean-god sat quietly on the surface eating plankton with both claws until disturbed. These miniature robots with human characters were perhaps the strangest creatures we met at sea. When their pivoting black eyes sighted the approaching enemy with the spaghetti sieve, they immediately took up a wrestler's defence position, arms flexed and pincers open. But seeing the size of the sieve they soon found it wiser to paddle away fast. This *Neptunus* did sideways, incredibly swiftly, and by a brilliant coordination of five pairs of legs. The forelimbs with the big claws were both turned to the left, right arm flexed at the elbow to reduce friction and left arm fully outstretched behind to serve as a regular steering-oar. The hindmost pair had the two outer joints flattened like oar-blades for fast paddling, with a complex system of hinges that ensured maximum effectiveness, and the other three pairs of slim limbs just scurried along. The way all the complicated segments of the hard-shelled robot pivoted and functioned, from its antennae to its propulsion and steering mechanism, was a masterpiece of engineering. Yet this is a trifle in an ocean where whales have always dived with sounding instruments operating like modern radar, and where regular jet propulsion is built into the body of squids so that they can shoot through the water with rocket-like speed behind their own smoke screen, or glide over the waves and climb aboard. In a city man may feel second to none. But alone in the immensity of the universe, among all the creatures that preceded man and built up the human species, even a most fervent atheist will wonder if Darwin found the visible road but not the invisible mechanism.

For two days we sailed among the reddish *Nautilus* in a calm sea; then the wind strengthened, we picked up speed, and sailed with our marine herd into another area dominated by the most beautiful sky-blue snail shells. Living snails were in them and floated upside down, hidden from above by plastic-like, segmented bubbles that helped them to sail about, but not fast enough to avoid recognition by the trigger-fish, whose turn it was to wriggle forth as fast as they could, to swallow the blue pearl-like flotsam, shell and sail.

The philosophers of the most ancient civilisations believed mankind to be the descendants of mother sea and father sky. Modern science has come to a somewhat similar conclusion. What else was there for the first living species to descend from? At night in the ocean even the stars seemed to come closer to the water and become part of man's world again, as they had once been to the people who first gave them names and used them as familiar landmarks when travelling in open spaces. Again we had this strong feeling that only life in the wilderness can give, of time fading away, and past and present becoming one. Time was not divided into ages, only into day and night.

When we stood night watch on the steering bridge, or lay on the cabin roof looking at the topmast circling among the ever more familiar constellations, we began to feel at home in the system up there, forgotten by other than astronomers and astronauts. The stars were no longer a chaos of sparks like the plankton of the sea. We recognised them and the time and direction of their paths, as they rotated over the bow and the sail, in the same order and with the same speed night after night. No wonder that the peoples of Mesopotamia and Egypt, with their wide open spaces, became master astronomers who knew the exact rotation period of all the main heavenly bodies, navigated by them, and gave our ancestors a proper calendar system.

There were nights on the bridge when I felt that *Tigris* was a sky rocket. The bundles were blue with phosphorescence and sent off dancing sparks, while behind each rudder-oar was a long bright light resembling the dim headlights of a vehicle; but since we drove away from them they became more like a burning exhaust, full of sparks, while we flew with our black sail in the opposite direction, amongst the stars.

On 1 March the diary reads: 'We started our journey in November and are still on board in March. It is incredible, but yesterday and today the sea on port side smells of fish! Is it possible that all these creatures packed just below the surface, which indeed smell above water, can send out an odour that reaches us when they are in sufficient numbers?'

A few days later we were suddenly torn away from the world of fish and early man. Norman had managed to establish good radio contact with the modern world, still far away. But we could clearly hear Frank at Bahrain Radio, and radio messages came with depressing regularity. *Tigris* was suddenly on a collision course with political events. Four months had gone since we launched our reed-ship in Iraq, and four weeks after we sailed from Pakistan when we ran into serious steering problems. Not because *Tigris* would no longer obey the rudder-oars, but because major areas in front of us had become forbidden territories.

We had hoped to sail from Meluhha to Punt, today Somalia, since Egyptian records speak of this fertile part of Africa as having been visited by their sailors and merchants. Requesting permission to land there, we got the first warning from London via Bahrain

Radio: 'No one in Somalia answering phones yesterday as major town had fallen to the Ethiopians. Very strongly urge you not to attempt to land on any of these territories.'

Somalia was at war. Yachts trespassing Somalian waters had been seized and their crews imprisoned. This meant that fifteen hundred miles of the African coast from the Gulf of Aden southwards was closed to us.

So we had to avoid the Horn of Africa on our way into the Red Sea. We must steer slightly more to the north in order to stay away from the African side of the Gulf of Aden. This meant that it was necessary to keep our course closer to the Arabian side.

Then came a warning from London that the Arabian side of the Gulf of Aden was also forbidden territory, for this was South Yemen, which had closed its borders to visitors. South Yemen filled the entire thousand mile stretch of the Arabian peninsula between Oman on the ocean side and North Yemen inside the Red Sea. South Yemen, with a communist government, had armed border clashes with both these capitalistic neighbours.

We now had to navigate with caution and try to reach and then follow the midline of the Gulf of Aden, 900 miles long, without touching the forbidden lands on either side. Our intentions caused renewed anxiety for our security among our consortium contacts in London:

Do you actually intend to navigate into the Gulf of Aden and from there into the Red Sea. Are you able to sail and navigate this course. Please beware of political situation in this area as previously advised. We have had no cooperation from either the South Yemen or Somalian Governments. Stop.

We were sure we were able to navigate that well. We aimed for the narrow Bab-el-Mandeb Strait leading into the Red Sea from the far end of the critical Gulf of Aden. Then the wind died down. Completely. There was not a gust from any direction, and we became a prey to the invisible ocean currents while there were still 1,400 miles left to the strait we had to hit. The sea became calmer than I have ever seen any ocean. Not a ripple except from us. The sail hung like a wall carpet, its beige pyramid and red sun mirrored perfectly on the surface. When we dived and rose again to the surface of this marine mirror there was no visible horizon anywhere; we became almost dizzy, floating about in all that blueness like spacemen beside a suspended *Tigris*, with sail and emblem duplicated like the figures on a playing card. Beautiful. But the two navigators checked sun and stars and told us we were drifting towards Socotra.

Socotra was a large island well in front of the Horn of Africa and equally far from Somalia and South Yemen. It now belonged to South Yemen. As a precaution we hurried to ask for permission to land on Socotra. The answer was a new disappointment:

Have approached the Embassy of South Yemen in London for permission for Tigris to land in Socotra if necessary. Stop. They are aware of Tigris's proximity to Socotra but stated that Tigris must not repeat not attempt to land on this island before such written permission has been given. Stop.

We drifted closer. London stressed: 'You risk arrest if you land on Socotra without permission. Stop.' And direct from a friendly west European Foreign Office came an independent warning: 'Do not go to Socotra now, you may get trouble.' Unconfirmed radio messages said it was believed that Russians were installing important military bases on this strategically located island which controlled the entrance to the Gulf of Aden and the Red Sea. No planes or ships were allowed to pass within sight. South Yemen simply answered that we could not approach the island unless we had advance written permission from the island itself.

Helpless without wind, we drifted still closer, while the well-meant warnings increased. We could be shot at. Court-martialled.

On 12 March the impressive mountain skyline of Socotra rose into sight, pale blue in the distant south-west, just where we were drifting. Carlo discovered the faint outlines just as the red sun set in the west, and we all shouted with mixed feelings of joy and concern. 'Hurrah, we see Socotra! We have reached Africa!'

This was great. We had crossed the Indian Ocean. There was now a very faint ripple on the sea, but not enough wind to lift the sail. Yet we must not come closer to that island. We struggled with sails and oars to keep away. The sun set and we saw nothing. We sent out new messages: we approached the forbidden island against our own will.

On 13 March the sun rose in a faint haze, with a lazy breeze still too faint to give us steering speed. By noon the haze gradually became so thin that I detected a sunlit formation of something bright but indistinct far over on port side. It most resembled an iceberg shaped like a seated polar bear. It seemed to be part of something bigger with darker outlines. We had clearly drifted much nearer the island, but could not quite make out what we saw in the haze. A seagull came out on a visit. The boys fished up clusters of floating algae.

Norman passed on a report of our unfortunate position, and got back the message that the London embassies of all nations represented on *Tigris* had approached South Yemen with negative result.

The sun set.

On 14 March, shortly after midnight, a strong wind sprang up and we started steering. No sooner were the two sails filled, when the wind turned from w to sw and s, and then suddenly back in the opposite direction. We ran about, turning the sail and helping with the rowing-oars all night, but never brought *Tigris* back on to course before the sail back-filled again. In the end we exhausted ourselves rowing in turns and circles, and in the dark I ran my forehead into the end of a thick bamboo rafter and got my hands and beard full of blood. At daybreak, shortly after six, I crawled out again and to my great amazement saw the entire rugged coastline of Socotra right before the bow. The seated polar bear was there too; it proved to be a colossal white sand dune running way up a hill and forming a headland in the central part of the island. Our bearings soon told us that we were looking at the north coast of Socotra, seventy miles long, from end to end, and when the sun rose we were only twenty miles off. As it set, the distance was only fourteen.

69–70. Waiting for wind. Inside the bamboo cabins, we sleep on asphalt-covered boxes containing our personal property; radio operator Norman hands the microphone to Yuri, who speaks to a Russian ham station, asking help from his Foreign Office to persuade South Yemen to give us a landing permit for the forbidden island of Socotra. *(opposite)*

71–72. Into forbidden waters off Socotra without adequate steering speed; within shooting range at the entrance to the capital port of Hadibu. *(p. 342)*

73. A birthday photo by self-exposure as the wind fills the sails again. On the cabin roof the four reed-ship veterans Norman, Thor, Yuri and Carlo, the man of the day; up the mast ladder, Norris, Toru, Detlef, Gherman, Rashad, Asbjörn and HP. *(p. 343)*

74. Into the African war zone; military aeroplanes and helicopters off the coast of Djibouti. *(p. 344)*

75. Into the final port, Djibouti. With all flags up and the captain at the helm, Asbjörn rushes to help lower the sails as we are welcomed by French warships. *(p. 344)*

In the meantime we had struggled non-stop all day to get further out, or at least hold our distance from shore. With our bow turning aimlessly in all directions, our school of dolphins swam bewildered around us in a ring. The island had peaks and pinnacles that seemed higher and wilder than any land we had sighted on the voyage. Norman was constantly on the radio, and learnt that an appeal on our behalf had been directed to the South Yemen Embassy at the United Nations, but to no avail. Permission could only be granted by the island itself. When Norman tried his amateur set a Russian radio ham named Valery was among those who called him back, and I asked Yuri to send a message to his superiors in Moscow explaining our unintentional drift towards Socotra. Soon afterwards we got a reply through Valery that the First Deputy Minister of the USSR Foreign Office had sent one dispatch to the South Yemen Embassy in Moscow and another to the Soviet Embassy in Aden. Evening came, and all sight of the island was lost again, while Detlef reported from the bridge that we seemed to hold our distance from land as we sailed westwards along the island against a tidal current of at least one knot. That night faint gusts of confused wind sometimes brought a marked smell of flowers or vegetation, and once we thought we scented roasted coffee. In the direction of land a single strong light was once turned on for seconds, then the night was as black as before.

It was 15 March. All wind died down at midnight, and we were a prey to the currents. We scouted in vain for lights from ship or shore. And as the black sky first grew bright green in the east and

76–77. Abandoning ship in Africa, Rashad, youngest crew member, walks ashore in Djibouti, as we are not permitted to land in any other area due to wars or for security reasons. With the African hills at the entrance to the Red Sea beside us we prepare *Tigris* for a proud end. *(p. 345)*

78. Farewell *Tigris*, you proved a good ship and we would not leave you here to rot. You ended as a flaming torch, with an appeal to all industrialised nations to stop unrestricted armament delivery to the part of the world that first gave us our civilisation. *(pp. 346–7)*

79. The end of *Tigris*, but we had the answers. You were still floating high after five months, and you had carried eleven men and all their necessities 4,200 miles, or 6,800 kilometres, from Mesopotamia (Iraq) by way of the Dilmun (Bahrain), Makan (Oman) and Meluhha (Indus Valley) of the Sumerians, across the Indian Ocean and past the Punt (Somalia) of the Egyptians, to Djibouti at the entrance to the Red Sea. *(opposite)*

then red, we scanned the horizon for land. There it was again, and only eight miles away. We had been carried back and were once more off that part of the island we had left the night before, but much nearer. The polar bear was now huge and close in, and we wondered how sand could have been sent that far up the hill. The sandy hill marked the eastern headland of the main island harbour with the capital of Hadibu. For the first time we could distinguish tall trees and a few houses on the west side of the bay. No sign of life. No smoke. Strangely, the big island seemed deserted. New kinds of fish came out to us, and we were surrounded by at least a hundred trigger-fish. Here again were plenty of *Neptunus* crabs. Birds too. But no people. The sea began boiling everywhere around us, far and near, in patches, with tiny silvery fish. Our dolphins were away hunting. There were numerous porpoises. There were even three big whales blowing water off the cape. But no boat came out from the main harbour, and nobody shot at us. We began to speculate if the island could have been abandoned due to political trouble or other reasons. Norman picked up a new message from Frank in Bahrain: 'The German Foreign Office advise Tigris not repeat not to land on Socotra due *severe difficulties* [underlined].'

We began to sail very slowly away with a faint breeze, made 1·2 knots over the surface, but the current was always against us. The surface for miles began to appear polluted, covered as if with soap-suds. The diary records:

This pollution increased in quantity as the afternoon and evening passed; it came in bands like the 'red tide' and could be seen from horizon to horizon, in places packed like sheets of snow. On inspection we found it full of tiny brown eroded oil clots as in the Atlantic. We have seen the clots without the suds for days. Also we saw oily slick mixed with the suds and drawn out in parallel strings like slime. Terrible. And the crabs run in this filth, also the gaping dolphins in pursuit of the billions of minute fish. All afternoon we have sailed here and there to escape the island, even tried to steer back and turn around its coast on the east cape, but all in vain. Now at 6 p.m. we steer for the same tall west cape we headed for last night; but our position is much worse, for we are almost within the three-mile limit. We are too close to feel comfortable. We see a few army tents on a cape where a trail runs uphill, and imagine seeing a truck beside them. But we have now also noted a considerable assemblage of huge modern concrete buildings mixed with some old Arab houses. Three of them lie side by side at the waterfront like apartment houses without windows, leaving the impression of ultra-modern plants contrasting the apparent wilderness around. Fabulous mountains. Never have we seen a more beautiful island. Norman and I agree that it resembles Tahiti, where we had first met, with the sky-piercing pinnacles of the Diademe rising in the interior. Carlo got the idea of trying to contact Socotra or South Yemen directly with our own radio. We have tried, but no answer. The large buildings are getting closer, some are at the water's edge. No life. No movement. No wind. A desperate situation. Risk of drifting on the rocks is great.

Night came as we were right up under the land. To our great surprise we suddenly saw sharp electric lights turned on in great numbers in the modern blocks, and a few scattered lights showed up elsewhere too. I closed the notes with the following words: 'We are indeed too close to feel comfortable. I have given orders to light extra kerosene lamps and turn all towards land so they know where we are and can interfere if we drift towards the perilous coast.'

On 16 March the spectacular island was still right there in all its impressive beauty. We were again immediately off the bay with the capital town of Hadibu, where the Sultan at least formerly resided. We were well inside the three-mile limit, with two blowing whales right beside us. I told everybody to put away their cameras. If somebody came out we would explain our problems. If not, we were close enough to go in with our dinghy and apologise.

At breakfast time a small engine-driven dhow turned up on the opposite horizon, coming from the direction of South Yemen and heading for the island. It adjusted course to come straight for us. A few hundred yards away the engine was turned off, and the four-man black crew stood gazing at *Tigris*. Their open boat was loaded to the brim with huge hammerhead sharks. I sent Rashad over with HP to get some information about this mysterious island. As they came back Rashad explained that the four black fishermen had been very friendly and spoke an Arab dialect. They had strongly recommended us to sail into this bay which had the only good landing beach. No problems on this island. People here were very friendly, the fishermen assured, and they had added that there were also 'Russians and Chinese' ashore. This unlikely combination caused a good deal of amusement. The fishermen left, went in a big curve and disappeared around the distant east cape of the island.

Encouraged by their lack of uneasiness, I wanted to go ashore, since we had long since trespassed territorial waters anyhow. I summoned all the men for an important pow-wow. I reminded them that I had promised to hear each man's voice, time permitting, before I took any major decision on this voyage. My vote would be decisive provided I had one man's support, which would ensure that I was not entirely out of my mind. Now I wanted a vote on my plan to land on Socotra. Rashad could speak Arabian and Yuri Russian. Nobody had shot at *Tigris*, so surely nobody would shoot at the innocent dinghy if we openly went in with our United Nations flag and explained our awkward position. Either they had to accept us, or they would have to tow us out again.

I did not overstress my proposal, for I was sure it was the only sensible thing to do and that everybody would be in accord as always so far. After all this island was incredibly beautiful; here we could certainly get fresh fruit, coconuts, poultry, steak, milk and good water. If there were Russians ashore they would surely give a friendly welcome to Yuri Alexandrovitch, who could even tell them of the messages on the way from their own Foreign Office. Prehistoric mariners must have left some vestiges on a big island with this position. Never had I seen a piece of land that more invited exploration. But I did not voice all this. I briefly suggested my plan, and pointed out that for half an hour now we had been favoured by a faint breeze blowing straight into the bay where we could beach or anchor.

Norman was the first to comment. I had never known him to be an orator, able to speak with so much enthusiasm and persuasion. To my surprise he pointed out that this faint breeze could also be taken in straight from starboard side and thus permit us to sail alongside the island and perhaps just carry us clear of the distant east cape with the triumph of continuing non-stop. It was certainly worth trying, he said. We should not cut the truly long leg of our trans-Indian Ocean voyage until absolutely necessary. We had a chance to beat all our own records in non-stop reed-ship sailing. We had plenty of food and water and needed nothing. If we sailed into this bay we would have to be towed out again or hire more men ashore to help us row the heavy vessel out. Besides, some of us always became ill ashore. It was surely in the interest of the expedition leader and everybody else not to interrupt our voyage here.

Carlo listened with an open mind. He, who had beamed with admiration when he caught sight of the splendid cluster of alpine pinnacles, was now completely uninterested in climbing. We had been at sea now for so long. It was better to take the risk of trying to clear the west cape and make straight for the Red Sea, where we could end the expedition.

Yuri was sure there would be no problem if we landed, but if Norman felt we might clear the west cape if the present wind lasted, then he, too, would vote for an attempt.

Norris was silent, but we heard the familiar hiccoughs of his baby as he filmed and registered all that was said. Without interference he let the multinational crew give their ballot on the most important decision of the whole expedition. Then he, too, voted against landing, as this could not possibly be an interesting place, otherwise Thor would have planned to go there from the beginning. One by one the votes ran contrary to my proposal: USA, Russia, Italy, Germany, Iraq, Japan, Mexico, Denmark. No support. Only my countryman HP agreed that we would never find a more interesting place to visit, so we ought to take a chance of going ashore.

I was amazed. Defeated. Perhaps I had been spoiled by my men always accepting my proposals, except when I wanted to interrupt the voyage of the first *Ra* and they all wanted to go on. That time I took the decision alone. There was no reason to take needless risks in a scientific experiment, and by then we had the answers anyhow. And we could build a second test raft. This time I had at least the necessary support of one vote. And I saw no point in setting a long-distance record. Besides, the risk of running on the reefs and wild rocks before we managed to clear the west cape of Socotra was perhaps fifty-fifty. A brief description of Socotra in the Indian Ocean sailing directory spoke of the west cape as treacherous, with fierce currents, rough sea and violent gusts of wind. Experience of the last four days had shown us that the wind might turn or die at any time, and then we would be lost. There was no other beach further west or a bay where a reed-ship could sail in and find security, only wild cliffs and a couple of impossible anchorages blocked by coral reefs. It was a desperate situation. To me it was totally meaningless to risk our lives in an on-shore wind against these extensive cliffs when we could stand straight into this fascinating bay, anchor and meet human beings to whom we could explain our case. But without support of my men and with all the warnings from the outside world I decided on a compromise.

The onshore wind increased sufficiently to whip up a few early white-caps. I told the men that I was willing to sail on a two-hour trial run with exact bearing on the west cape controlled from the bridge. If after two hours the bearing showed that we were able to maintain a course clear of the cape without leeway, then we should continue. If not, then there would still be time for us to take the wind in from the opposite tack and sail back into Hadibu port.

Everybody was happy, but at heart I was not at all pleased. This, to my mind, was hazardous.

Two hours passed as we sailed westwards along black inhospitable sea cliffs where the surf had begun to show up white against the rock as the steady onshore wind continued to blow. We were indeed taking a risk. After two hours Norman came crawling out of the cabin with two fresh radio messages, just as Detlef jumped down from the bridge with his bearing report. With Gherman on the tiller astern the rest of us gathered anxiously around the table as Norman read aloud: 'South Yemen Embassy at United Nations in touch with government to explain Tigris situation. Stop. Hope permission will follow. Stop.' The second message read: 'Norwegian Embassy in London advises South Yemen agrees OK for you to land Socotra.'

Some of the men shouted hurrah, and it was suggested that we should turn around at once and try to sail back to Hadibu Bay. I asked for Detlef's report. He said we made two knots over the bottom against a contrary current, and with leeway subtracted we were maintaining a course of 278°, which was 18° clear of the still very distant west cape.

Detlef was the first who wanted to run back to the bridge and reverse the course for Hadibu. 'Now we know we won't be shot at,' he cheered. But now I was firm. 'No,' I said. 'We decided unanimously to make a two-hour trial to see if we could clear the cape. We now have the result. It is affirmative. So now we sail on, otherwise this risky test-run has been pointless.'

Norman agreed. We ought to go ahead, as no Sumerians or Egyptians would have called at this island, knowing from earlier visits that there was neither gold nor other valuable products to pick up. But most of the others began to regret deeply the adventures we had missed by not sailing into Hadibu Bay and getting the answers to the still unsolved mysteries of Socotra. They gazed at the passing valleys and in our minds we all climbed the highlands and explored the mountain valleys up there between the wild peaks.

Carlo and Gherman dug up some bottles of red wine reserved for special occasions, and as the sun set in some fantastic cloud formations that we all for a while thought were more islands, we kept our bow to the right of the cape that was outlined against the golden sunset. The wild cliffs of Socotra were there at our side all the time, but the last cape was still very far ahead. We noticed a few primitive stone-house settlements on some level ground above the cliffs, but soon we saw nothing except our own bearded faces around the kerosene lamp. The grasshoppers were still singing in the reeds and little crabs crawled around our bare feet. After all, the eleven of us were still having a great time together on this Sumerian ma-gur. I began to forgive the men for their lack of support at the morning pow-wow. After all, the purpose of our voyage was to test our vessel, and they could not have given it a better certificate. This kind of prehistoric watercraft, deemed insecure outside Mesopotamia's river system, had carried us through the gulf from Bahrain to Oman, from Oman to Pakistan, from Pakistan to this African island, and it was still so seaworthy that the crew preferred to go on non-stop when they knew we next had to force ourselves another thousand miles and run the gauntlet of two long forbidden gulf coasts before we had a chance of another landing.

CHAPTER 11

Five Months for Us,
Five Millennia for Mankind

A NEW moon seemed to mark a new phase of our journey as we sailed away from Socotra, the first African island. We found rough seas and varying winds during the night, and both Norman and I at one time believed we sighted the black contours of the high west cape as we slid past in the sparse light of a slender sliver of moon. Farewell Socotra! For some time we all saw powerful search-lights far away in the direction we had come from. By sunrise we could see the whole of the tall west coast of the island behind us, and by late morning we were once again alone with the sea, surrounded by most of our former companions of the water.

Two days later we passed at good speed close to the uninhabited bird island Kal Farun, tall and shaped like a shark's tooth, glittering with guano. We noted that the drawing of Kal Farun and the one of Jazirat Subuniya had been interchanged through an error in the pilot book,[1] and assumed that few ships ventured here since the same book referred to strong tidal currents changing in opposite directions between these rock islands. But at that very moment, as the sun and sky turned red for sunset, we were overtaken by two ships that came on a converging course behind us, one of them clearly from the direction of Socotra. Both ships changed course to come straight for us side by side, and as they seemed to run up one on either side, we hoisted the United Nations flag. The two ships immediately responded by hoisting the Soviet colours, and we could now read their names: *Anapsky* and *Atchuievsky*. They were Russian trawlers that bore every sign of having been long at sea and in need of paint. Yuri climbed up on the cabin roof and rejoiced as he danced and waved his cap. Flocks of seabirds circled around us and dived for fish all the time. At first there was little reaction from the thirty men on each ship, although they were all lined up at the rails to gaze at us from the two trawlers which were now brilliantly lit, as darkness fell as soon as the sun was gone.

The two Russian ships slowed down and escorted us closely on either side astern, the one blowing its siren. Asbjörn hurriedly inflated the dinghy and he and Yuri jumped into it in the dancing black sea and rowed over to *Atchuievsky*, where Yuri climbed on board. Half an hour later he was back with a colossal deep-frozen red grouper, a twenty-pound bag of beheaded, neatly packed and deep-frozen prawns, a sack of potatoes, and a roll of old Russian maps of the gulf into which we were heading. For two days we ate delicious prawns in greater quantity than we would otherwise have dared, but we feared that they might turn bad in the heat. The two trawlers left with course for Aden.

Again we were alone, and for the last time we were free to lie on the roof and relive the days when man spread free commerce and dawning civilisation across these waters. The historian Pliny the Elder, in the first century after Christ, recorded[2] the truly impressive volume of trade carried on in his days by ships between Egypt and Ceylon, with further communication between Ceylon and 'the country of the Chinese'. He made it abundantly clear that the early Romans had learnt local sailing directions from the ancient Egyptians, who knew exactly where to steer and when to hoist sail in the right seasons. Thanks to Pliny and his informant, the leading Egyptian librarian and geographer Eratosthenes, we knew that *Tigris* was not the first reed-ship to have accomplished this easy voyage. He recorded that in earlier times the Egyptians, 'with vessels constructed of reeds and with the rigging used on the Nile', visited not only Ceylon, but also sailed on to mainland India, trading with the Prasii on the river Ganges. He gives the exact sailing route learnt by Eratosthenes from the Egyptian merchant mariners, and states that the voyage from the Red Sea ports begins in midsummer at the time the dogstar rises. Then, 'Travellers set sail from India on the return voyage at the beginning of the Egyptian month Tybis, which is our December, or at all events before the sixth day of the Egyptian Mechir, which works out at before 13 January in our calendar. . . .'

With *Tigris* we had not left the Indus Valley until 7 February; we knew it was too late and paid for it with a broken topmast. But we had reached African waters even so. Our problem, however, was not so much being late in the season as being late in historic time. We had made the crossing some centuries too late and were not allowed to land in Punt. We were now moving ahead dangerously near to that forbidden coast, and we must manoeuvre with utmost caution not to get within sight of Cape Guardafui, the projecting tip of the Horn of Africa, where the east coast turns abruptly west-southwest into the Gulf of Aden.

'I think we should stay south of centre of the Gulf of Aden,' said Norman, 'and one thing we'll have to do is be very accurate. We have only a fifteen mile gap between the continents of Asia and Africa, and we've got to hit that slot.'

The first political breakthrough came. Djibouti, a tiny new African republic in the innermost corner of the Gulf of Aden, just at the entrance to the Red Sea, had given permission for our landing. That little nation, not much more than a good port surrounded by a small piece of desert, was neutral.

We sailed past the cape of Punt far out of sight. In there, behind the horizon, the most developed nations of our civilisation were unloading the latest inventions for butchering people. Queen Hatshepsut's Egyptian fleet had come here three and a half thousand years ago to fetch living myrrh trees for planting at Thebes.

On a reed-ship sailing between the continents there is plenty of time to meditate. Except for human character, much has changed on all continents during the last five thousand years. Today the environmental changes accelerate around us. The road ahead is as unknown as is most of the road behind us, and the more we understand of the past the better we can predict the future. With a pedigree recently pushed back to over two million years we have much to learn from future excavations. The greatest discovery of recent years is how incredibly little we yet know of man's past, of the beginning. In the first decades after Darwin and the discovery of the unknown Sumerian civilisation, we thought we had all the answers: the jungle gave birth to man and two large and fertile river valleys gave birth to twin civilisations. Egypt and Mesopotamia. That made sense.

That two amazing civilisations suddenly arose side by side in the Middle East about 3000 BC was not surprising. The Garden of Eden was there, and Adam and Eve were born only a few millennia earlier. Then came the discoveries in the Indus Valley. First the well-preserved twin cities of Mohenjo-Daro and Harappa. But then field archaeologists found the ruins of the first civilised city builders here too, which dated roughly from 3000 BC as well. These three great civilisations surrounding the Arabian peninsula appeared as ready developed, organised dynasties at the same astonishingly high level and all three remarkably alike. The definite impression is as if related priest-kings at that time came from elsewhere with their respective entourages, and imposed their dynasties on areas formerly occupied by more primitive or at least culturally far less advanced tribes.

Why this impressive, seemingly overnight blossoming in three places simultaneously unless there was some link between them? If this question was pertinent before, it certainly becomes more so now, with the discovery that 3000 BC does not mark some sort of half-way point in human spread and evolution; the Mesopotamians, Egyptians and all other representatives of mankind had all had at least 2,000,000 years in which to move independently from palaeolithic barbarian to bronze age civilisation. Knowing this, how can we assume that three reed-boat building people began to travel in search of tin and copper at the same time, the two metals they needed to mix for shaping bronze in their wax-filled moulds? Nor is there any natural tie between bronze moulding and, say, the invention of script or the use of wheeled carts; yet these three civilisations suddenly shared all of man's major inventions and beliefs, as if they had inspired each other or suddenly had drawn from a common pool.

We are on firm historic ground when we admit that it was from the vast grain fields of Egypt and Sumer that all the arts of civilisation spread in Antiquity, first throughout the Middle East, then to Crete, Greece, Rome and finally the rest of Europe. The Indus civilisation, with elaborate ports on the coast, is known to have left its influence on distant parts of India. Since Pliny shows that Ceylon traded with China in prehistoric times, there would have been nothing to prevent the civilised Harappans from doing the same thing and spreading impulses important for the coming growth of the great Chinese civilisation which blossomed soon afterwards. In Ormara village we had seen the tiny, primitive dhows that regularly came to fetch shark cargo for Ceylon. The sea was their natural road, not the jungle of India. With Ceylon as a geographical intermediary in the east and Bahrain in the west, the Indus civilisation was not necessarily ignorant of any major nation along coastal Asia in the epoch when their ships began to plough the seas.

Chronologically, all the great civilisations of Antiquity known to us today appeared one by one in the centuries immediately after 3000 BC. They all followed the break-through in the three circum-Arabian river valleys. But however important, that break-through still does not mark the zero hour for civilised man, the real beginning. Established views collapsed again when a revision and adjustment of carbon datings disclosed that civilised man had been active on some of the most unlikely islands before he got around to founding the first dynasties in the three great river valleys. Important stone structures were built on islands around Great Britain, on Malta and Bahrain before they were built elsewhere. The long accepted teachings as to the beginning of civilisation are today found untenable.

Science itself is bewildered. There has been no time to revise the textbooks and they are now all heretical. Nor can the old texts be replaced until the majority of scholars agree on their replacement. At present some assert that civilisation must simply have started independently on various islands a millennium or more before it started in the continental river valleys. The stimulus for progress was perhaps peace and security through isolation rather than inspiration through contact, mass organisation and rich grain harvests. Others insist that such a theory conflicts with fundamental knowledge of social anthropology. Small islands, poor in soil and resources, offer none of the basic needs to produce civilisation. Besides, civilisation is not born overnight; if present on islands around Britain by 4000 BC, then civilised people must have reached there at that time or barbarians must have arrived there much earlier, with time to become civilised.

Whatever might be the answer, we are back with boats marking the beginnings. Whole families, civilised or primitive, were travelling together by sea. This confirms what we knew from the wall paintings in the Sahara and the rock carvings near the Red Sea: ship-building was old when pyramid-building started. Rivers and oceans were open when jungles were closed.

There is a glaring lack of knowledge about our own past in the two million years between the oldest hominid bones found under the silt in Africa and the evidence of seafaring inside and outside the Straits of Gibraltar some six or seven thousand years ago. Then another short gap until the sudden appearance of powerful, deified kings with whole entourages of skilled craftsmen, metal-workers, architects, astronomers and scribes at the mouth of the Mesopotamian rivers and on the banks of the Nile and the Indus around 3000 BC. These dates are milestones, nothing else. Most of the human past is totally lost. Buried or effaced. In the course of two million years of human activity, ice has come and gone. Land has emerged and submerged. Forest humus, desert sand, river silt and volcanic eruptions have hidden from our view large portions of the former surface of the earth. The sea level has altered, seventy per cent of our planet is below water and underwater archaeology has barely begun in coastal areas.

We are accustomed to finding sunken ships with old amphora and other cargo beneath the sea, but speculation as to the discovery of other human vestiges on the bottom of the ocean remains a subject for science fiction writers. Or almost; for flint arrowheads have been found on the bottom of the North Sea by trawlers, a reminder that Stone Age man hunted animals on foot in an area which changing water levels have hidden from easy access by an archaeologist. Until very recent years everyone derided Wegener's theory of continental drift; yet Wegener is today proved right and his theory generally accepted by geologists just when palaeontologists have discovered that at least another zero should be added to the hitherto accepted age of man.

Did the Atlantic basin sink to its present shape and depth before human times? It is still not known when the last of the Mid-Atlantic Ridge sank; the date can still fluctuate as much as the age of early man. But we at least know that there was some major geological catastrophe in the Atlantic in a period late enough to coincide with an identifiable stir among all known early civilisations. Its effects must have been worst on the founders of the island cultures around Britain, as the disturbance formed a lasting split in the Atlantic Ocean floor and right across the green countryside of Iceland. A tree fell into the rift and was imbedded in the lava that emerged; it has been dated by radio-carbon analysis to approximately 3000 BC.

Allowing for the usual radio-carbon margin of error, about a century plus or minus, this geological disturbance in the Atlantic of about 3000 BC coincides with the sudden blooming of civilisation in the three aforesaid river valleys. But not only there. Archaeology has disclosed that around 3000 BC a new epoch started even on the island of Cyprus; then the former occupation of neolithic sites came to an end, and it has been suspected that the cause may have been some unidentified natural disaster.[3] Correspondingly, 3000 BC has also been cited by archaeologists as the date marking the end of the neolithic phase and the beginning of a new cultural period on Malta.[4] Even on Crete archaeologists have found evidence of widespread dislocation and upheaval throughout the island about 3000 BC, with people taking refuge in caves and subsequently settling on high hills.[5]

Disastrous flood waves have struck civilised communities more than once, even in subsequent periods. Much attention has in recent years been paid to the violent waves that must have caused disaster on all Mediterranean island and mainland shores about 2000–1400 BC, when the volcano on Santorini exploded and buried the whole surface of the island with all its people and buildings. The discovery of a truly lost island civilisation beneath the volcanic ash on this island between Crete and Greece caused many serious scientists to revive the long discredited story of Atlantis. One may wonder at this revival of a much disputed Greek account of an allegedly Egyptian tradition, the more so since the island of Santorini never sank, nor is it in the Atlantic, the two basic points of the Atlantis myth. But the lost Atlantis has a grip on all our imaginations, for no other reason than that it was written down by a noted Greek almost 2400 years ago. It certainly reflects thoughts or notions concerning the 'beginnings', as put down in writing by men who cared about the past in that early period of documented history.

As to myself, I cracked many silly jokes about the possibility of rediscovering Atlantis when we strove to keep afloat on the papyrus ship *Ra*, built in Egypt, over the very locality where the Egyptians were said to believe their Atlantis had sunk. Hopefully, we said we were to discover how far early Mediterranean civilisation might have spread, not where the Egyptians said civilisation had originated.

As distinct from the *Ra* experiments, we had sailed in *Tigris* to trace the beginning of history, according to Sumerian writings. This had brought us to Dilmun, where the Sumerians said their forefathers had settled after the world catastrophe when most of mankind drowned. When listening to ancient man's opinion about our beginnings, we can nowhere get past the stories of a flood. Long before Christianity reached Hellas, the Greeks had three different versions of this disastrous flood: they had their own original deluge-myth, in which it was their own supreme god Zeus who had punished mankind; then, already in pre-Christian times, they received a Hebrew variant from the Hellenistic Jews; and independent of both of these they had received the Egyptian version following their intimate contact with the Nile country. If we care for the opinion of the ancient people whose cultural origin we seek, we have to bear with their flood stories which obstruct everything beyond. The Egyptians are no exception, if we are to trust the authority of Plato.

About four centuries before Christ, the thinker Plato wrote his dialogues *Timaeus* and *Critias*, in which he has Critias tell Socrates about Solon's interview with the learned priests of Sais, an Egyptian city at the head of the Nile delta. We are told that the story is a strange one, 'but Solon, the wisest of the seven wise men, once vouched its truth'. Solon started by telling the Egyptians about the beginning of history according to Greek memories, about how Deucalion and Pyrrha survived the deluge, and he tried to enumerate their descendants in order to work out how long ago the flood had occurred. He was then interrupted by a very old Egyptian priest, who told him that the Greeks were like children, with no ancient civilisation and no memories before the last flood. More than one flood had struck the Mediterranean and destroyed all growing civilisations, sweeping all scribes and learned men into the sea from Greece and surrounding territories. The only survivors in those parts had been unlettered and uncultivated herdsmen and shepherds in the mountains. Thus 'writings and other necessities of civilisation' had been destroyed, and the Greeks and their neighbours had to 'begin again like children', in complete ignorance of their own past achievements. But these disastrous flood waves had not struck Egypt in the same way. According to the old priest, written records from the earliest times had consequently been preserved in their temples. The oldest writings were said by the priest to describe the important events which he dated to a period nine thousand years before Solon's visit to Egypt:

Our records tell how your city checked a great power which arrogantly advanced from its base in the Atlantic ocean to attack the cities of Europe and Asia. For in those days the Atlantic was navigable. There was an island opposite the strait which you call (so you say) the Pillars of Heracles, an island larger than Libya and Asia combined; from it travellers could in those days reach the other islands, and from them the whole opposite continent which surrounds what can truly be called the ocean. For the sea within the strait we were talking about is like a lake with a narrow entrance; the outer ocean is the real ocean and the land which entirely surrounds it is properly termed continent. On this island of Atlantis had arisen a powerful and remarkable dynasty of kings, who ruled the whole island, and many other islands as well and parts of the continent; in addition it controlled, within the strait, Libya up to the borders of Egypt and Europe as far as Tyrrhenia.

The cities, temples and canals of Atlantis as described in this story can match the most impressive structures of the Pharaohs and are full of Egyptian flavour, but the reference to the port is at least remarkable: '. . . the canal and large harbour were crowded with vast numbers of merchant ships from all quarters, from which rose a constant din of shouting and noise day and night.'

It was the detailed description of this Atlantic island and the greatness of its culture and power that interested the Egyptian priesthood and the Greek narrator, whereas the dramatic details of its submergence are greatly underplayed: 'At a later time there were earthquakes and floods of extraordinary violence, and in a single dreadful day and night all your fighting men were swallowed up by the earth, and the island Atlantis was similarly swallowed up by the sea and vanished; this is why the sea in that area is to this day impassable to navigation, which is hindered by mud just below the surface, the remains of the sunken island.'[6]

On the other side of the Atlantic, the priesthood of the Aztecs and Mayas also had their records written in hieroglyphics, most of which were burnt by the Spaniards, who nevertheless recorded that these Mexican aborigines believed in a great deluge and a land sunk in the Atlantic. The Aztecs took their own national name from that island, which they in their tongue referred to as Aztlán, saying that it had been their former fatherland. The whole foundation for their religious beliefs was the assertion that their own royal families descended from certain white and bearded men resembling the Spaniards who had come from that sunken land and instructed their savage ancestors in the rites of sun-worship and all the arts of civilisation: writing, cotton cultivation, calendar system, and architecture, including the building of cities and pyramids. The amazingly accurate Maya calendar, more exact than ours today by half a day in every 5,000 years, began with a zero year of 4 *Ahau* 2 *Cumhu*, which converted into our calendar system becomes 12 August 3113 BC. The Maya astronomical clock was more exact than our approximate radio-carbon dating. With a human background of two million years, we may again wonder at this close coincidence in time with the catastrophe that split Iceland and the beginning of new cultural eras on Crete, Cyprus, Malta, in Egypt, Mesopotamia, and the Indus Valley. No satisfactory explanation has ever been found as to why the Maya chose the date of 12 August 3113 BC for their beginning of time reckoning. All other calendar systems have chosen a zero year to coincide with some event in the life of the personage who founded their religion: those of the Buddhists, Hebrews, Christians and Moslems. Maya religion was founded by Kukulcan, the sacred priest-king who arrived from across the Atlantic, and claimed descent from the sun.[7]

Could it be, I thought, that all these sun-worshippers had been chased away in August 3113 BC from a former unidentified habitat by some natural catastrophe not yet known to us?

'Look at the moon!' It was Gherman who shouted in surprise from the steering bridge. I closed my books and put away my notes to crawl out and see what he found so strange. HP was there in a bound and grabbed Norman's astronomical almanac. The sky was clear but the moon was fading; it ceased to be a flat and shining disc and became globular and pale like a lost balloon. Then it began to disappear. It looked frightening. '24 March 1978, total moon eclipse over part of Asia and the Indian Ocean,' HP reported. This was the classical sight that would have been interpreted by prehistoric sky-watchers as an ill omen. We were in no way superstitious, but we shared the sombre feelings of earlier man that night, as we sailed along the coast of Punt at a good speed but with maximum caution so as not to stray within the war zone.

Three days later a beautiful bird, able to raise a majestic crest of feathers on its head, came from Africa and landed in the forestay. This was an upupa (called in Britain the hoopoe), known to the ancient Vikings as the 'army-bird' *(hærfugl)* and regarded by them as a sign of war. Carlo could confirm that the upupa in Italy was known as the 'cock-of-Mars' *(gallo-di-Marte)*, named after the Roman god of war. This would have been too many evil omens for ancient voyagers, and we could easily see how it would have impressed them if they next discovered what we already knew, that a war was really raging just beyond the horizon.

That same night Toru shouted from the bridge: 'Did you hear that?' We did. Inside the cabin we too had heard the distant rumble of gun-fire on the port side. And we heard it again.

Next day we even heard the growing drone of an aeroplane. 'By gosh, he's coming straight for us!' Detlef shouted from the bridge. We all rushed out into the burning sun of the Gulf of Aden just as a twin-engined military plane dived down over the oar in our mast top, so low that the sails flapped back with the wind-pressure. I was just in time to stop Detlef, who already had one leg over the bridge railing ready to jump from the ship. We hoisted the United Nations flag. The plane turned and, very low, came straight back again. 'Here they come again!' I yelled as I saw it turning. Rashad shouted that they were customs control men coming to bomb us, thinking we carried contraband. Asbjörn suggested they had chosen us as a training target. 'Hurrah, they are American marines!' Norman cheered and danced and waved on the cabin roof as they again droned low overhead. Someone up there waved back from the cockpit. We were all relieved.

Norris had filmed and tape-recorded this episode, and as a moment later he replayed his tape on the cabin roof, everyone but me rushed out once more as they heard my voice shouting: 'Here they come again' followed by the recorded droning. We laughed, but not for long, as Norris's tape had not ended when the droning came back louder than ever and of a different kind. We looked up, and there was a military helicopter coming straight for *Tigris*; Norman yelled that it was not American.

We were all ready to jump as the heavy war-bird turned so close over the mast that we could see the uniformed men inside salute and point to their colours: they were French! No sooner had they disappeared into the blue before two other helicopters appeared, one on either horizon. They came towards us from different directions. This looked worse. They met above us and we were safe, for one was American and the other French. Friendly pilots photographed us from the air. Friendly to us but not to everybody in this surrounding area. We had reached the inner end of the gulf. On port side was the last of Somalia and a front line which was guarded on either side with the support of the great powers. On starboard was South Yemen, also with border feuds, and the old and famous port of Aden, also closed.

Norman was strongly in favour of taking advantage of our present favourable position to head straight through the narrow Bab-el-Mandeb strait and continue up the Red Sea. But since neither of the two nations flanking the Red Sea inside the strait had responded to our request for landing permits, we unanimously agreed on altering course for Djibouti, the tiny nation that had welcomed us to land on the African side of the strait.

On 28 March we saw the blue mountains of Africa, and that night we steered by the lights from the shore. Long before daybreak we passed the lighthouses outside Djibouti harbour, and dropped anchor. As day broke we found ourselves riding with a huge battleship of some sort as our nearest neighbour. The sun rose and a small yacht came out of port and guided us in, under full sail, past the flagship and other units of the French Indian Ocean fleet. The ancient harbour seemed packed with warships that were there to protect neutral Djibouti from intrusion by belligerent nations fighting each other all around. This mini-republic had just been granted its independence from France. Officers and men on all the naval vessels were lined up to welcome peaceful *Tigris* as I shouted 'sails down' and turned the tiller for anchoring. Norman climbed the mast and made a masterly performance of riding the yardarm down to deck all alone, as the rest of us worked at rudder-oars, punt-poles and anchors.

To receive us in Djibouti and collect our films were Bruce Norman and Roy Davies from the BBC. They brought with them from London the news that under no circumstances could we land in Ethiopia on the African side, inside the Red Sea. Massawa, where I had loaded the papyrus from Lake Tana for *Ra I* and *Ra II*, and where for this reason I had hoped to end the *Tigris* voyage after an estimated five more days of sailing, was in a state of siege. The city and the port were in the hands of the Ethiopians, supported by Russians, but the entrance and the surrounding coasts were held by Eritrean liberation forces and any trespassers would be shot. But Roy triumphantly handed me two letters from the North Yemen authorities. One was from Mohamed Abdulla Al-Eryani, Ambassador of the Yemen Arab Republic in London. The other was from the Minister Plenipotentiary, Mohamed Al-Makhadhi. The first confirmed North Yemen's interest in our expedition, expressed the warmest wishes for its success, and referred to the Minister. The Minister wrote: 'I can assure you of our fullest cooperation at all times since Dr T. Heyerdahl's expedition is a very remarkable and praiseworthy one. Yours sincerely . . .'

There was every reason to celebrate. We could now rest a few days and then continue through the strait to land in North Yemen. It was on the opposite side from Massawa, but that meant nothing.

We jumped ashore among friendly, black Africans, and checked in at Siesta Hotel. I shall never forget the big juicy pepper steak that was put in front of my nose just as the telephone rang. Counter-orders from London! North Yemen had withdrawn permission for us to sail into their national waters 'for security reasons'. Since we did not have as much as a pistol on board, nobody could be afraid of us. The very friendly previous messages indicated that the concern was perhaps for our security, not their own. With the UN flag astern and men from east and west on board, a slow-moving Sumerian ma-gur would be a tempting prey for modern hijackers. This was a hot corner of our twentieth century planet. Nobody knew it then, but possibly fear was in the air: for in the following year the

Presidents of both North and South Yemen were assassinated on two successive days.

There was suddenly nowhere to sail in any direction. Scientifically it did not matter a bit that we were not allowed to add another five days to an experiment that had gone on for five months. But what hurt all of us was that we had come back to our own world, our own contemporaries, and met again the results of twenty centuries of progress since the time of Christ, the peace-loving moralist whose birth marks our own zero year. And here, around us on all sides, wonderful people were taught to kill each other by our own experts, and were helped to do so by the most advanced methods man had invented at the end of five millennia of known history.

I did not tell my celebrating companions the bad news. I sneaked away from the party and spent all night on my back on board *Tigris*, gazing into the cane roof of the cosy cabin and wondering what we could do. We had to abandon ship and end the expedition here, that was certain. So far we had never given a thought to what to do with *Tigris*; in fact we had boldly promised to stay on board as long as it would float. It so happened that it still floated high, the distance from deck to water was as on *Ra I* and *Ra II* when we started. The outer mats began to tear and were marked by pollution, but the forty-four inner bundles made by the Marsh Arabs were as good as ever, and so was the palm-stem repair of the bow. Both *Kon-Tiki* and *Ra II* were taken to Oslo after their expeditions, and were on exhibition, with sails up, in the Kon-Tiki Museum. But with the new hall for *Ra II* there was no possible room for further extension. If we left *Tigris* in the polluted harbour of Djibouti, the ropes would quickly rot and the beautiful reed-ship would fall apart and disintegrate in a few months. Business people in different parts were ready to buy *Tigris*, the last offer being from the man who towed us out from Karachi. But I hated the thought of our proud vessel travelling about stage-managed by some speculator. In addition, I was upset by the unbelievable nightmare of modern war and the suffering of the refugees around us. I was sure the rest of the world was as ignorant as we had been of what was going on; to them, as to us, war in some distant part, away from our own doorsteps, was unreal, merely part of the daily news.

I took a hard decision. Instead of being left to rot, *Tigris* should have a proud end, as a torch that would call to men of reason to resume the cause of peace in a corner of the world where civilisation first took foothold. We should set the reed bundles ablaze at the entrance to the Red Sea as a fiery protest against the accelerating arms race and the fighting in Africa and Asia.

The others were informed of my decision as they came on board next morning and we all gathered around the breakfast table. They were shocked at first, but everybody gave wholehearted support to the plan.

That day I was received at the Djibouti Palace by President Hassan Gouled Aptidon, an elderly leader of a young nation, and a wise, friendly and human representative of black Africa. I asked him for permission to dissolve the expedition and abandon ship in his country, and explained to him how precisely we had been forced to navigate to reach his neutral territory. 'You were lucky,' he said with a calm smile; 'your vessel was able to sail away from the war. But my little nation is forced to remain here, with war on all sides, and with constant fear of invasion.' He added that we were welcome to leave our ship and move ashore, but we should know that his nation was full of refugees, all roads to the outside world were blocked and the only railway out, to Addis Ababa, had been blown up. Some meat was flown in from Nairobi, but all other food came by air from Paris, and his people suffered because they could not afford the prices. Apart from the port, the Republic of Djibouti had no source of income, as the limited country around the city was pure desert.

A very cordial reception by the French Rear-Admiral Darrieus on board the floating navy dock-ship TCD *Ouragan* followed, and a buffet dinner hosted by the charming American couple Chantal and Walter Clarke, who were as depressed at the local situation as we were. As Chargé d'Affairs Walter Clarke had not yet had time to open an American Embassy, so young was the republic.

Only one person had to know about our plan: the harbour master. Otherwise the port fire brigade and navy helicopters would come out the moment they thought they detected an accident. We wanted to be alone at the end.

Tigris was towed out of the harbour with usual clearance papers and anchored, sails up, off the lighthouse on Musha island, a small coral isle outside the port. Before we lowered the United Nations flag I wrote a telegram to the man who had granted *Tigris* the right to sail symbolically under this flag. The message was passed to everyone on board for approval or disproval:

Secretary-General Kurt Waldheim,
United Nations.

As the multinational crew of the experimental reed-ship *Tigris* brings the test voyage to its conclusion today we are grateful to the Secretary-General for the permission to sail under United Nations flag and we are proud to report that the double objectives of the expedition have been achieved to our complete satisfaction.

Ours has been a voyage into the past to study the qualities of a prehistoric type of vessel built after ancient Sumerian principles. But it has also been a voyage into the future to demonstrate that no space is too restricted for peaceful coexistence of men who work for common survival. We are eleven men from countries governed by different political systems. And we have sailed together on a small raft-ship of tender reeds and rope a distance of over six thousand kilometres from the Republic of Iraq by way of the Emirate of Bahrain, the Sultanate of Oman and the Republic of Pakistan to the recently born African nation of Djibouti. We are able to report that in spite of different political views we have lived and struggled together in perfect understanding and friendship shoulder to shoulder in cramped quarters during calms and storms, always according to the ideals of the United Nations: cooperation for joint survival.

When we embarked last November on our reed-ship *Tigris* we knew we would sink or survive together, and this knowledge united us in friendship. When we now, in April, disperse to our respective homelands we sincerely respect and feel sympathy for each other's nation, and our joint message is not directed to any one country but to modern man everywhere. We have shown that the ancient people in Mesopotamia, the Indus Valley and Egypt could have built man's earliest civilisations through the benefit of mutual contact with the primitive vessels at their disposal five thousand years ago. Culture arose through intelligent and profitable exchange of thoughts and products. Today we burn our proud ship with sails up and rigging and vessel in perfect shape to protest against the inhuman elements in the world of 1978 to which we have come back as we reach land from the open sea. We are forced to stop at the entrance to the Red Sea. Surrounded by military aeroplanes and warships from the world's most civilised and developed nations we are denied permission by friendly governments, for security reasons, to land anywhere but in the tiny and still neutral republic of Djibouti, because elsewhere around us brothers and neighbours are engaged in homicide with means made available to them by those who lead humanity on our joint road into the third millennium.

To the innocent masses in all industrialised countries we direct our appeal. We must wake up to the insane reality of our time which to all of us has been reduced to mere unpleasant headlines in the news. We are all irresponsible unless we demand from the responsible decision makers that modern armaments must no longer be made available to the people whose former battle axes and swords our ancestors condemned. Our planet is bigger than the reed bundles that have carried us across the seas and yet small enough to run the same risks unless those of us still alive open our eyes and minds to the desperate need of intelligent collaboration to save ourselves and our common civilisation from what we are about to convert into a sinking ship.

The Republic of Djibouti, 3 April, 1978*

Everybody signed. Thor, Norman, Yuri, Carlo, Toru, Detlef, Gherman, Asbjörn, Rashad, HP, Norris. All eleven. Then we ate a last meal at the plank table between the two cabins: Yuri's dried rainbow-runner, Rashad's pickled flying-fish, biscuits. We had great memories from around this table. Norman remarked that we had sailed 6,800 kilometres together; 4,200 miles. *Tigris* had now been afloat 143 days, or twenty weeks and three days, that is a good five months.

Norris looked at his watch and pointed at his camera. The sun was getting low. It would soon set behind the blue mountains of Africa, which fell off in a blunt cape at the entrance to the Red Sea. Everybody but HP, Asbjörn and I were set ashore on the low coral banks with the dinghy. We had chartered a little yacht to bring us back to port. The captain and his mate brought it into safety behind the island when they realised what we were up to. HP had been a peace-time demolition sergeant in the Norwegian army and had bought an innocent clock-like time-keeper in a Djibouti photo-shop. It was zero hour for *Tigris*. Asbjörn had been in charge of our kerosene lamps on board and knew where to find the fuel. HP where to pour it. I looked at the empty table as I jumped into the

---

* Editorial note: The Secretary-General responded with a long and extremely positive message, extending his warmest congratulations on the successful outcome of the experiment and assuring us that the appeal would not go unheeded at the United Nations.

dinghy after the others. Nobody had troubled to clean the table tonight. Provisions for eleven men for another month, blankets and everything else serviceable had been carried ashore to the refugees.

We lined up ashore and none of us could say much. 'Take off your hats,' I said at last as the flames licked out of the main cabin door. The sail caught fire in a rain of sparks, accompanied by sharp shotlike reports of splitting bamboo and the crackling of burning reeds. Nobody else spoke, and I barely heard myself mumble:

'She was a fine ship.'

REFERENCES

CHAPTER 1

1. Amiet, P., *La Glyptique Mesopotamienne Archaique,* Paris, 1961.
2. Salonen, A., 'Die Wasserfahrzeuge in Babylonien', Ed. K. Tallqvist, *Studia Orientalia edidit Societas Orientalis Fennica*, Vol. 8, 4, p. 70, 1939.
3. *The New English Bible*, Genesis 15, 18–21, 1970.
4. Tallqvist, K., *Gilgameš-eposet*, p. 92, Stockholm, 1977.
5. *The New English Bible*, Genesis 6, 14–16, 1970.
6. Tallqvist, op. cit., p. 125.
7. Glob, P. V., *Al-Bahrain*, p. 214, Copenhagen, 1968.

CHAPTER 4

1. Bibby, G., *Looking for Dilmun*, p. 253, New York, 1969.
2. Kramer, S. N., *Sumerian Mythology. A study of spiritual and literary achievements in the third millennium BC*, pp. 37–8, 60. Philadelphia, 1944.

CHAPTER 5

1. Oppenheim, A. L., 'The Seafaring Merchants of Ur', *Journ. Amer. Oriental Soc.*, Vol. 74, No. 1, pp. 6–17, 1954.
2. Gordon, E. I., 'The Sumerian Proverb Collections: A Preliminary Report', *Journ. Amer. Oriental Soc.*, Vol. 74, No. 2, pp. 82–5, 1954.
3. Salonen, A., op. cit., pp. 12–14, 49, 66, 70.
4. Woolley, C. L., *The Sumerians*, pp. 7–8, 192–4, New York, 1965. By permission of Oxford University Press.
5. Kramer, S. N., op. cit., p. 60.
6. Woolley, C. L., op. cit., pp. 35–45.
7. Bibby, G., op. cit., p. 80.
8. Ibid.
9. Danish archaeologists resumed excavations at this Dilmun port shortly after *Tigris* left Bahrain. They now discovered a large and deep basin with embankment walls inside the proper city walls. At high tide shallow vessels could sail straight into a sheltered dock cut into the bedrock inside the seaward wall of the city, which also served as a protective mole against the sea. (Bibby, G., 'Gensyn med Bahrain', *Sfinx*, No. 4, pp. 99–103.)
10. Bibby, G., op. cit., pp. 186–9.
11. Heyerdahl, T., *Aku-Aku. The Secrets of Easter Island*, London, 1958; *The Art of Easter Island,* London, 1976.
12. Mallowan, M. E. L., *Nimrud and its remains*, Vol. 1, pp. 78–81, 323, New York, 1966.
13. Bibby, G., op. cit., pp. 69–77, 160–1.

CHAPTER 7

1. *Journal of Oman Studies*, 1976.
2. Bibby, G., op. cit., pp. 191, 219–20.
3. Woolley, C. L., op. cit., pp. 45–6.

4. Bibby, G., op. cit., p. 220.
5. Plinius Gaius Secundus (AD 77), *Naturalis Historia*, Book VI, 98, The Loeb Classical Library.

CHAPTER 8

1. Bibby, G., op. cit., p. 219.
2. Kramer, S. N., op. cit., p. 112.
3. Bibby, G., op. cit., pp. 221–2.
4. Bibby, G., op. cit., p. 192.
5. Dales, G. F., 'Harappan Outposts on the Makran Coast,' *Antiquity* (1962), Vol. 36, No. 142, pp. 86–92.
6. Plinius, op. cit., Book VI, 97–8.

CHAPTER 9

1. Khan, F. A., 'Indus Valley Civilization', *Cultural Heritage of Pakistan*, p. 11 (Karachi, 1966).
2. Raikes, R. L., 'The End of the Ancient Cities of the Indus', *American Anthropologist*, Vol. 66, No. 2, p. 291 (1964).
3. Rao, S. R., 'Shipping and Maritime Trade of the Indus People', *Expedition*, University of Penn., Philadelphia, Vol. 7, No. 3, pp. 30–37 (1965).
4. Rao, S. R., 'A "Persian Gulf" Seal from Lothal', *Antiquity*, Vol. 37, No. 146, pp. 96–9 (1963).
5. Mackay, E. J. H., *Further Excavations at Mohenjo-Daro*, New Delhi Government Press, 1938, Vol. I, pp. 340–1.
6. LeBaron Bowen, R. Jr., 'Boats of the Indus Civilization', *The Mariner's Mirror*, 41–2, pp. 279–90 (1956).
7. Fairservis, W. A., *The Roots of Ancient India*, George Allen & Unwin, London, 1971, pp. 277–8.
8. Plinius, op. cit., Book VI, p. 205.
9. MS. 26 February 1977.

CHAPTER 10

1. Heyerdahl, T., *Early Man and the Ocean*, Ch. 2 with map, George Allen & Unwin, London, 1978.

CHAPTER 11

1. *Red Sea and Gulf of Aden Pilot*, pp. 557 and 559, London, 1967.
2. Plinius, op. cit., Book VI, pp. 398–421.
3. Karageorgis, V., *The Ancient Civilization of Cyprus*, p. 37, New York, 1969.
4. Evans, J. D., *The Prehistoric Antiquities of the Maltese Islands*, pp. 212–14, London, 1971.
5. Hood, S., *The Minoans: Crete in the Bronze Age*, p. 29, London, 1971.
6. Plato (ca. 427–347 BC), *Timaeus and Critias*, Penguin Classics, pp. 33–8, 141, London, 1965.
7. This topic and the arguments for trans-Atlantic contact in pre-European times are dealt with in detail in Heyerdahl, T., *Early Man and the Ocean*, London, 1978.

INDEX

Index

377

Now published in Unwin Paperbacks for the first time. A new paperback edition of Thor Heyerdahl's first great ocean adventure:

THE KON-TIKI EXPEDITION

THOR HEYERDAHL

tells the enthralling story of crossing the Pacific on a balsa-wood raft. **Kon-Tiki** became an international classic with sales counted in millions.

'an incredible adventure which happens to be true'
SOMERSET MAUGHAM

'an enthralling account of an experience without parallel'
RICHARD HUGHES

'a bizarre and adventurous enterprise, excitingly and modestly recounted'
MALCOLM MUGGERIDGE